Fluid Mechanics with Civil Engineering Applications

E. John Finnemore
Professor Emeritus of Civil Engineering
Santa Clara University, Santa Clara, CA

Edwin Maurer
Professor of Civil, Environmental, and Sustainable Engineering
Santa Clara University, Santa Clara, CA

Eleventh Edition

New York   Chicago   San Francisco
Athens   London   Madrid
Mexico City   Milan   New Delhi
Singapore   Sydney   Toronto

Fluid Mechanics with Civil Engineering Applications, Eleventh Edition

1 2 3 4 5 6 7 8 9   LKV   29 28 27 26 25 24 23

Library of Congress Control Number: 2023945694

ISBN 978-1-264-78729-6
MHID 1-264-78729-4

This book is printed on acid-free paper.

Sponsoring Editors
Ania Levinson, Robin Najar

Editorial Supervisor
Janet Walden

Project Manager
Poonam Bisht, MPS Limited

Acquisitions Coordinator
Olivia Higgins

Copy Editor
Subashree Baskaran, MPS Limited

Proofreader
MPS Limited

Indexer
MPS Limited

Production Supervisor
Lynn M. Messina

Composition
MPS Limited

Illustration
MPS Limited

Art Director, Cover
Jeff Weeks

Cover images from Shutterstock: Background photo: Landscape overlooking Kerr Dam with wide open gates spilling water waterfall, by *Valley Journal*. Foreground, lower left photo: Engineers inspect the quality of a concrete pipe, by Chirawut8878.

About the Authors

E. John Finnemore is Professor Emeritus of Civil Engineering at Santa Clara University, California. Born in London, England, he received a B.Sc. (Eng.) degree from London University in 1960, and M.S. and Ph.D. degrees from Stanford University in 1966 and 1970, all in civil engineering. Finnemore worked with consulting civil engineers in England and Canada for five years before starting graduate studies, and for another seven years in California after completing his doctorate. He served one year on the faculty of Pahlavi University in Shiraz, Iran, and he has been a member of the faculty at Santa Clara University since 1979. He has taught courses in fluid mechanics, hydraulic engineering, hydrology, and water resources engineering, and has authored numerous technical articles and reports in several related fields. His research has often involved environmental protection, such as stormwater management and onsite wastewater disposal. Professor Finnemore has served on governmental review boards and as a consultant to various private concerns. He is a Fellow of the American Society of Civil Engineers and a registered civil engineer in California. He lives with his wife Gulshan in Cupertino, California.

Edwin Maurer is Professor of Civil, Environmental, and Sustainable Engineering at Santa Clara University, where he joined the faculty in 2003. He received a B.S. from the University of Rhode Island, an M.S. from the University of California, Berkeley, and a Ph.D. from the University of Washington. Prior to entering academia, he worked for seven years in a variety of consulting engineering engagements and served in the Peruvian altiplano for four years designing small community water systems and teaching hydrology at a local university. At Santa Clara University, he has taught classes in fluid mechanics and hydraulics, surface and groundwater hydrology, geographical information systems, and sustainable engineering. He has published extensively on hydrologic modeling and the impacts of climate change on water.

**To that great love which
encourages humanity in all
its noble endeavors
and**

**to Gulshan for her loving
support**

**To Deidre, Frances, and
Lucie for their patience,
humor, and love**

Contents at a Glance

Contents

Preface

Philosophy and History

This eleventh edition of the classic textbook, *Fluid Mechanics with Civil Engineering Applications,* continues and improves on its tradition of explaining the physical phenomena of fluid mechanics and applying its basic principles in the simplest and clearest possible manner without the use of complicated mathematics. It focuses on civil, environmental, and agricultural engineering problems, although mechanical engineering topics are also strongly represented. The book is written as a text for a first course in fluid mechanics for civil and related engineering students, with sufficient breadth of coverage that it can serve in a number of ways for a second course if desired.

Thousands of engineering students and practitioners throughout the world have used this book for over 100 years; it is widely distributed as an International Edition, and translations into Spanish and Korean are available. The book is now in its fourth generation of authorship. Though this edition is very different from the first edition, it retains the same basic philosophy and presentation of fluid mechanics as an engineering subject that Robert L. Daugherty originally developed over his many years of teaching at Cornell University, Rensselaer Polytechnic Institute, and the California Institute of Technology. The first edition that Professor Daugherty authored was published in 1916 with the title *Hydraulics.* He revised the book four times. On the fifth edition (fourth revision) Dr. Alfred C. Ingersoll assisted him, and they changed the title of the book to *Fluid Mechanics with Engineering Applications.* The sixth and seventh editions were entirely the work of Professor Franzini. A student of Daugherty's at Caltech, Franzini had received his first exposure to the subject of fluid mechanics from the fourth edition of the book. Professor Franzini enlisted the services of Professor Finnemore, a former student of Franzini's at Stanford, to assist him with the eighth, ninth, and tenth editions. The tenth edition was chosen for special treatment and promotion by McGraw Hill. This eleventh edition is the work of Dr. Finnemore and Dr. Maurer, a new contributor for this book.

The Book, Its Organization

We feel it is most important that the engineering student clearly visualize the physical situation under consideration. Throughout the book, therefore, we place considerable emphasis on physical phenomena of fluid mechanics. We stress the governing principles, the assumptions made in their development, and their limits of applicability, and we show how we can apply the principles to the solution of practical engineering problems. The emphasis is on teachability for the instructor and on clarity for both the instructor and the student, so that they can readily grasp basic principles and applications. Numerous worked sample problems are presented to demonstrate the application of basic principles. These sample problems also help to clarify the text. Drill exercises with answers are available online at www.mhprofessional.com/FMWCEA11-9781264787296 for most sections to help students rapidly reinforce their understanding of the subjects and concepts. Also, for each chapter, problems are available at the same Web site without solutions for assignment purposes. These were carefully selected to provide the student with a thorough workout in the application of basic principles; their solutions are available only to instructors. Only by working numerous exercises and problems will students experience the evolution so necessary to the learning process. We recommend ways to study fluid mechanics and to approach problem solving in Chapter 1.

The book is essentially "self-contained." The treatment is such that an instructor generally need not resort to another reference to answer any question that a student might normally be expected to ask. This has required more detailed discussion than that needed for a more superficial presentation of certain topics. A list of selected references is provided at the end of the book to serve as a guide for those students who wish to probe deeper into the various fields of fluid mechanics. The appendix section contains information on physical properties of fluids and other useful tables. Chapter 1 contains information on dimensions and units, and, for convenient reference, the appendices contain conversion factors and important quantities and definitions.

Even though we use British Gravitational (BG) units (feet, slugs, seconds, pounds) as the primary system of units, we give the corresponding SI units in the text. We provide sample problems, and exercises and problems in BG and in SI units in near equal numbers. We have made every effort to ease the changeover from BG units to SI units; Chapter 1 includes a discussion of unit systems and conversion of units. We encourage instructors to assign problems in each system so that students become conversant with both.

Improvements to This Edition

Probably the most noticeable change made throughout this edition will be our rearrangement of chapters and more abbreviated content. We have focused on the content most important to a first course in civil engineering fluid mechanics, and included additional explanation and new sample problems to guide students toward more advanced topics with a stronger background. By removing chapters and sections not commonly covered in a first course, we hope to make the book more accessible to students.

In this revision, we have given special attention to the chapters on hydrostatics, pipe flow solutions, and open channel flow. Our own experience and the many comments generously provided by other faculty using the text pointed toward improvements needed with the presentation of materials, especially providing additional or revised sample problems demonstrating solution methods for simpler problems.

The prior edition included an appendix demonstrating the use of specific software or calculator models for solving implicit equations or systems of equations, since these are commonly encountered in both pipe and open channel flow. With rapid changes in technology, open-source programming languages with powerful equation solvers, and the wide availability or graphing calculators with equation solvers, the new edition avoids demonstrating solution methods with specific methods, since they become out-dated very rapidly. Instead, we encourage students to explore applications using their preferred language or tool.[1]

Use of the Book, Course Planning

Schools having stringent requirements in fluid mechanics might wish to cover the entire text in their course or courses required of all civil, environmental, and agricultural engineers. At other schools only partial coverage of the text might suffice for the course required of all such engineers, and they might cover other portions of the text in a second course for students in a particular branch or branches of engineering. In this way, civil, environmental, and agricultural engineers might emphasize Chapter 8 and perhaps Chapter 10 in a second course. For more advanced material on some topics, the tenth edition of this text can be a resource, as well as the many references listed in Appendix C.

As noted above, students (and instructors) have online access to exercise statements and solutions for most sections of the text, providing ample practice material for students. Students will also have access to the problems that accompany each chapter. For instructors only, a companion Solutions Manual is available from McGraw Hill that contains carefully explained solutions to all the problems prepared for each chapter; for convenience, the problem statements and problem figures are repeated with the solutions. The manual contains suggestions on how to use it most effectively to select problems for assignment, and a Problem Selection Guide for each chapter categorizes the problems by their difficulty, length, units used, and any special features. The problems for students and the instructor's Solutions Manual may be downloaded from www.mhprofessional.com/FMWCEA11-9781264787296.

---

[1] As one example, see E. Maurer, 2023. *Hydraulics and Water Resources: Examples Using R*, https://doi.org/10.5281/zenodo.7576843.

Acknowledgments

We appreciate the many comments and suggestions that we have received from users of the book throughout the years, and from numerous anonymous in-depth reviews arranged by McGraw Hill. Further comments and suggestions for future editions of the book are always welcome.

We are very grateful for the care, assistance, and guidance that many people at McGraw Hill and its subcontractors have provided to us in the preparation of this edition.

E. John Finnemore
Edwin Maurer

List of Symbols

The following table lists the letter symbols generally used throughout the text. Because there are so many more concepts than there are English and suitable Greek letters, certain conflicts are unavoidable. However, where we have used the same letter for different concepts, the topics are so far removed from each other that no confusion should result. Occasionally we will use a particular letter in one special case only, but we will clearly indicate this local deviation from the table, and will not use it elsewhere. We give the customary units of measurement for each item in the British Gravitational (BG) system, with the corresponding SI unit in parentheses or brackets.

For the most part, we have attempted to adhere to generally accepted symbols, but not always.

A = any area, ft^2 (m^2)

= cross-sectional area of a stream normal to the velocity, ft^2 (m^2)

= area in turbines or pumps normal to the direction of the absolute velocity of the fluid, ft^2 (m^2)

A_s = area of a liquid surface as in a tank or reservoir, ft^2 or acre (m^2 or hectare)

a = linear acceleration, ft/sec^2 (m/s^2)

B = any width, ft (m)

= width of open channel at water surface, ft (m)

= width of pump impeller at periphery, ft (m)

b = bottom width of open channel, ft (m)

$\mathbf{C}$ = cavitation number = $(p - p_v)/(\frac{1}{2}\rho V^2)$ [dimensionless]

C = any coefficient [dimensionless]

= Chézy coefficient [ft$^{1/2}$sec^{-1} (m$^{1/2}$s^{-1})]

C_c = coefficient of contraction ⎫

C_d = coefficient of discharge ⎬ for orifices, tubes, and nozzles [all dimensionless]

C_v = coefficient of velocity ⎭

C_{HW} = Hazen-Williams pipe roughness coefficient, ft$^{0.37}$/sec (m$^{0.37}$/s)

C_p = pressure coefficient = $\Delta p/(\frac{1}{2}\rho V^2)$. [dimensionless]

c = specific heat of liquid, Btu/(slug·°R) [cal/(g·K) or N·m/(kg·K)]

= wave velocity (celerity), fps (m/s)

= sonic (i.e., acoustic) velocity (celerity), fps (m/s)

c_j = velocity (celerity) of pressure wave in elastic fluid inside an elastic pipe, ft/sec (m/s)

c_p = specific heat of gas at constant pressure, ft·lb/(slug·°R) [N·m/(kg·K)]

c_v = specific heat of gas at constant volume, ft·lb/(slug·°R) [N·m/(kg·K)]

$D =$ diameter of pipe or pump impeller, ft or in (m or mm)

$D''V =$ product of pipe diameter in inches and mean flow velocity in fps

$E =$ specific energy in open channels $= y + V^2/2g$, ft (m)

$\quad =$ linear modulus of elasticity, psi (N/m^2)

$E_j =$ "joint" volume modulus of elasticity for elastic fluid in an elastic pipe, psi (N/m^2)

$E_v =$ volume modulus of elasticity, psi (N/m^2)

$e =$ height of surface roughness projections, ft (mm)

$\quad = 2.718\ 281\ 828\ 46\ldots$

$\mathbf{F} =$ Froude number $= V/\sqrt{gL}$ [dimensionless]

$F =$ any force, lb (N)

$f =$ friction factor for pipe flow [dimensionless]

$G =$ weight flow rate $= dW/dt = \dot{m}g = \gamma Q$, lb/sec (N/s)

$g =$ acceleration due to gravity $= 32.1740$ ft/sec^2 (9.806 65 m/s^2) (standard)

$\quad = 32.2$ ft/sec^2 (9.81 m/s^2) for usual computation

$H =$ total energy head $= p/\gamma + z + V^2/2g$, ft (m)

$\quad =$ head on weir or spillway, ft (m)

$h =$ any head, ft (m)

$\quad =$ enthalpy (energy) per unit mass of gas $= i + p/\rho$, ft·lb/slug (N·m/kg)

$h' =$ minor head loss, ft (m)

$\quad =$ effective head, ft (m) in the context of reaction turbines

$h_a =$ accelerative head $= (L/g)(dV/dt)$, ft (m)

$h_c =$ depth to centroid of area, ft (m)

$h_f =$ head loss due to wall or pipe friction, ft (m)

$h_L =$ total head loss due to all causes, ft (m)

$h_M =$ energy *added* to a flow by a machine per unit weight of flowing fluid, ft·lb/lb (N·m/N)

$h_O =$ stagnation (or total) enthalpy of a gas $= h + \frac{1}{2}V^2$, ft·lb/slug (N·m/kg)

$h_p =$ depth to center of pressure, ft (m)

$\quad =$ head added to a flow by a pump, ft (m)

$h_t =$ head removed from a flow by a turbine, ft (m)

$I =$ moment of inertia of area, ft^4 or in^4 (m^4 or mm^4)

$\quad =$ internal thermal energy per unit weight $= i/g$, ft·lb/lb (N·m/N)

$I_c =$ moment of inertia about centroidal axis, ft^4 or in^4 (m^4 or mm^4)

$i =$ internal thermal energy per unit mass $= gI$, ft·lb/slug (N·m/kg)

$K =$ any constant [dimensionless]

$k =$ any loss coefficient [dimensionless]

$\quad =$ specific heat ratio $= c_p/c_v$ [dimensionless]

$L =$ length, ft (m)

$\ell =$ mixing length, ft or in (m or mm)

$\mathbf{M} =$ Mach number $= V/c$ [dimensionless]

$M =$ molar mass, slugs/slug-mol (kg/kg-mol)

$m =$ mass $= W/g$, slugs (kg)

$\dot{m} =$ mass flow rate $= dm/dt = \rho Q$, slugs/sec (kg/s)

$\mathbf{N} =$ any dimensionless number

$N_s =$ specific speed $= n_e\sqrt{\text{gpm}}/h^{3/4}$ for pumps [dimensionless]

NPSH $=$ net positive suction head, ft (m)

NPSH$_R =$ required net positive suction head, ft (m)

$n =$ an exponent or any number in general

$\quad =$ Manning coefficient of roughness, sec/ft$^{1/3}$ (s/m$^{1/3}$)

$\quad\quad$ = revolutions per minute, min^{-1}

n_e = rotative speed of hydraulic machine at maximum efficiency, rev/min

P = power, ft·lb/sec (N·m/s)

$\quad\quad$ = height of weir or spillway crest above channel bottom, ft (m)

$\quad\quad$ = wetted perimeter, ft (m)

p = fluid pressure, lb/ft^2 or psi (N/m^2 = Pa)

p_{atm} = atmospheric pressure, psia (N/m^2 abs)

p_O = stagnation pressure, psf or psi (Pa)

p_v = vapor pressure, psia (N/m^2 abs)

Q = volume rate of flow (discharge rate), cfs (m^3/s)

Q_H = heat *added* to a flow per unit weight of fluid, ft·lb/lb (N·m/N)

q = volume rate of flow per unit width of rectangular channel, cfs/ft = ft^2/sec (m^2/s)

q_H = heat transferred per unit mass of fluid, ft·lb/slug (N·m/kg)

$\mathbf{R}$ = Reynolds number = $LV\rho/\mu = LV/\nu$ [dimensionless]

R = gas constant, ft·lb/(slug·°R) or N·m/(kg·K)

R_h = hydraulic radius = A/P, ft (m)

R_m = manometer reading, ft or in (m or mm)

R_0 = universal gas constant = 49,709 ft·lb/(slug-mol·°R) [8312 N·m/(kg-mol·K)]

r = any radius, ft or in (m or mm)

r_0 = radius of pipe, ft or in (m or mm)

S = slope of energy grade line = h_L/L

S_c = critical slope of open channel flow

S_0 = slope of channel bed $\quad\quad$ [dimensionless]

S_w = slope of water surface

$\quad s$ = specific gravity of a fluid = ratio of its density to that of a standard fluid (water, air, or hydrogen) [dimensionless]

T = temperature, °F or °R (°C or K)

$\quad\quad$ = period of time for travel of a pressure wave, sec (s)

$\quad\quad$ = torque, ft·lb (N·m)

T_O = stagnation temperature of a gas = $T + \frac{1}{2}V^2/c_p$, °F or °R (°C or K)

T_r = travel time (pulse interval) of a pressure wave, sec (s)

t = time, sec (s)

$\quad\quad$ = thickness, ft or in (m or mm)

t_c = time for complete or partial closure of a valve, sec (s)

U, U_0 = uniform velocity of fluid, fps (m/s)

u = velocity of a solid body, fps (m/s)

$\quad\quad$ = tangential velocity of a point on a rotating body = $r\omega$, fps (m/s)

$\quad\quad$ = local velocity of fluid, fps (m/s)

u' = turbulent velocity fluctuation in the direction of flow, fps (m/s)

u_* = shear stress velocity or friction velocity = $\sqrt{\tau_0/\rho}$, ft/sec (m/s)

V = mean velocity of fluid, fps (m/s)

$\quad\quad$ = absolute velocity of fluid in hydraulic machines, fps (m/s)

V_c = critical mean velocity of open channel flow, fps (m/s)

V_j = jet velocity, fps (m/s)

$\forall$ = any volume, ft^3 (m^3)

v = specific volume = $1/\rho$, ft^3/slug (m^3/kg)

u, v, w = components of velocity in x, y, z, directions, fps (m/s)

$W =$ Weber number $= V\sqrt{\rho L/\sigma}$ [dimensionless]
$W =$ total weight, lb (N)
$x =$ a distance, usually parallel to flow, ft (m)
$x_c =$ distance from leading edge to point where boundary layer becomes turbulent, ft (m)
$y =$ a distance along a plane in hydrostatics, ft (m)
$\quad =$ total depth of open channel flow, ft (m)
$y_c =$ critical depth of open channel flow, ft (m)
$\quad =$ distance to centroid, ft (m)
$y_h =$ hydraulic (mean) depth $= A/B$, ft (m)
$y_0 =$ depth for uniform flow in open channel (normal depth), ft (m)
$y_p =$ distance to center of pressure, ft (m)
$z =$ elevation above any arbitrary datum plane, ft (m)
α (alpha) $=$ an angle
$\quad =$ kinetic energy correction factor [dimensionless]
β (beta) $=$ an angle
$\quad =$ momentum correction factor [dimensionless]
γ (gamma) $=$ specific weight, lb/ft³ (N/m³)
δ (delta) $=$ thickness of boundary layer, in (mm)
$\quad \delta_v =$ thickness of viscous sublayer in turbulent flow, in (mm)
$\quad \delta_t =$ thickness of transition boundary layer in turbulent flow, in (mm)
ε (epsilon) $=$ kinematic eddy viscosity, ft²/sec (m²/s)
η (eta) $=$ eddy viscosity, lb·sec/ft² (N·s/m²)
$\quad =$ efficiency of hydraulic machine
θ (theta) $=$ any angle
μ (mu) $=$ absolute or dynamic viscosity, lb·sec/ft² (N·s/m²)
ν (nu) $=$ kinematic viscosity $= \mu/\rho$, ft²/sec (m²/s)
ξ (xi) $=$ vorticity, sec⁻¹ (s⁻¹)
Π (pi) $=$ dimensionless parameter
$\quad \pi = 3.141\ 592\ 653\ 59\ldots$
ρ (rho) $=$ density, mass per unit volume $= \gamma/g$, slug/ft³ (kg/m³)
Σ (sigma) $=$ summation
σ (sigma) $=$ surface tension, lb/ft (N/m)
$\quad =$ cavitation parameter in turbomachines [dimensionless]
$\quad =$ submergence of weir $= h_d/h_u$ [dimensionless]
$\quad \sigma_c =$ critical cavitation parameter in turbomachines [dimensionless]
τ (tau) $=$ shear stress, lb/ft² (N/m²)
$\quad \tau_0 =$ shear stress at a wall or boundary, lb/ft² (N/m²)
ϕ (phi) $=$ any function
$\quad =$ peripheral-velocity factor $= u_{\text{periph}}/\sqrt{2gh}$ [dimensionless]
$\quad \phi_e =$ peripheral-velocity factor at point of maximum efficiency [dimensionless]
ω (omega) $=$ angular velocity $= u/r = 2\pi n/60$, rad/sec (rad/s)

Suitable subscripts on variables often indicate their values at specific points. We always assume fluid flows from subscript 1 to subscript 2.

List of Abbreviations

abs = absolute
atm = atmospheric, atmospheres
avg = average
bhp = brake (or shaft) horsepower
Btu = British Thermal Unit
°C = degree celsius
cal = calorie
cfm = cubic feet per minute
cfs = cubic feet per second
cm = centimeter
d = day or days (SI)
°F = degree fahrenheit
fpm = feet per minute
fps = feet per second
ft = foot or feet
g = gram or grams
gal = gallon
gpd = (U.S.) gallons per day
gpm = (U.S.) gallons per minute
h = hour or hours (SI)
ha = hectare
hp = horsepower
hr = hour or hours (BG)
Hz = hertz (cycles per second)
in = inch or inches
J = joules = N·m = W·s
K = kelvin (unit of temperature)
kg = kilograms = 10^3 grams
kgf = kilogram force
kgm = kilogram mass
km = kilometer
L = liter
lb = pounds of force (*not* lbs)
lbf = pound force

$$
\begin{aligned}
\text{lbm} &= \text{pound mass} \\
\text{ln} &= \log_e \\
\text{log} &= \log_{10} \\
\text{m} &= \text{meter or meters} \\
\text{mb} &= \text{millibars} = 10^{-3}\,\text{bar} \\
\text{mb abs} &= \text{millibars, absolute} \\
\text{mgd} &= \text{million (U.S.) gallons per day} \\
\text{min} &= \text{minute or minutes (BG and SI)} \\
\text{mL} &= \text{milliliter} \\
\text{mm} &= \text{millimeters} = 10^{-3}\,\text{meter} \\
\text{mol} &= \text{mole} \\
\text{mph} &= \text{miles per hour} \\
\text{N} &= \text{newton or newtons} = \text{kg·m/s}^2 \\
\text{N/m}^2\,\text{abs} &= \text{newtons per square meter, absolute} \\
\text{oz} &= \text{ounce} \\
\text{P} &= \text{poise} = 0.10\,\text{N·s/m}^2 \\
\text{Pa} &= \text{pascal} = \text{N/m}^2 \\
\text{pcf} &= \text{pounds per cubic foot} \\
\text{psf} &= \text{pounds per square foot} \\
\text{psfa} &= \text{pounds per square foot, absolute} \\
\text{psfg} &= \text{pounds per square foot, gage} \\
\text{psi} &= \text{pounds per square inch} \\
\text{psia} &= \text{pounds per square inch, absolute} \\
\text{psig} &= \text{pounds per square inch, gage} \\
°\text{R} &= \text{degree rankine} \\
\text{rad} &= \text{radians} \\
\text{rev} &= \text{revolutions} \\
\text{rpm} &= \text{revolutions per minute} \\
\text{rps} &= \text{revolutions per second} \\
\text{s} &= \text{second or seconds (SI)} \\
\text{sec} &= \text{second or seconds (BG)} \\
\text{St} &= \text{stoke} = \text{cm}^2/\text{s} \\
\text{W} &= \text{watt or watts} = \text{J/s} \\
\text{y} &= \text{year or years (SI)} \\
\text{yr} &= \text{year or years (BG)}
\end{aligned}
$$

CHAPTER 1

Introduction

In preparing this eleventh edition of *Fluid Mechanics with Civil Engineering Applications*, we added the qualifier "Civil" to the title not only because the concepts and solutions are more pertinent to civil engineers, but to highlight the perspective of the authors as civil engineers, using these concepts in professional practice, research, and teaching throughout our careers. We have strived to present the material in such a way that you, the student, can readily learn the fundamentals of fluid mechanics and see how those fundamentals can be applied to practical engineering problems. Only by understanding the text and working many problems can students master the application of the fundamentals.

1.1   Scope of Fluid Mechanics

Undoubtedly you have observed the movement of clouds in the atmosphere, the flight of birds through the air, the flow of water in streams, and the breaking of waves at the seashore. Fluid mechanics phenomena are involved in all of these. Fluids include gases and liquids, with air and water as the most prevalent. Some of the many other aspects of our lives that involve fluid mechanics are flow in pipelines and channels, movements of air and blood in the body, air resistance or drag, wind loading on buildings, motion of projectiles, jets, shock waves, lubrication, combustion, irrigation, sedimentation, and meteorology and oceanography. The motions of moisture through soils and oil through geologic formations are other applications. A knowledge of fluid mechanics is required to properly design water supply systems, wastewater treatment facilities, dam spillways, valves, flow meters, hydraulic shock absorbers and brakes, automatic transmissions, aircraft, ships, submarines, breakwaters, marinas, rockets, computer disk drives, windmills, turbines, pumps, heating and air conditioning systems, bearings, artificial organs, and even sports items like golf balls, yachts, efficient vehicles, and hang gliders. It is clear that everybody's life is affected by fluid mechanics in a variety of ways. All engineers should have at least a basic knowledge of fluid phenomena.

Fluid mechanics is the science of the mechanics of liquids and gases, and is based on the same fundamental principles that are employed in the mechanics of solids. However, the mechanics of fluids is a more complicated subject than the mechanics of solids, because with solids one deals with separate and tangible elements, while with fluids there are no separate elements to be distinguished.

Fluid mechanics can be divided into three branches: *fluid statics* is the study of the mechanics of fluids at rest; *kinematics* deals with velocities and streamlines without considering forces or energy; and *fluid dynamics* is concerned with the relations between velocities and accelerations and the forces exerted by or upon fluids in motion.

1

Classical *hydrodynamics* is largely a subject in mathematics, since it deals with an imaginary ideal fluid that is completely frictionless. The results of such studies, without consideration of all the properties of real fluids, are of limited practical value. Consequently, in the past, engineers turned to experiments, and from these developed empirical formulas that supplied answers to practical problems. When dealing with liquids, this subject is called *hydraulics.*

Empirical hydraulics was confined largely to water and was limited in scope. With developments in aeronautics, chemical engineering, and the petroleum industry, the need arose for a broader treatment. This has led to the combining of classical hydrodynamics (ideal fluids) with the study of real fluids, both liquids (hydraulics) and gases, and this combination we call *fluid mechanics.* In modern fluid mechanics the basic principles of hydrodynamics are combined with experimental data. The experimental data can be used to verify theory or to provide information supplementary to mathematical analysis. The end product is a unified body of basic principles of fluid mechanics that we can apply to the solution of fluid-flow problems of engineering significance. With the advent of the computer the field of *computational fluid dynamics* developed. Various numerical methods such as finite differences, finite elements, boundary elements, and analytic elements are now used to solve advanced problems in fluid mechanics.

1.2 Historical Sketch of the Development of Fluid Mechanics[1]

From time to time we discover more about the knowledge that ancient civilizations had about fluids, particularly in the areas of irrigation channels and sailing ships. The Romans are well known for their aqueducts and baths, many of which were built in the fourth century B.C.E. In addition, civilizations around the world had developed sophisticated, engineered water storage and conveyance systems many centuries ago, some examples being the Teotihuacan (in what is now Mexico), the Wari (Peru), and Mesopotamian (Iran). Some of these still operate today, and others are being revived for their potential to help current water management respond in a sustainable way to intensified water stress in a disrupted climate.

The Greeks are known to have made quantified measurements, the best known being those of Archimedes who discovered and formulated the principles of buoyancy in the third century B.C.E. We know of no basic improvements to the understanding of flow until Leonardo da Vinci (1452–1519), who performed experiments, investigated, and speculated on waves and jets, eddies and streamlining, and even on flying. He contributed to the one-dimensional equation for conservation of mass.

Isaac Newton (1642–1727), by formulating his laws of motion and his law of viscosity, in addition to developing the calculus, paved the way for many great developments in fluid mechanics. Using Newton's laws of motion, numerous 18th-century mathematicians solved many frictionless (zero-viscosity) flow problems. However, most flows are dominated by viscous effects, so engineers of the 17th and 18th centuries found the inviscid flow solutions unsuitable, and by experimentation they developed empirical equations, thus establishing the science of hydraulics.

Late in the 19th century the importance of dimensionless numbers and their relationship to turbulence was recognized, and dimensional analysis was born. In 1904

---

[1] See also Rouse, H., and S. Ince, *History of Hydraulics*, Dover, New York, 1963.

Ludwig Prandtl published a key paper, proposing that the flow fields of low-viscosity fluids be divided into two zones, namely a thin, viscosity-dominated *boundary layer* near solid surfaces, and an effectively inviscid outer zone away from the boundaries. This concept explained many former paradoxes and enabled subsequent engineers to analyze far more complex flows. However, we still have no complete theory for the nature of turbulence, and so modern fluid mechanics continues to be a combination of experimental results and theory.

While the history of modern fluid mechanics education dates back over a century (e.g., the first edition of this textbook was published in 1916[2]), it took several decades for materials other than print media to be used. In the 1960s many films were created to aid in visualizing fluid phenomena. Fortunately, many of these have been preserved and archived. Two prominent archives are:

- The videos of Dr. Hunter Rouse, Iowa Institute of Hydraulic Research (https:// www.iihr.uiowa.edu/rouse-educational-films/); and

- The videos of Dr. Ascher Shapiro, National Committee for Fluid Mechanics Films. (http://web.mit.edu/hml/ncfmf.html)

1.3   The Book, Its Contents, and How to Study Fluid Mechanics

In this introductory chapter we attempt to give you some insight into what fluid mechanics is all about. In the previous sections we discussed the scope of fluid mechanics and the historical development of the subject. This and the next section explain how to best use this book to study fluid mechanics. The last section of this chapter discusses the importance of dimensions and units.

You can get a feel for the contents of the book and the variety of topics it covers by reviewing the Contents at the front of the text. Most of the subject titles are self-explanatory.

Because problem solving is such an important part of the study of fluid mechanics, before beginning you should make yourself very familiar with the supporting resources available. You will often be expected to know where to find such information, without any direct reference. For convenience, many unit conversions and related data have been collected in the appendices of this book. Many data on material properties, often needed, are collected into the figures and tables of Appendix A; but some are also in the chapters, such as Fig. 2.1. The lists of symbols and abbreviations preceding this chapter are also a useful resource. Appendix B summarizes important information about equations, which form a key part of the language of fluid mechanics. As you progress you will increasingly realize how helpful programming procedures can be in solving many fluid mechanics problems, such as flow in pipes and pipe networks, water surface profiles in open channels and culverts, and unsteady flow problems. The most convenient of these procedures are in mathematics software packages such as Mathcad, in spreadsheets like Excel, and in equation solvers on programmable scientific calculators or in most programming languages (e.g., Python, R, MATLAB); equation solving methods, using tools like these, are mentioned particularly in Chaps. 7 and 8, where the benefit of

---

[2] Daugherty, R.L., *Hydraulics*, McGraw-Hill, 1916. With the fifth edition in 1954 the title became *Fluid Mechanics with Engineering Applications*.

using these tools to solve implicit equations is discussed. Sample problems for equation solvers are helpful and are indicated by the abacus symbol "⚏" preceding the sample problem number. Note, however, that in all instances the use of programming and computers is optional. To help you broaden your horizons by reading books on subjects related to those in this text, a list of such references is provided in Appendix C.

Throughout this book we strive to develop basic concepts in a logical manner so that you can readily read the material and understand it. Material is divided into "building blocks" within separate sections of the chapters. Once the basic concepts are developed, we often provide sample problems to illustrate applications of the concepts; then we usually provide online exercises, which you should perform as needed to reinforce your understanding. The exercises normally address only material in a specific section and are generally quite straightforward. They are drill exercises, to familiarize you with the subject and concepts, and can be accessed at https://www.mhprofessional.com/FMWCEA11-9781264787296.

For each chapter we have a set of summary problems. These are intended to be more like real-world or examination problems, where it is not indicated which section(s) they address. In some instances they may require the application of concepts from a number of sections or even chapters. You will find it a great advantage to have developed your familiarity with the concepts by doing drill exercises before tackling the end-of-chapter problems. Although answers to the exercises are given, answers to the end-of-chapter problems are not. One reason is that many problems in fluid mechanics require trial-and-error solution methods, and having answers reduces learning of such methods. Another is that as you progress in competence, you need to rely on yourself more and learn ways to check yourself; real-world problems do not come with accompanying answers.

As we stressed at the outset, there are two major keys to success in mastering fluid mechanics. The first of these is learning the fundamentals, and this requires that you read and understand the text. There are many phenomena and situations that must be described in words and figures, and that equations and numbers alone cannot explain. So be sure you adequately read (and reread) the text.

The second key to success is working many problems. In this text we stress the application of basic principles to the solution of practical engineering problems. *Only by working many problems can you truly understand the basic principles and how to apply them.* This includes working through the sample problems and exercises, *without referencing the solutions,* to gain confidence with the material. We feel this is very important! Because of this importance, we next include some suggestions that will aid you in solving problems.

1.4   Approach to Problem Solving

Research on successful methods for problem solving in engineering has produced a variety of strategies, many of which follow the general structure illustrated in Fig. 1.1.[3]

Figure 1.1 illustrates the importance of first considering methods, including reviewing pertinent concepts and examining past problems you have solved, before implementing a solution. The last step is essential; your result should be of a magnitude that it is a possible solution. For example, if your calculated flow through a small water pipe exceeds that of the Mississippi River, a flag should be raised. Also, check that any assumptions you made initially are satisfied and appropriate.

---

[3] Woods, D.R., "An Evidence-Based Strategy for Problem Solving," *J. of Engineering Education,* 89(4), 443–459, 2000.

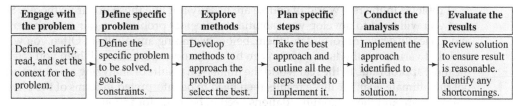

Engage with the problem	Define specific problem	Explore methods	Plan specific steps	Conduct the analysis	Evaluate the results
Define, clarify, read, and set the context for the problem.	Define the specific problem to be solved, goals, constraints.	Develop methods to approach the problem and select the best.	Take the best approach and outline all the steps needed to implement it.	Implement the approach identified to obtain a solution.	Review solution to ensure result is reasonable. Identify any shortcomings.

FIGURE 1.1 A systematic approach to solving engineering problems.

In planning and executing solutions to problems ("Plan specific steps" and "Conduct the analysis" in Fig. 1.1), additional details are often recommended:

 a. Draw a neat figure or figures, fully labeled, of the situation to be analyzed.

 b. State all assumptions you consider necessary.

 c. Reference all principles, equations, tables, etc., that you will use. (Remember all the available supporting resources, mentioned in Sec. 1.3.)

 d. Solve the problem as far as possible algebraically (in terms of symbols) before inserting numbers.

 e. Check the dimensions of the various terms for consistency (per Sec. 1.5).

 f. Insert numerical values[4] for the variables at the last possible stage, using a consistent (SI or BG) set of units (per Sec. 1.5). Evaluate a numerical answer, with units, and report it to an appropriate precision. (This should be no more precise, as a percentage, than that of the least precise inserted value; and however precise the inserted values may be, a common practical rule in engineering is to report results to three significant figures, or to four figures if they begin with a "1," which yields a maximum possible error of 0.25%. Do not round off values in intermediate steps, but only do so when presenting your answer.)

Not only do you need to learn and understand the material, but also you need to *know how and when to use it!* Seek and build understanding of *applications* for your knowledge, particularly to problems that are not straightforward. It is for non-straightforward problems that we need well-trained engineers. Understanding is particularly demonstrated by successful application of the principles to situations different from those you have met before.

Although the preceding emphasizes analysis, which can involve algebra, trial-and-error methods, graphical methods, and calculus, other problem-solving methods such as computer and experimental techniques can be used and should be mastered to a reasonable extent. Become familiar with the use of computers to solve problems by iterative procedures, to perform repetitive numerical evaluations, to perform numerical integration, etc. Also, programmable calculators often have root finders to solve implicit equations, and have integration and graphing capabilities. Familiarity with these will greatly add to your effectiveness in fluid mechanics and as an engineer in general. Take every opportunity to learn about practical issues in the laboratory and on field trips; never forget, as the title of this book reminds us, that all this theory and analysis is for application to the real world.

---

[4] Assume that given values (only) are fully accurate, regardless of the number of significant figures.

Problems in the real world of course are usually not like those in our textbooks. So next you will need to develop your abilities to *recognize* problems in our environment, and to clearly *define* (or formulate) them, before beginning any analysis. Often you will find that various methods of solution can be used, and experience will help you select the most appropriate. In the real world the numerical results of analyzing a problem are not the ultimate goal; for those results then need to be interpreted in terms of the physical problem, and then recommendations need to be made for action.

Remain conscious of your goal, to become a capable and responsible engineer, and remain conscious of your path to that goal, which involves the many steps we have outlined here.

1.5 Dimensions and Units

To properly define a physical property or a fluid phenomenon, one must express the property or phenomenon in terms of some set of units. For example, the diameter of a pipe might be 160 millimeters and the average flow velocity 8 meters per second.[5] A different set of units might have been used, such as a diameter of 0.16 meter and a velocity of 800 centimeters per second. Or, the diameter and velocity might have been expressed in English (U.S. Customary) or other units.

In this book we use two systems of units: the British Gravitational (BG) system when dealing with English units, and the SI (Système Internationale d'Unités) when dealing with metric units. The SI was adopted in 1960 at the Eleventh General International Conference on Weights and Measures, at which the United States was represented. As of the year 2000, nearly every major country in the world, except the United States, was using the SI. Because of the intransigence on this issue in the United States, and because English units have been used in the technical literature for so many years it is necessary that engineers be familiar with *both* the systems, BG and SI, used in this book.

In fluid mechanics the basic dimensions are length (L), mass (M), time (T), force (F), and temperature (θ). In order to satisfy Newton's second law, $F = ma = MLT^{-2}$, where acceleration a is expressed by its basic dimensions as LT^{-2}, we note that units for only three of the first four of these dimensions can be assigned arbitrarily; the fourth unit must agree with the other three, and is therefore known as a **derived unit**. In the two systems of units used in this book, the commonly used units for the five basic dimensions mentioned are:

Dimension	BG Unit	SI Unit
Length (L)	Foot (ft)	Meter (m)
Mass (M)	Slug* ($= \text{lb·sec}^2/\text{ft}$)	Kilogram (kg)
Time (T)	Second (sec)	Second (s)
Force (F)	Pound (lb)	Newton* (N) ($= \text{kg·m/s}^2$)
Temperature (θ)		
Absolute	Rankine (°R)	Kelvin (K)
Ordinary	Fahrenheit (°F)	Celsius (°C)

*Derived units.

---

[5] This book uses the American spelling *meter*, although the official spelling is *metre*.

The SI employs L, M, and T and derives F from MLT^{-2}. Force in the SI is defined by the **newton,** the force required to accelerate one kilogram of mass at a rate of one meter per second per second; that is,

$$1 \text{ N} = (1 \text{ kg})(1 \text{ m/s}^2)$$

On the other hand, the BG system, also sometimes known as the U.S. Customary (USC) system, employs L, F, and T, and derives M from $F/a = FL^{-1}T^2$. The BG unit of mass, the **slug,** is therefore defined as that mass that accelerates at one foot per second per second when acted upon by a force of one pound; that is,

$$1 \text{ slug} = (1 \text{ lb})/(1 \text{ ft/sec}^2) = 1 \text{ lb·sec}^2/\text{ft}$$

or $$1 \text{ lb} = (1 \text{ slug})(1 \text{ ft/sec}^2)$$

When working in the BG system, it often pays to keep mass expressed in basic units (lb·sec²/ft or weight/gravitational acceleration) for as long as possible.

We see that the definition of mass in the BG system depends on the definition of one **pound,**[6] which is the force of gravity acting on (or weight of) a platinum standard whose mass is 0.453 592 43 kg. **Weight** is the gravitational attraction force F between two bodies, of masses m_1 and m_2, given by Newton's Law of Gravitation as

$$F = G \frac{m_1 m_2}{r^2}$$

where G is the universal constant of gravitation and r is the distance between the centers of the two masses. If m_1 is the mass m of an object on the earth's surface and m_2 is the mass M of the earth then r is the radius of the earth, so that

$$F = m \left(\frac{GM}{r^2} \right)$$

and the weight of the object is

$$W = mg$$

where the gravitational acceleration $g = GM/r^2$. Clearly g varies slightly with altitude and latitude on earth, since the earth is not truly spherical, while in space and on other planets it is much different. Furthermore, the preceding does not take into account the earth's rotation, which by centrifugal action reduces the apparent weight of an object by at most 0.35% at the equator. Because the force (weight) depends on the value of g, which in turn varies with location, a system such as the BG system based on length (L), force (F), and time (T) is referred to as a **gravitational system.** On the other hand, systems like the SI, which are based on length (L), mass (M), and time (T), are **absolute** because they are *independent* of the gravitational acceleration g.

---

[6] We are using the common avoirdupois pound (7000 grains), not the Troy pound (5760 grains) that is mostly used for precious metals and by pharmacies.

A partial list of derived quantities encountered in fluid mechanics and their commonly used dimensions in terms of L, M, T, and F is:

Quantity	Commonly Used Dimensions	BG Unit	SI Unit
Acceleration (a)	LT^{-2}	ft/sec²	m/s²
Area (A)	L^2	ft²	m²
Density (ρ)	ML^{-3}	slug/ft³	kg/m³
Energy, work, or quantity of heat	FL	ft·lb	N·m = J
Flowrate (Q)	L^3T^{-1}	ft³/sec (cfs)	m³/s
Frequency	T^{-1}	cycle/sec (sec⁻¹)	Hz (hertz, s⁻¹)
Kinematic viscosity (ν)	L^2T^{-1}	ft²/sec	m²/s
Power (P)	FLT^{-1}	ft·lb/sec	N·m/s = W
Pressure (p)	FL^{-2}	lb/in² (psi)	N/m² = Pa
Specific weight (γ)	FL^{-3}	lb/ft³ (pcf)	N/m³
Velocity (V)	LT^{-1}	ft/sec (fps)	m/s
Viscosity (μ)	FTL^{-2}	lb·sec/ft²	N·s/m²
Volume ($\forall$)	L^3	ft³	m³

Using the identity $F = MLT^{-2}$, all dimensions containing an F could have been expressed using an M instead, and vice versa. Other derived quantities will be dealt with when they are encountered in the text. Radians do not have dimensions, because they are defined as an arc length divided by a radius.

On the earth's surface the variation in g is small, and, by international agreement, standard gravitational acceleration at sea level is 32.1740 ft/sec² or 9.806 65 m/s² (for problem solving we usually use 32.2 ft/sec² or 9.81 m/s²). So variations in g are generally not considered in this text as long as we are analyzing problems on the earth's surface. Fluid problems for other locations, such as on the moon, where g is quite different from that on earth, can be handled by the methods presented in this text if proper consideration is given to the value of g.

For unit mass (1 slug or 1 kg) on the earth's surface, we note that:

- In BG units: $W = mg = (1 \text{ slug})(32.2 \text{ ft/sec}^2) = 32.2 \text{ lb}$;
- In SI units: $W = mg = (1 \text{ kg})(9.81 \text{ m/s}^2) = 9.81 \text{ N}$.

Other systems of units used elsewhere include the English Engineering (EE) system, the Absolute Metric (cgs) system, and the mks metric system. The EE system uses pound force (lbf) and pound mass (lbm), and the mks metric system uses kilogram force (kgf) and kilogram mass (kgm). As a result, both of these are said to be *inconsistent systems,* because unit force does not cause unit mass to undergo unit acceleration; they require an additional proportionality constant or conversion factor. The SI and BG systems used in this book are *consistent systems* having conversion factors with a magnitude of one. Although the cgs metric system is both consistent and nongravitational, it is little used for engineering applications because its unit of force, the dyne, is so small; 1 dyne = (1 g)(1 cm/s²) = 10^{-5} N.

Do not be confused by popular usage of kilograms to measure weight (force). When European shoppers buy a kilo of sugar, say, in our terms they are buying sugar with a mass of 1 kg, in effect defining a force of 1 kg (1 kgf) = (1 kgm)(9.81 m/s^2), which is equivalent to 9.81 N. Because a 1-lb weight has a mass of about 0.4536 kg, the shoppers' conversion factor is 1.0/0.4536 = 2.205 lb/kgf. In engineering we are careful to distinguish between mass and weight, reserving kg for mass and using newtons for force in the SI system.

In this book we shall use the abbreviation kg for kilogram mass, and lb for pound force. The abbreviation lb for pound is taken from the Latin *libra*, plural *librae*, so the correct plural abbreviation is lb not lbs. The units second, minute, hour, day, and year are correctly abbreviated as s, min, h, d, and y in the SI system, and although in the BG system they should be abbreviated as sec, min, hr, day, and yr, it is common to use the SI abbreviations for both systems. There are many "nonstandard" or traditional abbreviations used by engineers, such as fps for ft/sec, gpm for gal/min, and cfs for ft^3/sec (also sometimes referred to as the second-foot and the cusec). The more common of these are included in the list just preceding this chapter. Acres, tons, and slugs are not abbreviated. When units are named after people, like the newton (N), joule (J), and pascal (Pa), they are capitalized when abbreviated but not capitalized when spelled out. The abbreviation capital L for liter is a special case, used to avoid confusion with one (1). Also note that in the SI the unit for absolute temperature measurement is the degree kelvin, which is abbreviated K *without* a degree (°) symbol.

The British or imperial gallon is, within 0.1%, equal to 1.2 U.S. gallons. Where the kind of gallons is not specified, in this book assume them to be U.S. gallons.

When dealing with unusually large or very small numbers, a series of prefixes has been adopted for use with SI units. The most commonly used prefixes are given for convenient reference in Appendix D. Hence Mg (megagram) represents 10^6 grams, mm (millimeter) represents 10^{-3} meters, and kN (kilonewton) represents 10^3 newtons, for example. Note that multiples of 10^3 are preferred in engineering usage; other multiples like cm are to be avoided if possible. Also, in the SI it is conventional to separate sequences of digits into groups of three by spaces rather than by commas, as was done earlier for the mass of the standard pound. Thus, 10 cubic meters of water weigh 98 100 N or 98.1 kN.

Often we need to convert quantities from BG units into SI units, and vice versa. Because time units are the same in both systems, we only need to convert units of length, and force or mass, from which all other units can then be derived. For length, by definition, one foot is *exactly* 0.3048 meters, and so an inch is *exactly* 25.4 mm. For force, using $W = mg$ and definitions given earlier, 1 lb = (0.453 592 43 kg)(9.806 65 m/s^2), or about 4.448 N. For mass, 1 slug = (1 lb)/(1 ft/sec^2) is about equal to (4.448 N)/ (0.3048 m/s^2) = 14.59 kg. Conversion factors for many other units, derived from these three basic ones, are given for convenience in tables in Appendix D of this book; *exact* conversion factors are indicated by an asterisk (*). These tables include conversions of units *within* the BG system and *within* the SI. They include some definitions, other useful conversions, and relations between the four principal temperature scales.

In the SI, lengths are commonly expressed in millimeters (mm), centimeters (cm; try to avoid), meters (m), or kilometers (km), depending on the distance being measured. A meter is about 39 inches or 3.3 ft and a kilometer is approximately five-eighths of a mile. Areas are usually expressed in square centimeters (cm^2), square meters (m^2), or *hectares* (100 m × 100 m = 10^4 m^2), depending on the area being measured. The hectare, used for measuring large areas, is equivalent to about 2.5 acres. A newton is equivalent to almost 0.225 lb. The SI unit of stress (or pressure), newton per square meter (N/m²), is known as the **pascal** (Pa), and is equivalent to about 0.021 lb/ft² or 0.00015 lb/in². In SI units, energy, work, or quantity of heat are ordinarily expressed in joules (J). A *joule*[7] is equal to a newton-meter, i.e., J = N·m. The unit of power is the **watt** (W), which is equivalent to a joule per second, i.e., W = J/s = N·m/s.

When we have to work with less usual units, like centipoise (for viscosity) or ergs (for energy), it is best to convert them into SI or BG units as soon as possible.

Sample Problem 1.1

Bernoulli's equation for the flow of an ideal fluid, which is discussed in Chap. 5, can be written

$$\frac{p}{\gamma} + z + \frac{V^2}{2g} = \text{constant} \qquad (5.7)$$

where p = pressure, γ = specific weight, z = elevation, V = mean flow velocity, and g = acceleration of gravity. Demonstrate that this equation is dimensionally homogeneous, i.e., that all terms have the same dimensions.

Solution

Term 1:   Dimensions of $\dfrac{p}{\gamma} = \dfrac{F/L^2}{F/L^3} = L$

Term 2:   Dimensions of $z = L$

Term 3:   Dimensions of $\dfrac{V^2}{2g} = \dfrac{(L/T)^2}{L/T^2} = L$

So all the terms have the same dimensions, L, which must also be the dimensions of the constant at the right-hand side of Eq. (5.7).   *ANS*

---

[7] Joule is pronounced (jo͞ol) to rhyme with cool.

Sample Problem 1.2

Convert 200 Btu to (*a*) BG, (*b*) SI, and (*c*) cgs metric units of energy.

Solution

From Appendix D:

$$1 \text{ Btu} = 778 \text{ ft·lb}, \quad 1 \text{ ft·lb} = 1.356 \text{ N·m} = 1.356 \text{ J}, \quad 1 \text{ N} = 10^5 \text{ dyne}.$$

(*a*) For BG units: $200 \text{ Btu} = 200 \text{ Btu}\left(\dfrac{778 \text{ ft·lb}}{1 \text{ Btu}}\right) = 155{,}600 \text{ ft·lb}.$ *ANS*

(*b*) For SI units: $155{,}600 \text{ ft·lb} = 155{,}600 \text{ ft·lb}\left(\dfrac{1.356 \text{ N·m}}{1 \text{ ft·lb}}\right)$

$$= 210\,994 \text{ N·m} = 211 \text{ kN·m} = 211 \text{ kJ}. \quad \textit{ANS}$$

(*c*) For cgs units: $210\,994 \text{ N·m} = 211 \times 10^3 \text{ N·m}\left(\dfrac{10^5 \text{ dyne}}{1 \text{ N}}\right)\left(\dfrac{10^2 \text{ cm}}{\text{m}}\right)$

$$= 211 \times 10^{10} \text{ dyne·cm} = 211 \times 10^{10} \text{ erg.} \quad \textit{ANS}$$

CHAPTER 2

Properties of Fluids

In this chapter we discuss a number of fundamental properties of fluids. An understanding of these properties is essential for us to apply basic principles of fluid mechanics to the solution of practical problems.

2.1 Distinction Between a Solid and a Fluid

The molecules of a *solid* are usually closer together than those of a *fluid.* The attractive forces between the molecules of a solid are so large that a solid tends to retain its shape. This is not the case for a fluid, where the attractive forces between the molecules are smaller. An ideal elastic solid will deform under load and, once the load is removed, will return to its original state. Some solids are plastic. These deform under the action of a sufficient load and the deformation continues as long as a load is applied, provided the material does not rupture. Deformation ceases when the load is removed, but the plastic solid does not return to its original state.

The intermolecular cohesive forces in a fluid are not great enough to hold the various elements of the fluid together. Hence a fluid will flow under the action of the slightest stress and flow will continue as long as the stress is present.

2.2 Distinction Between a Gas and a Liquid

A fluid may be either a *gas* or a *liquid.* The molecules of a gas are much farther apart than those of a liquid. Hence a gas is very compressible, and when all external pressure is removed, it tends to expand indefinitely. A gas is therefore in equilibrium only when it is completely enclosed. A liquid is relatively incompressible, and if all pressure, except that of its own vapor pressure, is removed, the cohesion between molecules holds them together, so that the liquid does not expand indefinitely. Therefore, a liquid may have a free surface, i.e., a surface from which all pressure is removed, except that of its own vapor.

A *vapor* is a gas whose temperature and pressure are such that it is very near the liquid phase. Thus steam is considered a vapor because its state is normally not far from that of water. A gas may be defined as a highly superheated vapor; that is, its state is far removed from the liquid phase. Thus air is considered a gas because its state is normally very far from that of liquid air.

The volume of a gas or vapor is greatly affected by changes in pressure or temperature or both. It is usually necessary, therefore, to take account of changes in volume and temperature in dealing with gases or vapors. Whenever significant temperature or phase changes are involved in dealing with vapors and gases, the subject is largely

13

dependent on heat phenomena (***thermodynamics***). Thus fluid mechanics and thermodynamics are interrelated.

2.3   Density, Specific Weight, Specific Volume, and Specific Gravity

The ***density*** ρ (rho),[1] or more strictly, ***mass density***, of a fluid is its *mass* per unit volume, while the ***specific weight*** γ (gamma) is its *weight* per unit volume. In the British Gravitational (BG) system (Sec. 1.5) density ρ will be in slugs per cubic foot [kg/m^3 in SI units (Système Internationale d'Unités)], which can also be expressed as units of $lb \cdot sec^2/ft^4$ ($N \cdot s^2/m^4$ in SI units) (Sec. 1.5).

Specific weight γ represents the force exerted by gravity on a unit volume of fluid, and therefore must have the units of force per unit volume, such as pounds per cubic foot (N/m^3 in SI units).

Density and specific weight of a fluid are related as:

$$\rho = \frac{\gamma}{g} \quad \text{or} \quad \gamma = \rho g \tag{2.1}$$

Since the physical equations are dimensionally homogeneous, the dimensions of density are

$$\text{Dimensions of } \rho = \frac{\text{dimensions of } \gamma}{\text{dimensions of } g} = \frac{lb/ft^3}{ft/sec^2} = \frac{lb \cdot sec^2}{ft^4} = \frac{mass}{volume} = \frac{slugs}{ft^3}$$

In SI units

$$\text{Dimensions of } \rho = \frac{\text{dimensions of } \gamma}{\text{dimensions of } g} = \frac{N/m^3}{m/s^2} = \frac{N \cdot s^2}{m^4} = \frac{mass}{volume} = \frac{kg}{m^3}$$

Note that density ρ is absolute, since it depends on mass, which is independent of location. Specific weight γ, on the other hand, is not absolute, since it depends on the value of the gravitational acceleration g, which varies with location, primarily latitude and elevation above mean sea level.

Densities and specific weights of fluids vary with temperature. Appendix A provides commonly needed temperature variations of these quantities for water and air. It also contains densities and specific weights of common gases at standard atmospheric pressure and temperature. We shall discuss the specific weight of liquids further in Sec. 2.6.

Specific volume v is the volume occupied by a unit mass of fluid.[2] We commonly apply it to gases, and usually express it in cubic feet per slug (m^3/kg in SI units). Specific volume is the reciprocal of density. Thus

$$v = \frac{1}{\rho} \tag{2.2}$$

---

[1] The names of Greek letters are given in the List of Symbols.
[2] Note that in this book we use a "rounded" lower case v (vee), to help distinguish it from a capital V and from the Greek ν (nu).

Specific gravity s of a liquid is the dimensionless ratio

$$s_{\text{liquid}} = \frac{\rho_{\text{liquid}}}{\rho_{\text{water at standard temprature}}}$$

Physicists use 4°C (39.2°F) as the standard, but engineers often use 60°F (15.56°C). In the metric system the density of water at 4°C is 1.00 g/cm³ (or 1.00 g/mL),[3] equivalent to 1000 kg/m³, and hence the specific gravity (which is dimensionless) of a liquid has the same numerical value as its density expressed in g/mL or Mg/m³. Appendix A contains information on specific gravities and densities of various liquids at standard atmospheric pressure.

The specific gravity of a gas is the ratio of its density to that of either hydrogen or air at some specified temperature and pressure, but there is no general agreement on these standards, and so we must explicitly state them in any given case.

Since the density of a fluid varies with temperature, we must determine and specify specific gravities at particular temperatures.

Sample Problem 2.1

The specific weight of water at ordinary pressure and temperature is 62.4 lb/ft³. The specific gravity of mercury is 13.56. Compute the density of water and the specific weight and density of mercury.

Solution

$$\rho_{\text{water}} = \frac{\gamma_{\text{water}}}{g} = \frac{62.4 \text{ lb/ft}^3}{32.2 \text{ ft/sec}^2} = 1.938 \text{ slugs/ft}^3 \qquad \textit{ANS}$$

$$\gamma_{\text{mercury}} = s_{\text{mercury}}\gamma_{\text{water}} = 13.56(62.4) = 846 \text{ lb/ft}^3 \qquad \textit{ANS}$$

$$\rho_{\text{mercury}} = s_{\text{mercury}}\rho_{\text{water}} = 13.56(1.938) = 26.3 \text{ slugs/ft}^3 \qquad \textit{ANS}$$

Sample Problem 2.2

The specific weight of water at ordinary pressure and temperature is 9.81 kN/m³. The specific gravity of mercury is 13.56. Compute the density of water and the specific weight and density of mercury.

Solution

$$\rho_{\text{water}} = \frac{9.81 \text{ kN/m}^3}{9.81 \text{ m/s}^2} = 1.00 \text{ Mg/m}^3 = 1.00 \text{ g/mL} \qquad \textit{ANS}$$

$$\gamma_{\text{mercury}} = s_{\text{mercury}}\gamma_{\text{water}} = 13.56(9.81) = 133.0 \text{ kN/m}^3 \qquad \textit{ANS}$$

$$\rho_{\text{mercury}} = s_{\text{mercury}}\rho_{\text{water}} = 13.56(1.00) = 13.56 \text{ Mg/m}^3 \qquad \textit{ANS}$$

---

[3] One cubic centimeter (cm³) is equivalent to one milliliter (mL).

2.4   Compressible and Incompressible Fluids

Fluid mechanics deals with both incompressible and compressible fluids, that is, with liquids and gases of either constant or variable density. Although there is no such thing in reality as an incompressible fluid, we use this term where the change in density with pressure is so small as to be negligible. This is usually the case with liquids. We may also consider gases to be incompressible when the pressure variation is small compared with the absolute pressure.

Ordinarily we consider liquids to be incompressible fluids, yet sound waves, which are really pressure waves, travel through them. This is evidence of the elasticity of liquids. In problems involving water hammer (Sec. 10.6) we must consider the compressibility of the liquid.

The flow of air in a ventilating system is a case where we may treat a gas as incompressible, for the pressure variation is so small that the change in density is of no importance. But for a gas or steam flowing at high velocity through a long pipeline, the drop in pressure may be so great that we cannot ignore the change in density. For an airplane flying at speeds below 250 mph (100 m/s), we may consider the air to be of constant density. But as an object moving through the air approaches the velocity of sound, which is of the order of 760 mph (1200 km/h) depending on temperature, the pressure and the density of the air adjacent to the body become materially different from those of the air at some distance away, and we must then treat the air as a compressible fluid.

2.5   Compressibility of Liquids

The compressibility (change in volume due to change in pressure) of a liquid is inversely proportional to its **volume modulus of elasticity**, also known as the **bulk modulus**. This modulus is defined as

$$E_v = -v\frac{dp}{dv} = -\left(\frac{v}{dv}\right)dp$$

where v = specific volume and p = pressure. As v/dv is a dimensionless ratio, the units of E_v and p are identical. The bulk modulus is analogous to the modulus of elasticity for solids; however, for fluids it is defined on a volume basis rather than in terms of the familiar one-dimensional stress–strain relation for solid bodies.

In most engineering problems, the bulk modulus at or near atmospheric pressure is the one of interest. The bulk modulus is a property of the fluid and for liquids is a function of temperature and pressure. A few values of the bulk modulus for water are given in Table 2.1. At any temperature we see that the value of E_v increases continuously with pressure, but at any one pressure the value of E_v is a maximum at about 120°F (50°C). Thus water has a minimum compressibility at about 120°F (50°C).

Note that we often specify applied pressures, such as those in Table 2.1, in absolute terms, because atmospheric pressure varies. The unit psia or kN/m² abs indicates absolute pressure, which is the actual pressure on the fluid, relative to absolute zero. The standard atmospheric pressure at sea level is about 14.7 psia or 101.3 kN/m² abs

Pressure, psia	Temperature, °F				
	32°	68°	120°	200°	300°
15	293,000	320,000	333,000	308,000	
1,500	300,000	330,000	342,000	319,000	248,000
4,500	317,000	348,000	362,000	338,000	271,000
15,000	380,000	410,000	426,000	405,000	350,000

[a] To transform these values to meganewtons per square meter, multiply them by 0.006 895. The values in the first line are for conditions close to normal atmospheric pressure; for a more complete set of values at normal atmospheric pressure, see Table A.1 in Appendix A. The five temperatures are equal to 0, 20, 48.9, 93.3, and 148.9°C, respectively.

TABLE 2.1 Bulk modulus of water E_v, psi[a]

(1013 mb abs) (see Sec. 2.9 and Table A.3). Bars and millibars were previously used in metric systems to express pressure; 1 mb = 100 N/m². We measure most pressures relative to the atmosphere and call them gage pressures. We explain this more fully in Sec. 3.4.

The volume modulus, E_v, of mild steel is about 26,000,000 psi (170 000 MN/m²). Taking a typical value for the volume modulus of cold water to be 320,000 psi (2200 MN/m²), we see that water is about 80 times as compressible as steel. The compressibility of liquids covers a wide range. Mercury, for example, is approximately 8% as compressible as water, while the compressibility of nitric acid is nearly six times greater than that of water.

In Table 2.1 we see that at any one temperature the bulk modulus of water does not vary a great deal for a moderate range in pressure. By rearranging the definition of E_v, as an approximation we may use for the case of a fixed mass of liquid at constant temperature

$$\frac{\Delta v}{v} \approx -\frac{\Delta p}{E_v} \tag{2.3a}$$

or

$$\frac{v_2 - v_1}{v_1} \approx -\frac{p_2 - p_1}{E_v} \tag{2.3b}$$

where E_v is the mean value of the modulus for the pressure range and the subscripts 1 and 2 refer to the before and after conditions respectively.

Assuming E_v to have a value of 320,000 psi, we see that increasing the pressure of water by 1000 psi will compress it only $\frac{1}{320}$, or about 0.3%, of its original volume. Therefore we find that the usual assumption regarding water as being incompressible is justified.

Sample Problem 2.3

At a depth of 8 km in the ocean the pressure is 81.8 MPa. Assume that the specific weight of seawater at the surface is 10.05 kN/m³ and that the average volume modulus is 2.34×10^9 N/m² for that pressure range. (*a*) What will be the change in specific volume between that at the surface and at that depth? (*b*) What will be the specific volume at that depth? (*c*) What will be the specific weight at that depth?

Solution

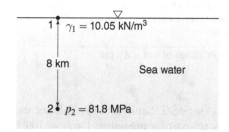

(*a*) Eq. (2.2): $v_1 = 1/\rho_1 = g/\gamma_1 = 9.81/10\,050 = 0.000\,976$ m³/kg

Eq. (2.3*a*): $\Delta v = -0.000\,976(81.8 \times 10^6 - 0)/(2.34 \times 10^9)$

$$= -34.1 \times 10^{-6} \text{ m}^3/\text{kg} \qquad\qquad ANS$$

(*b*) Eq. (2.3*b*): $v_2 = v_1 + \Delta v = 0.000\,942$ m³/kg   *ANS*

(*c*) $\gamma_2 = g/v_2 = 9.81/0.000\,942 = 10\,410$ N/m³   *ANS*

2.6   Specific Weight of Liquids

The specific weights γ of some common liquids at 68°F (20°C) and standard sea-level atmospheric pressure[4] with $g = 32.2$ ft/sec² (9.81 m/s²) are given in Table 2.2. The specific weight of a *liquid* varies only slightly with pressure, depending on the bulk modulus of the liquid (Sec. 2.5); it also depends on temperature, and the variation may be considerable. Since specific weight γ is equal to ρg, the specific weight of a *fluid* depends on the local value of the acceleration of gravity in addition to the variations with temperature and pressure. The variation of the specific weight of water with temperature and pressure, where $g = 32.2$ ft/sec² (9.81 m/s²), is shown in Fig. 2.1. The presence of dissolved air, salts in solution, and suspended matter will increase these values a very slight amount. Ordinarily we assume ocean water to weigh 64.0 lb/ft³ (10.1 kN/m³). Unless otherwise specified or implied by a given temperature, the value to use for water in the problems in this book is $\gamma = 62.4$ lb/ft³ (9.81 kN/m³). Under extreme conditions the specific weight of water is quite different. For example, at 500°F (260°C) and 6000 psi (42 MN/m²) the specific weight of water is 51 lb/ft³ (8.0 kN/m³).

---

[4] See Secs. 2.9 and 3.5.

	lb/ft³	**kN/m³**
Carbon tetrachloride	99.4	15.6
Ethyl alcohol	49.3	7.76
Gasoline	42	6.6
Glycerin	78.7	12.3
Kerosene	50	7.9
Motor oil	54	8.5
Seawater	63.9	10.03
Water	62.3	9.79

TABLE 2.2   Specific weights γ of common liquids at 68°F (20°C), 14.7 psia (1013 mb abs) with $g = 32.2$ ft/sec² (9.81 m/s²)

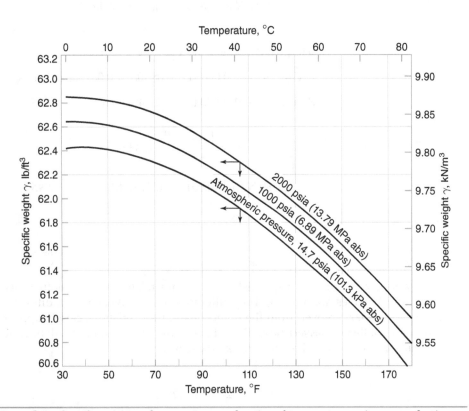

FIGURE 2.1   Specific weight γ of pure water as a function of temperature and pressure for the condition where $g = 32.2$ ft/sec² (9.81 m/s²).

Sample Problem 2.4

A vessel contains 85 L of water at 10°C and atmospheric pressure. If the water is heated to 70°C, what will be the percentage change in its volume? What weight of water must be removed to maintain the volume at its original value? Use Appendix A.

Solution

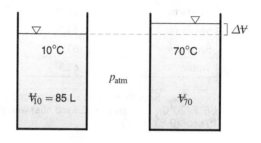

Volume,
$$\mathcal{V}_{10} = 85\,L = 0.085\ \text{m}^3$$

Table A.1:
$$\gamma_{10} = 9.804\ \text{kN/m}^3, \quad \gamma_{70} = 9.589\ \text{kN/m}^3$$

Weight of water, $W = \gamma \mathcal{V} = \gamma_{10}\mathcal{V}_{10} = \gamma_{70}\mathcal{V}_{70}$

i.e.,
$$9.804(0.085)\text{kN} = 9.589\ \mathcal{V}_{70}; \quad \mathcal{V}_{70} = 0.08691\ \text{m}^3$$

$$\Delta\mathcal{V} = \mathcal{V}_{70} - \mathcal{V}_{10} = 0.08691 - 0.08500 = 0.001906\ \text{m}^3 \text{ at } \gamma_{70}$$

$$\Delta\mathcal{V}/\mathcal{V}_{10} = 0.001906/0.085 = 2.24\% \text{ increase} \qquad \textit{ANS}$$

Must remove (at γ_{70}):
$$W\left(\frac{\Delta\mathcal{V}}{\mathcal{V}_{70}}\right) = \gamma_{70}\Delta\mathcal{V}$$

$$= (9589\ \text{N/m}^3)(0.001906\ \text{m}^3) = 18.27\ \text{N} \qquad \textit{ANS}$$

2.7   Property Relations for Perfect Gases

The various properties of a gas, listed below, are related to one another (see, e.g., Appendix A, Tables A.2 and A.5). They differ for each gas. When the conditions of most real gases are far removed from the liquid phase, these relations closely approximate those of hypothetical **perfect gases.** Perfect gases are, here (and often), defined to have constant specific heats and to obey the **perfect-gas law,**

$$\frac{p}{\rho} = pv = RT \tag{2.4a}$$

or

$$p\mathcal{V} = mRT \tag{2.4b}$$

where p = absolute pressure (Sec. 3.4)

ρ = density (mass per unit volume, total mass/total volume, $m/V\!\!\!\!-$)

v = specific volume (volume per unit mass, $= 1/\rho$)

R = a gas constant, which depends on the gas (see Table A.5)

T = absolute temperature in degrees Rankine or Kelvin[5]

For air, the value of R is 1715 ft·lb/(slug·°R) or 287 N·m/(kg·K) (Appendix A, Table A.5); making use of the definitions of a slug and a newton (Sec. 1.5), these units are sometimes given as $ft^2/(sec^2·°R)$ and $m^2/(s^2·K)$, respectively. Since $\gamma = \rho g$, Eq. (2.4) can also be written

$$\gamma = \frac{gp}{RT} \tag{2.5}$$

from which the specific weight of any gas at any temperature and pressure can be computed if R and g are known. Because Eqs. (2.4) and (2.5) relate the various gas properties at a particular state, they are known as *equations of state* and as *property relations.*

In this book we shall assume that all gases are perfect. Perfect gases are sometimes also called *ideal gases.* Do not confuse a perfect (ideal) gas with an ideal fluid (Sec. 2.10).

Avogadro's law states that all gases at the same temperature and pressure under the action of a given value of g have the same number of molecules per unit of volume, from which it follows that the specific weight of a gas[6] is proportional to its molar mass. Thus, if M denotes *molar mass* (formerly called *molecular weight*), $\gamma_2/\gamma_1 = M_2/M_1$ and, from Eq. (2.5), $\gamma_2/\gamma_1 = R_1/R_2$ for the same temperature, pressure, and value of g. Hence for a perfect gas

$$M_1 R_1 = M_2 R_2 = \text{constant} = R_0$$

R_0 is known as the *universal gas constant,* and has a value of 49,709 ft·lb/(slug-mol·°R) or 8312 N·m/(kg-mol·K). Rewriting the preceding equation in the form

$$R = \frac{R_0}{M}$$

enables us to obtain any gas constant R required for Eq. (2.4) or (2.5).

For real (nonperfect) gases, the specific heats may vary over large temperature ranges, and the right-hand side of Eq. (2.4) is replaced by zRT, so that $R_0 = MzR$, where z is a compressibility factor that varies with pressure and temperature. Values of z and R are given in thermodynamics texts and in handbooks. However, for normally encountered monatomic and diatomic gases, z varies from unity by less than 3%, so the perfect-gas idealizations yield good approximations, and Eqs. (2.4) and (2.5) will give good results.

When various gases exist as a mixture, as in air, *Dalton's law of partial pressures* states that each gas exerts its own pressure as if the other(s) were not present. Hence it is the

---

[5] Absolute temperature is measured above absolute zero. This occurs on the Fahrenheit scale at −459.67°F (0° Rankine) and on the Celsius scale at −273.15°C (0 Kelvin). Except for low-temperature work, these values are usually taken as −460°F and −273°C. Remember that no degree symbol is used with Kelvin.
[6] The specific weight of air (molar mass ≈ 29.0) at 68°F (20°C) and 14.7 psia (1013 mb abs) with $g = 32.2$ ft/sec² (9.81 m/s²) is 0.0752 lb/ft³ (11.82 N/m³).

partial pressure of each that we must use in Eqs. (2.4) and (2.5) (see Sample Problem 2.5). Water vapor as it naturally occurs in the atmosphere has a low partial pressure, so we may treat it as a perfect gas with $R = 49{,}709/18 = 2760$ ft·lb/(slug·°R) [462 N·m/(kg·K)]. But for steam at higher pressures this value is not applicable.

As we increase the pressure and simultaneously lower the temperature, a gas becomes a vapor, and as gases depart more and more from the gas phase and approach the liquid phase, the property relations become much more complicated than Eq. (2.4), and we must then obtain specific weight and other properties from vapor tables or charts. Such tables and charts exist for steam, ammonia, sulfur dioxide, freon, and other vapors in common engineering use.

Another fundamental equation for a perfect gas is

$$pv^n = p_1 v_1^n = \text{constant} \tag{2.6a}$$

or

$$\frac{p}{p_1} = \left(\frac{\rho}{\rho_1}\right)^n = \text{constant} \tag{2.6b}$$

where p is absolute pressure, $v\,(= 1/\rho)$ is specific volume, ρ is density, and n may have any nonnegative value from zero to infinity, depending on the process to which the gas is subjected. Since this equation describes the change of the gas properties from one state to another for a particular process, we call it a *process equation.* For a process of change for a given mass of a gas (thus, m and R remain constant), Eq. (2.4) becomes

$$\frac{p V}{T} = \frac{p_1 V_1}{T_1} = \text{constant} \tag{2.7}$$

If the process of change is at a constant temperature (*isothermal*), $n = 1$ in Eq. (2.6a), and T is constant in Eq. (2.7), resulting in Eq. (2.8).

Isothermal case: $pV = p_1 V_1 = \text{constant}$ (2.8)

If there is no heat transfer to or from the gas, the process is *adiabatic.* A frictionless (and reversible) adiabatic process is an *isentropic* process, for which we denote n by k, where $k = c_p/c_v$, the ratio of specific heat at constant pressure to that at constant volume. This *specific heat ratio* k is also called the *adiabatic exponent.* For expansion with friction n is less than k, and for compression with friction n is greater than k. Values for k are given in Appendix A, Table A.5, and in thermodynamics texts and handbooks. For air and diatomic gases at usual temperatures, we can take k as 1.4 (which for isentropic conditions also means $n = 1.4$).

By combining Eqs. (2.4a) and (2.6a), we can obtain other useful relations such as

$$\frac{T_2}{T_1} = \left(\frac{v_1}{v_2}\right)^{n-1} = \left(\frac{\rho_2}{\rho_1}\right)^{n-1} = \left(\frac{p_2}{p_1}\right)^{(n-1)/n} \tag{2.9}$$

Sample Problem 2.5
If an artificial atmosphere consists of 20% oxygen and 80% nitrogen by volume, at 14.7 psia and 60°F, what are (a) the specific weight and partial pressure of the oxygen and (b) the specific weight of the mixture?

Solution

Table A.5:
$$R(\text{oxygen}) = 1554 \text{ ft}^2/(\text{sec}^2 \cdot °R),$$
$$R(\text{nitrogen}) = 1773 \text{ ft}^2/(\text{sec}^2 \cdot °R)$$

Eq. (2.5):   100% O_2: $\gamma = \dfrac{32.2(14.7 \times 144)}{1554(460 + 60)} = 0.0843 \text{ lb/ft}^3$

Eq. (2.5):   100% N_2: $\gamma = \dfrac{32.2(14.7 \times 144)}{1773(520)} = 0.0739 \text{ lb/ft}^3$

(a) Each ft^3 of mixture contains 0.2 ft^3 of O_2 and 0.8 ft^3 of N_2.

So for 20% O_2, $\gamma = 0.20(0.0843) = 0.01687 \text{ lb/ft}^3$ *ANS*

From Eq. (2.5), for 20% O_2, $p = \dfrac{\gamma RT}{g} = \dfrac{0.01687(1554)520}{32.2}$
$$= 423 \text{ lb/ft}^2\text{abs} = 2.94 \text{ psia}   \text{\textit{ANS}}$$

Note that this $= 20\%(14.7 \text{ psia})$.

(b) For 80% N_2, $\gamma = 0.80(0.0739) = 0.0591 \text{ lb/ft}^3$.

Mixture: $\gamma = 0.01687 + 0.0591 = 0.0760 \text{ lb/ft}^3$.   *ANS*

2.8 Compressibility of Perfect Gases

Differentiating Eq. (2.6) gives $npv^{n-1}dv + v^n dp = 0$. Inserting the value of dp from this into $E_v = -(v/dv)\,dp$ from Sec. 2.5 yields

$$E_v = np \qquad\qquad (2.10)$$

So for an isothermal process of a gas $E_v = p$, and for an isentropic process $E_v = kp$.

Thus, at a pressure of 15 psia, the isothermal modulus of elasticity for a gas is 15 psi, and for air in an isentropic process it is 1.4(15 psi) = 21 psi. Assuming from Table 2.1 a typical value of the modulus of elasticity of cold water to be 320,000 psi, we see that air at 15 psia is 320,000/15 = 21,000 times as compressible as cold water isothermally, or 320,000/21 = 15,000 times as compressible isentropically. This emphasizes the great difference between the compressibility of normal atmospheric air and that of water.

Sample Problem 2.6

(*a*) Calculate the density, specific weight, and specific volume of air at 59°F and 14.7 psia (pounds per square inch absolute; see Sec. 2.7). (*b*) An air parcel at the conditions of (*a*) at the earth's surface rises isentropically to 10,000 ft, where pressure is 10.1 psia. What would be its temperature and by what percent would it expand? (*c*) A bicycle tire (volume = 0.05 ft³) at the conditions in (*a*) is inflated to the very low pressure of 20 psia. A pump with a piston volume of 0.007 ft³ is used to inflate the tire. After one stroke of the pump (adding the entire piston contents to the tire), how much air mass is added, and what is the final tire pressure (assume tire volume and temperature are constant)?

Solution

Table A.5 for air: Molar mass $M = 28.96$, $k = 1.40$

(*a*) Sec. 2.7: $R \approx \dfrac{R_0}{M} = \dfrac{49,709}{28.96} = 1716$ ft·lb/(slug·°R)   (as in Table A.5)

From Eq. (2.4): $\rho = \dfrac{p}{RT} = \dfrac{14.7 \times 144 \text{ lb/ft}^2}{[1716 \text{ ft·lb/(slug·°R)}][(460 + 59)°R]}$

$$= 0.00238 \text{ slug/ft}^3 \qquad\qquad ANS$$

With $g = 32.2$ ft/sec², $\gamma = \rho g = 0.00238(32.2) = 0.0765$ lb/ft³   *ANS*

Eq. (2.2): $v = \dfrac{1}{\rho} = \dfrac{1}{0.00238} = 421$ ft³/slug   *ANS*

(*b*) Eq. (2.9): $T_2 = (59 + 460°R)\left(\dfrac{10.1}{14.7}\right)^{(1.4-1)/1.4} = 466.23 \ °R = 6.23 \ °F$ *ANS*

Eq. (2.7): $\dfrac{\Psi_2}{\Psi_1} = \dfrac{14.7}{10.1}\dfrac{466.23}{(460 + 59)} = 1.307$ thus, expansion is 30.7%   *ANS*

(c) Eq. (2.4): $m_{pumped} = \dfrac{p V_{pump}}{RT} = \dfrac{(14.7 \times 144)(0.007)}{1716(460 + 59)} = 1.663 \times 10^{-5}$ slug   *ANS*

The original mass of air in the tire:

$$m_{tire} = \frac{p V_{tire}}{RT} = \frac{(15 \times 144)(0.05)}{1716(460 + 59)} = 1.212 \times 10^{-4} \text{slug}$$

The final tire pressure:

Eq. (2.4): $p = \dfrac{m_{total}RT}{V} = \dfrac{(1.212 \times 10^{-4} + 1.663 \times 10^{-5}) \times (1716) \times (460 + 59)}{0.05}$

$$= 2456 \text{ lb/ft}^2 = 17.06 \text{ psi} \qquad\qquad\qquad\qquad \textit{ANS}$$

Sample Problem 2.7

Calculate the density, specific weight, and specific volume of chlorine gas at 25°C and pressure of 600 kN/m² abs (kilonewtons per square meter absolute; see Sec. 2.7). Given the molar mass of chlorine $(Cl_2) = 71$.

Solution

Sec. 2.7: $\qquad\qquad\qquad R = \dfrac{R_0}{M} = \dfrac{8312}{71} = 117.1 \text{ N·m/(kg·K)}$

From Eq. (2.4a): $\qquad\qquad \rho = \dfrac{P}{RT} = \dfrac{600\,000 \text{ N/m}^2}{[117.1 \text{ N·m/(kg·K)}][(273 + 25)K]}$

$$= 17.20 \text{ kg/m}^3 \qquad \textit{ANS}$$

With $g = 9.81 \text{ m/s}^2$, $\gamma = \rho g = 17.20(9.81) = 168.7 \text{ N/m}^3 \qquad \textit{ANS}$

Eq. (2.2): $\qquad\qquad\qquad v = \dfrac{1}{\rho} = \dfrac{1}{17.20} = 0.0581 \text{ m}^3/\text{kg} \qquad \textit{ANS}$

2.9 Standard Atmosphere

Standard atmospheres were first adopted in the 1920s in the United States and in Europe to satisfy a need for standardization of aircraft instruments and aircraft performance. As knowledge of the atmosphere increased, and man's activities in it rose to ever greater altitudes, such standards have been frequently extended and improved.

The International Civil Aviation Organization (ICAO) adopted its latest ***ICAO Standard Atmosphere*** in 1964, which extends up to 32 km (105,000 ft). The International Standards Organization (ISO) adopted an ***ISO Standard Atmosphere*** to 50 km (164,000 ft) in 1973, which incorporates the ICAO standard. The United States has adopted the ***U.S. Standard Atmosphere,*** last revised in 1976. This incorporates the ICAO and ISO standards, and extends to at least 86 km (282,000 ft or 53.4 mi); for some quantities it extends as far as 1000 km (621 mi).

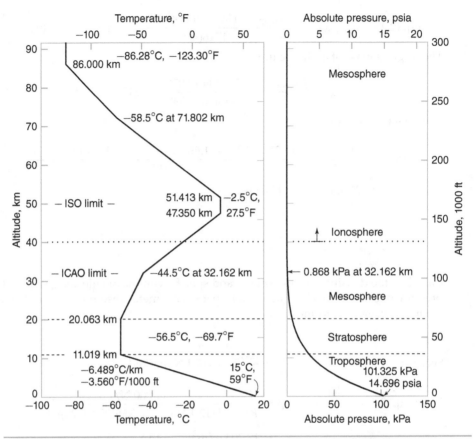

Figure 2.2   The U.S. Standard Atmosphere, temperature, and pressure distributions.

Figure 2.2 graphically presents variations of temperature and pressure in the U.S. Standard Atmosphere. In the lowest 11.02 km (36,200 ft), called the *troposphere*, the temperature decreases rapidly and linearly at a *lapse rate* of −6.489°C/km (−3.560°F/1000 ft). This lapse rate is lower in magnitude than that calculated for isentropic changes in Sample Problem 2.6; the lapse rate in the standard atmosphere reflects the heat imparted to rising air by condensing water vapor, which partially offsets the cooling predicted by isentropic assumptions. In the next layer, called the *stratosphere*, about 9 km (30,000 ft) thick, the temperature remains constant at −56.5°C (−69.7°F). Next, in the *mesosphere*, it increases, first slowly and then more rapidly, to a maximum of −2.5°C (27.5°F) at an altitude around 50 km (165,000 ft or 31 mi). Above this, in the upper part of the mesosphere known as the *ionosphere*, the temperature again decreases.

The standard absolute pressure[7] behaves very differently from temperature (Fig. 2.2), decreasing quite rapidly and smoothly to almost zero at an altitude of 30 km (98,000 ft). The pressure profile was computed from the standard temperatures using methods of fluid statics (Sec. 3.2). The representation of the standard temperature

---

[7] Absolute pressure is discussed in Secs. 2.7 and 3.4.

profile by a number of linear functions of elevation (Fig. 2.2) greatly facilitates such computations (see Sample Problem 3.1*b*).

Temperature, pressure, and other variables from the ICAO Standard Atmosphere, including density and viscosity, are tabulated together with gravitational acceleration out to 30 km and 100,000 ft in Appendix A, Table A.3. Engineers generally use such data in design calculations where the performance of high-altitude aircraft is of interest. The standard atmosphere serves as a good approximation of conditions in the atmosphere; of course the actual conditions vary somewhat with the weather, the seasons, and the latitude.

2.10 Ideal Fluid

An *ideal* fluid is usually defined as a fluid in which there is no *friction*; it is *inviscid* (its viscosity is zero). Thus the internal forces at any section within it are always normal to the section, even during motion. So these forces are purely pressure forces. Although such a fluid does not exist in reality, many fluids approximate frictionless flow at sufficient distances from solid boundaries, and so we can often conveniently analyze their behaviors by assuming an ideal fluid. As noted in Sec. 2.7, take care to not confuse an ideal fluid with a perfect (ideal) gas.

In a *real* fluid, either liquid or gas, tangential or shearing forces always develop whenever there is motion relative to a body, thus creating fluid friction, because these forces oppose the motion of one particle past another. These friction forces give rise to a fluid property called viscosity.

2.11 Viscosity

The *viscosity* of a fluid is a measure of its resistance to shear or angular deformation. Motor oil, for example, has high viscosity and resistance to shear, is cohesive, and feels "sticky," whereas gasoline has low viscosity. The friction forces in flowing fluid result from the cohesion and momentum interchange between molecules. Figure 2.3 indicates

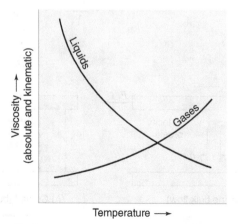

FIGURE 2.3 Trends in viscosity variation with temperature.

how the viscosities of typical fluids depend on temperature. As the temperature increases, the viscosities of all liquids *decrease*, while the viscosities of all gases *increase*. This is because the force of cohesion, which diminishes with temperature, predominates with liquids, while with gases the predominating factor is the interchange of molecules between the layers of different velocities. Thus a rapidly-moving gas molecule shifting into a slower-moving layer tends to speed up the latter. And a slow-moving molecule entering a faster-moving layer tends to slow down the faster-moving layer. This molecular interchange sets up a shear, so that it produces a friction force between adjacent layers. At higher temperatures molecular activity increases, so causing the viscosity of gases to increase with temperature.

Figures A.1 and A.2 in Appendix A graphically present numerical values of absolute and kinematic viscosities for a variety of liquids and gases, and show how they vary with temperature.

Consider the classic case of two parallel plates (Fig. 2.4), sufficiently large that we can neglect edge conditions, a small distance Y apart, with fluid filling the space between. The lower plate is stationary, while the upper one moves parallel to it with a velocity U due to a force F corresponding to some area A of the moving plate.

At boundaries, particles of fluid adhere to the walls, and so their velocities are zero relative to the wall. This so-called **no-slip condition** occurs with all viscous fluids. Thus in Fig. 2.4 the fluid velocities must be U where in contact with the plate at the upper boundary and zero at the lower boundary. We call the form of the velocity variation with distance between these two extremes, as depicted in Fig. 2.4, a **velocity profile**. If the separation distance Y is not too great, if the velocity U is not too high, and if there is no net flow of fluid through the space, the velocity profile will be linear, as in Fig. 2.4a. If, in addition, there is a small amount of bulk fluid transport between the plates, as could result from pressure-fed lubrication for example, the velocity profile becomes the sum of the previous linear profile plus a parabolic profile (Fig. 2.4b); the parabolic additions to (or subtractions from) the linear profile are zero at the walls (plates) and maximum at the centerline. The behavior of the fluid is much as if it consisted of a series of thin layers, each of which slips a little relative to the next.

For a large class of fluids under the conditions of Fig. 2.4a, experiments have shown that

$$F \propto \frac{AU}{Y}$$

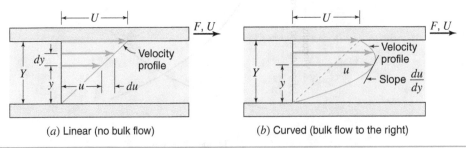

(a) Linear (no bulk flow)   (b) Curved (bulk flow to the right)

FIGURE 2.4   Velocity profiles.

We see from similar triangles that we can replace U/Y by the velocity gradient du/dy. If we now introduce a constant of proportionality μ (mu), we can express the shearing stress τ (tau) between any two thin sheets of fluid by

No bulk flow:
$$\tau = \frac{F}{A} = \mu \frac{U}{Y} = \mu \frac{du}{dy} \qquad (2.11)$$

We call Eq. (2.11) *Newton's equation of viscosity,* since Sir Isaac Newton (1642–1727) first suggested it. Although better known for his formulation of the fundamental laws of motion and gravity and for the development of differential calculus, Newton, an English mathematician and natural philosopher, also made many pioneering studies in fluid mechanics. In transposed form, Eq. (2.11) defines the proportionality constant

$$\mu = \frac{\tau}{du/dy} \qquad (2.12)$$

known as the *coefficient of viscosity,* the *absolute viscosity,* the *dynamic viscosity* (since it involves force), or simply the *viscosity* of the fluid. We shall use "absolute viscosity" to help differentiate it from another viscosity that we will discuss shortly.

We noted in Sec. 2.1 that the distinction between a solid and a fluid lies in the manner in which each can resist shearing stresses. We will clarify a further distinction among various kinds of fluids and solids by referring to Fig. 2.5. For a solid, shear stress depends on the *magnitude* of the deformation; but Eq. (2.11) shows that in many fluids the shear stress is proportional to the *slope* of the velocity profile (Fig. 2.4)

A fluid for which the constant of proportionality (i.e., the absolute viscosity) does not change with rate of deformation is called a *Newtonian fluid,* and this plots as a straight line in Fig. 2.5. The slope of this line is the absolute viscosity, μ. The ideal fluid, with no viscosity (Sec. 2.10), falls on the horizontal axis, while the true elastic solid plots along the vertical axis. A plastic that sustains a certain amount of stress before suffering

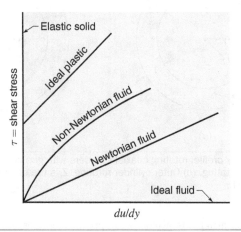

FIGURE 2.5   Relationship between shear stress and the rate of deformation for different fluids.

a plastic flow corresponds to a straight line intersecting the vertical axis at the yield stress. There are certain non-Newtonian fluids[8] in which μ varies with the rate of deformation. These are relatively uncommon in engineering usage, so we will restrict the remainder of this text to the common fluids that under normal conditions obey Newton's equation of viscosity.

In a *journal bearing*, lubricating fluid fills the small annular space between a shaft and its surrounding support. This fluid layer is very similar to the layer between the two parallel plates, except it is curved. There is another more subtle difference, however. For coaxial cylinders (Fig. 2.6) with constant rotative speed ω (omega), the resisting and driving torques are equal, so $\tau_1(2\pi r_1 Z)r_1 = \tau_2(2\pi r_2 Z)r_2$. But because the radii at the inner and outer walls are different, it follows that the shear stresses and velocity gradients there must also be different [Eq. (2.9) and Fig. 2.6]. The shear stress and velocity gradient must vary continuously across the gap, and so the velocity profile must curve.

$$\left(\frac{du}{dy}\right)_1 = \left(\frac{du}{dy}\right)_2 \frac{r_2^2}{r_1^2}$$

However, as the gap distance $Y \rightarrow 0$, $du/dy \rightarrow U/Y = $ constant. So, when the gap is very small, we can assume the velocity profile to be a straight line, and we can solve problems in a similar manner as for flat plates.

The dimensions of absolute viscosity are force per unit area divided by velocity gradient. In the BG system the dimensions of absolute viscosity are as follows:

$$\text{Dimensions of } \mu = \frac{\text{dimensions of } \tau}{\text{dimensions of } du/dy} = \frac{\text{lb/ft}^2}{\text{fps/ft}} = \text{lb·sec/ft}^2$$

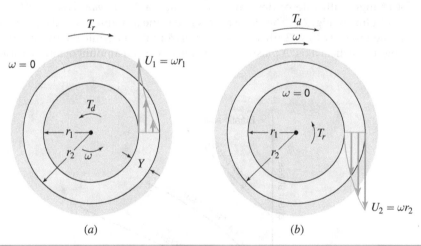

(a)   (b)

FIGURE 2.6   Velocity profile, rotating coaxial cylinders with gap completely filled with liquid. (a) Inner cylinder rotating. (b) Outer cylinder rotating. Z is the dimension at right angles to the plane of the sketch.

---

[8] Typical non-Newtonian fluids include paints, printer's ink, gels and emulsions, sludges and slurries, and certain plastics. An excellent treatment of the subject is given by W. L. Wilkinson in *Non-Newtonian Fluids*, Pergamon Press, New York, 1960.

In SI units

$$\text{Dimensions of } \mu = \frac{N/m^2}{s^{-1}} = N{\cdot}s/m^2$$

A widely used unit for viscosity in the metric system is the *poise* (P), named after Jean Louis Poiseuille (1799–1869). A French anatomist, Poiseuille was one of the first investigators of viscosity. The poise = 0.10 N·s/m². The *centipoise* (cP) (= 0.01 P = 1 mN·s/m²) is frequently a more convenient unit. It has a further advantage in that the viscosity of water at 68.4°F is 1 cP. Thus the value of the viscosity in centipoises is an indication of the viscosity of the fluid relative to that of water at 68.4°F.

In many problems involving viscosity the absolute viscosity is divided by density. This ratio defines the *kinematic viscosity* ν (nu), so-called because force is not involved, the only dimensions being length and time, as in kinematics (Sec. 1.1). Thus

$$\nu = \frac{\mu}{\rho} \tag{2.13}$$

We usually measure kinematic viscosity ν in ft²/sec in the BG system, and in m²/s in the SI. Previously, in the metric system the common units were cm²/s, also called the *stoke* (St), after Sir George Stokes (1819–1903), an English physicist and pioneering investigator of viscosity. Many found the *centistoke* (cSt) (0.01 St = 10^{-6} m²/s) a more convenient unit to work with.

An important practical distinction between the two viscosities is the following. The absolute viscosity μ of most fluids is virtually independent of pressure for the range that is ordinarily encountered in engineering work; for extremely high pressures, the values are a little higher than those shown in Fig. A.1. The kinematic viscosity ν of gases, however, varies strongly with pressure because of changes in density. Therefore, if we need to determine the kinematic viscosity ν at a nonstandard pressure, we can look up the (pressure-independent) value of μ and calculate ν from Eq. (2.13). This will require knowing the gas density, ρ, which, if necessary, we can calculate using Eq. (2.4).

Sample Problem 2.8
A 1-in-wide space between two horizontal plane surfaces is filled with SAE 30 Western lubricating oil at 80°F. What force is required to drag a very thin plate of 4-ft² area through the oil at a velocity of 20 ft/min if the plate is 0.33 in from one surface?

Solution

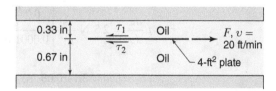

Fig. A.1: $\qquad\qquad \mu = 0.0063 \text{ lb·sec}/\text{ft}^2$

Eq. (2.9): $\qquad\qquad \tau_1 = 0.0063 \times (20/60)/(0.33/12) = 0.0764 \text{ lb}/\text{ft}^2$

Eq. (2.9): $\qquad\qquad \tau_2 = 0.0063 \times (20/60)/(0.67/12) = 0.0394 \text{ lb}/\text{ft}^2$

From Eq. (2.9): $\qquad\qquad F_1 = \tau_1 A = 0.0764 \times 4 = 0.305 \text{ lb}$

From Eq. (2.9): $\qquad\qquad F_2 = \tau_2 A = 0.0394 \times 4 = 0.158 \text{ lb}$

$$\text{Force} = F_1 + F_2 = 0.463 \text{ lb} \qquad \textbf{\textit{ANS}}$$

Sample Problem 2.9

The piston in Fig. S2.9 has a diameter of 8 cm and operates in a cylinder of diameter 8.02 cm, with the 0.01 cm gap around it filled with SAE 30 (Eastern) oil. When the piston is traveling upward at 0.2 m/s what is the vertical force exerted by the piston rod when the engine is cold ($T = 20°C$)?

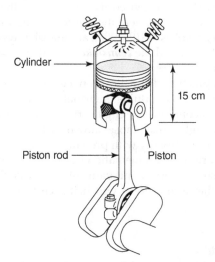

Cylinder

15 cm

Piston rod   Piston

FIGURE S2.9   Schematic of a piston (original by Pearson Scott Foresman).

Solution

From Fig. A1: $\qquad\qquad \mu = 2.5 \text{ N·s}/\text{m}^2$

Eq. (2.11) $\qquad\qquad \tau = \mu \dfrac{du}{dy} = 2.5 \text{ N·}\dfrac{\text{s}}{\text{m}^2} \left(\dfrac{0.2\frac{\text{m}}{\text{s}}}{0.0001 \text{ m}} \right) = 5000 \text{ N}/\text{m}^2$

$$A_{\text{contact}} = \pi Dh = 3.14(0.08 \text{ m})(0.15 \text{ m}) = 0.0377 \text{ m}^2$$

Eq. (2.11) $\qquad\qquad F = \tau A = 5000 \times 0.0377 = 188.5 \text{ N} \qquad \textbf{\textit{ANS}}$

Sample Problem 2.10

In Fig. S2.10 oil of absolute viscosity μ fills the small gap of thickness Y. (a) Neglecting fluid stress exerted on the circular underside and on the annular top surface, obtain an expression for the torque T required to rotate the truncated cone at constant speed ω. (b) What is the rate of heat generation, in joules per second, if the oil's absolute viscosity is 0.20 N·s/m², $\alpha = 45°$, $a = 45$ mm, $b = 60$ mm, $Y = 0.2$ mm, and the speed of rotation is 90 rpm?

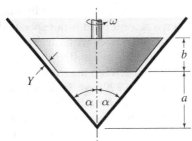

FIGURE S2.10

Solution

(a) $U = \omega r$;   for small gap Y, $\dfrac{du}{dy} = \dfrac{U}{Y} = \dfrac{\omega r}{Y}$

Eq. (2.11): $\tau = \mu\dfrac{du}{dy} = \dfrac{\mu\omega r}{Y}$; $dA = 2\pi r ds = \dfrac{2\pi r dy}{\cos\alpha}$

From Eq. (2.11): $dF = \tau dA = \dfrac{\mu\omega r}{Y}\left(\dfrac{2\pi r dy}{\cos\alpha}\right)$

$$dT = rdF = \dfrac{2\pi\mu\omega}{Y\cos\alpha}r^3 dy;\quad r = y\tan\alpha$$

$$dT = \dfrac{2\pi\mu\omega\tan^3\alpha}{Y\cos\alpha}y^3 dy$$

$$T = \dfrac{2\pi\mu\omega\tan^3\alpha}{Y\cos\alpha}\int_a^{a+b}y^3 dy;\quad \left.\dfrac{y^4}{4}\right|_a^{a+b} = \left[\dfrac{(a+b)^4}{4} - \dfrac{a^4}{4}\right]$$

$$T = \dfrac{2\pi\mu\omega\tan^3\alpha}{4Y\cos\alpha}[(a+b)^4 - a^4]\quad \textit{ANS}$$

(b) $[(a+b)^4 - a^4] = (0.105\text{ m})^4 - (0.045\text{ m})^4 = 0.0001175\text{ m}^4$

$$\omega = \left(90\dfrac{\text{rev}}{\text{min}}\right)\left(2\pi\dfrac{\text{rad}}{\text{rev}}\right)\left(\dfrac{1\text{ min}}{60\text{ s}}\right) = 3\pi\text{ rad/s} = 3\pi\text{ s}^{-1}$$

Heat generation rate = power = $T\omega = \dfrac{2\pi\mu\omega^2\tan^3\alpha}{4Y\cos\alpha}[(a+b)^4 - a^4]$

$$= \dfrac{2\pi(0.20\text{ N·s/m}^2)(3\pi\text{ s}^{-1})^2(1)^3[0.0001175\text{ m}^4]}{4(2\times10^{-4}\text{ m})\cos 45°}$$

$$= 23.2\text{ N·m/s} = 23.2\text{ J/s}\quad \textit{ANS}$$

2.12   Surface Tension

Liquids have cohesion and adhesion, both of which are forms of molecular attraction. *Cohesion* enables a liquid to resist tensile stress, while *adhesion* enables it to adhere to another body.[9] At the interface between a liquid and a gas, i.e., at the liquid surface, and at the interface between two *immiscible* (not mixable) liquids, the out-of-balance attraction force between molecules forms an imaginary surface film which exerts a tension force in the surface. This liquid property is known as *surface tension.* Because this tension acts in a surface, we compare such forces by measuring the tension force per unit length of surface. When a second fluid is not specified at the interface, it is understood that the liquid surface is in contact with air. The surface tensions of various liquids cover a wide range, and they decrease slightly with increasing temperature. Values of the surface tension for water between the freezing and boiling points vary from 0.005 18 to 0.004 04 lb/ft (0.0756 to 0.0589 N/m); Table A.1 of Appendix A contains more typical values. Table A.4 includes values for other liquids. *Capillarity* is the property of exerting forces on fluids by fine tubes or porous media; it is due to both cohesion and adhesion. When the cohesion is of less effect than the adhesion, the liquid will wet a solid surface it touches and rise at the point of contact; if cohesion predominates, the liquid surface will depress at the point of contact. For example, capillarity makes water rise in a glass tube, while mercury depresses below the true level, as shown in the insert in Fig. 2.7, which is drawn to scale and reproduced actual size. We call the curved liquid surface that develops in a tube a *meniscus.*

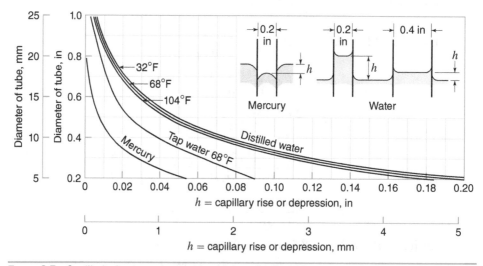

FIGURE 2.7   Capillarity in clean circular glass tubes, for liquid in contact with air.

***

[9] In 1877 Osborne Reynolds demonstrated that a ¼-in-diameter column of mercury could withstand a tensile stress (negative pressure, below atmospheric) of 3 atm (44 psi or 304 kPa) for a time, but that it would separate upon external jarring of the tube. Liquid tensile stress (said to be as high as 400 atm) accounts for the rise of water in the very small channels of xylem tissue in tall trees. For practical engineering purposes, however, we assume liquids are incapable of resisting any direct tensile stress.

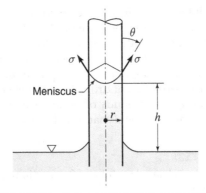

FIGURE 2.8   Capillary rise.

A cross section through capillary rise in a tube looks like Fig. 2.8. From free-body considerations, equating the (capillary) lifting force created by surface tension to the gravity force,

$$\text{Capillary Force} = L\sigma\cos\theta$$

$$2\pi r\sigma\cos\theta = \pi r^2 h\gamma$$

So

$$h = \frac{2\sigma\cos\theta}{\gamma r} \tag{2.14}$$

where σ = surface tension (sigma) in units of force per unit length
L = length of the interface
θ = wetting angle (theta)
γ = specific weight of liquid
r = radius of tube
h = capillary rise[10]

We can use this expression to compute the *approximate* capillary rise or depression in a tube. If the tube is clean, $\theta = 0°$ for water and about 140° for mercury. Note that the meniscus (Figs. 2.7 and 2.8) lifts a small volume of liquid, near the tube walls, in addition to the volume $\pi r^2 h$ used to obtain Eq. (2.14). For larger tube diameters, with smaller capillary rise heights, this small additional volume can become a large fraction of $\pi r^2 h$. So Eq. (2.14) overestimates the amount of capillary rise or depression, particularly for larger diameter tubes. The curves of Fig. 2.7 are for water or mercury in contact with air; if mercury is in contact with water, the surface tension effect is slightly less than when in contact with air. For tube diameters larger than ½ in (12 mm), capillary effects are negligible.

Surface tension effects are generally negligible in most engineering situations. However, they can be important in problems involving capillary rise, such as in the soil water zone; without capillarity most forms of vegetable life would perish. Surface tension allows tall trees to lift water to heights exceeding 100 m (328 ft), due to very small

---

[10] Measurements to a meniscus are usually taken to the point on the centerline.

spacing between cellulose microfibrils, effectively introducing a low value for r in Eq. (2.14). When we use small tubes to measure fluid properties, such as pressures, we must take the readings while aware of the surface tension effects; a true reading would occur if surface tension effects were zero. These effects are also important in hydraulic model studies when the model is small, in the breakup of liquid jets, and in the formation of drops and bubbles. The formation of drops is extremely complex to analyze, but is, for example, of critical concern in the study of cloud formation, pharmaceutical production using sprays, and the transmission dynamics of viruses.

Sample Problem 2.11
Water at 10°C stands in a clean glass tube of 2-mm diameter at a height of 35 mm. What is the true static height?

Solution
Table A.1 at 10°C: $\qquad\qquad \gamma = 9804 \text{ N/m}^3, \sigma = 0.0742 \text{ N/m}.$

Sec. 2.12 for clean glass tube: $\qquad\qquad \theta = 0°.$

Eq. (2.14): $\qquad h = \dfrac{2\sigma\cos\theta}{\gamma r} = \dfrac{2(0.0742 \text{ N/m})(1)}{(9804 \text{ N/m}^3)0.001 \text{ m}}$

$$= 0.01514 \text{ m} = 15.14 \text{ mm}$$

Sec. 2.12:   True static height $= 35.00 - 15.14 = 19.86$ mm $\textit{ANS}$

Sample Problem 2.12
A metal needle or thin rod can float on water, which, if ferrous, can be used as a compass. (a) Find the maximum diameter of a mild steel rod ($\rho = 7.85$ g/cm³) that will float on 15°C water, ignoring the buoyant force (B). (b) Repeat part (a) including the buoyant force (see Sec. 3.9). Assume the rod is at the point of sinking, half submerged, so forces act vertically. Neglect end effects.

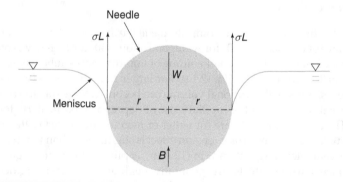

FIGURE S2.12

Solution
(a) The mass of the needle:

Sec. 1.5: $\qquad m = \rho_{st}\forall_{st} = 7.85\dfrac{\text{g}}{\text{cm}^3}\dfrac{10^6 \text{cm}^3}{\text{m}^3}\dfrac{\text{kg}}{10^3\text{g}}(\pi r^2 L) = 24\,662 \; r^2 L \text{ (kg)}$

Convert to weight (Sec. 1.5):

$$W = mg = 24\,661r^2L\left(9.81\ \frac{\text{m}}{\text{s}^2}\right) = 241\,929\ r^2L\ (\text{N})$$

Table A.1 at 15°C: $\qquad\qquad\qquad\qquad \sigma = 0.0735\ \text{N/m}.$

Total vertical force due to surface tension, [for vertical forces, $\theta = 0°$ in Eq. (2.14)]:

$$2\sigma L = 2(0.0735\ \text{N/m})L = 0.1470L\ (\text{N})$$

To float, W must not exceed the upward surface tension force:

$$241\,929\ r^2L \le 0.147L \quad\text{or}\quad r^2 \le 6.076 \times 10^{-7} \quad\text{or}\quad r \le 7.795 \times 10^{-4}\ \text{m} = 0.779\ \text{mm}$$

Thus a needle will float as long as its diameter $D = 2r \le 2(0.779) = 1.559\ \text{mm}$ $\qquad$ **ANS**

Note that the result is independent of the length of the needle.

(b) A more complete solution accounts for the buoyant force, B, of the submerged portion of the needle (Sec. 3.9). Using Table A.1, $\gamma_w = 9798\ \text{N/m}^3$. Then using the submerged volume of the needle, V_B:

$$B = \gamma_W V_B = 9798\ \frac{\text{N}}{\text{m}^3}\left(\frac{\pi r^2 L}{2}\right) = 15\,396\ r^2L\ (\text{N})$$

To float, $W - B$ must not exceed the force of surface tension:

$(241\,929 - 15\,396)\ r^2L \le 0.147L$ or $r^2 \le 6.49 \times 10^{-7}$ or $r \le 8.06 \times 10^{-4}$ m $= 0.806$ mm, thus $D = 2r \le 2(0.806) = 1.612$ mm $\qquad$ **ANS**

2.13 Vapor Pressure of Liquids

All liquids tend to evaporate or vaporize, which they do by projecting molecules into the space above their surfaces. If this is a confined space, the partial pressure exerted by the molecules increases until the rate at which molecules reenter the liquid is equal to the rate at which they leave. For this equilibrium condition, we call the vapor pressure the *saturation pressure*. For atmospheric water vapor, the ratio of actual to saturated vapor pressure is called *relative humidity* (RH), so actual vapor pressure $= RH(p_v)$. The maximum amount of water vapor air can hold is a function of temperature; when that level is reached, the relative humidity is 100 percent and no more net evaporation will occur.

Molecular activity increases with increasing temperature and decreasing pressure, and so the saturation pressure does the same. At any given temperature, if the pressure on the liquid surface falls below the saturation pressure, a rapid rate of evaporation results, known as *boiling*. Thus we can refer to the saturation pressure as the *boiling pressure* for a given temperature, and it is of practical importance for liquids.[11] The relationship between temperature, pressure, and state are illustrated by a phase diagram, an example of which is shown in Fig. 2.9.

---

[11] Values of the saturation pressure for water for temperatures from 32°F to 705.4°F are given in J. H. Keenan, *Thermodynamic Properties of Water including Vapor, Liquid and Solid States*, John Wiley & Sons, Inc., New York, 1969, and in other steam tables. There are similar vapor tables published for ammonia, carbon dioxide, sulfur dioxide, and other vapors of engineering interest.

38   Chapter Two

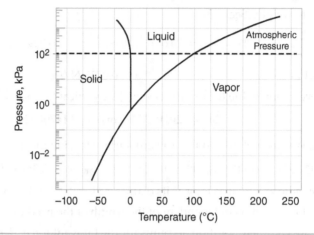

FIGURE 2.9   A phase diagram for water.

	psia	N/m² abs	mb abs
Mercury	0.000 025	0.17	0.0017
Water	0.34	2 340	23.4
Carbon tetrachloride	1.90	13 100	131
Gasoline	8.0	55 200	552

TABLE 2.3   Saturation vapor pressure of selected liquids at 68°F (20°C)

We call the rapid vaporization and recondensation of liquid as it briefly passes through a region of low absolute pressure *cavitation.* This phenomenon is often very damaging, and so we must avoid it; we shall discuss it in more detail in Sec. 5.10.

Table 2.3 calls attention to the wide variation in saturation vapor pressure of various liquids; Appendix A, Table A.4 contains more values. The very low vapor pressure of mercury makes it particularly suitable for use in barometers. Values for the vapor pressure of water at different temperatures are in Appendix A, Table A.1.

Sample Problem 2.13
At approximately what temperature will water boil if the elevation is 10,000 ft?

Solution
From Appendix A, Table A.3, the pressure of the standard atmosphere at 10,000-ft elevation is 10.11 psia. From Appendix A, Table A.1, the saturation vapor pressure p_v of water is 10.11 psia at about 193°F (by interpolation). Hence the water at 10,000 ft will boil at about 193°F.   *ANS*

Compared with the boiling temperature of 212°F at sea level, this explains why it takes longer to cook at high elevations.

Sample Problem 2.14
Air at standard atmospheric pressure has a temperature of 35°C and a relative humidity of 40%. To what temperature would it need to cool to reach saturation, so that condensation would be possible?

Solution
Table A.1: Saturation vapor pressure at 35°C, $p_v = 5.81$ kPa by interpolation.

Actual vapor pressure of the air: $0.4(5.81) = 2.32$ kPa

The temperature at which the actual vapor pressure becomes saturated is interpolated between 15°C and 20°C from Table A.1:

$$T = 15 + (20 - 15)\left(\frac{2.32 - 1.71}{2.34 - 1.71}\right) = 19.8 \ ^\circ C \qquad \textbf{\textit{ANS}}$$

Sample Problem 2.15
A hand pump installed at an elevation of 1000 m is designed to draw 15°C water up from a well by suction. What is the theoretical maximum suction pressure (or vacuum) that can be used to lift water?

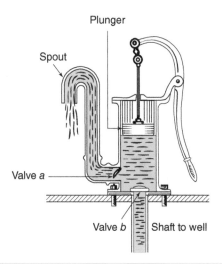

FIGURE S2.15   Schematic of a hand pump (original from the archives of Pearson Scott Foresman).

Solution
Table A.3: Atmospheric pressure at 1000 m = 89.876 kPa.

Table A.1: For water at 15 °C saturation vapor pressure, $p_v = 1.710$ kPa.

When pressure in the piston is reduced from atmospheric to p_v the liquid will vaporize so no water can be pumped. Thus, $89.876 - 1.710 = 88.166$ kPa represents the largest vacuum possible. Relationships in Sec. 3.3 allow this to be converted to a height of water column, h (using the specific weight of water $\gamma = 9.798$ kN/m³):

$$h = \frac{p}{\gamma} = \frac{88.166 \text{ kPa}}{9.798 \text{ kN/m}^3} = 9.00 \text{ m} \qquad ANS$$

This means water at these conditions that is deeper than 9 m could not be lifted by this type of pump.

CHAPTER 3

Fluid Statics

I n fluids at rest there are no shear stresses; hence only normal forces due to pressure are present. Normal forces produced by static fluids are often very important. For example, they tend to overturn concrete dams, burst pressure vessels, and break lock gates on canals. To design such facilities, we need to be able to compute the magnitudes and locations of normal pressure forces. Understanding them, we can also develop instruments to measure pressures, and systems that transfer pressures, such as for automobile brakes and hoists.

Note that normal pressure forces alone can occur in a moving fluid if the fluid is moving in bulk without deformation, i.e., as if it were solid or rigid. For such an example, see Sec. 3.10. However, this is relatively rare.

The *average pressure intensity*, p, is the force exerted on a unit area. If F represents the total normal pressure force on some finite area A, while dF represents the force on an infinitesimal area dA, the pressure is

$$p = \frac{dF}{dA} \tag{3.1}$$

If the pressure is uniform over the total area, then $p = F/A$. In the British Gravitational (BG) system we generally express pressure in pounds per square inch (psi) or pounds per square foot ($\text{lb}/\text{ft}^2 = \text{psf}$), while in SI units we commonly use the pascal ($\text{Pa} = \text{N}/\text{m}^2$) or kPa ($\text{kN}/\text{m}^2$). Previously, bars and millibars were used in metric systems to express pressure; 1 mb = 100 Pa.

3.1 Pressure at a Point the Same in All Directions

In a solid, because of the possibility of tangential stresses between adjacent particles, the stresses at a given point may be different in different directions. But no tangential stresses can exist in a fluid at rest, and the only forces between adjacent surfaces are pressure forces normal to the surfaces. Therefore the pressure at any point in a fluid at rest is the same in every direction.

We can prove this by referring to Fig. 3.1, which represents a very small wedge-shaped element of fluid at rest whose thickness perpendicular to the plane of the paper is constant and equal to dy. Let p be the average pressure in any direction in the plane of the paper, let α be as shown, and let p_x and p_z be the average pressures in the horizontal and vertical directions.[1] The forces acting on the element of fluid, with the exception of those in the

---

[1] Note that the axes are arranged differently from those usually used in solid mechanics. They are chosen to retain a right-handed coordinate system, and to make z vertical, because z is traditionally used for elevation in fluid mechanics.

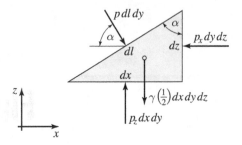

FIGURE 3.1 Forces on a fluid element at rest.

y direction on the two faces parallel to the plane of the paper, are shown in the diagram. For our purpose, forces in the y direction need not be considered because they cancel. Since the fluid is at rest, no tangential forces are involved. As this is a condition of equilibrium, the sum of the force components on the element in any direction must be equal to zero. Writing such an equation for the components in the x direction, $p \, dl \, dy \cos \alpha - p_x dy \, dz = 0$. Since $dz = dl \cos \alpha$ it follows that $p = p_x$. Similarly, summing forces in the z direction gives $p_z \, dx \, dy - p \, dl \, dy \sin \alpha - \frac{1}{2}\gamma \, dx \, dy \, dz = 0$. The third term is of higher order than the other two terms and so may be neglected. It follows from this that $p = p_z$. We can also prove that $p = p_y$ by considering a three-dimensional case. The results are independent of α; thus the pressure at any point in a fluid at rest is the same in all directions.

3.2　Variation of Pressure in a Static Fluid

Consider the differential element (or control volume) of static fluid shown in Fig. 3.2. Since the element is very small, we can assume that the density of the fluid within the element is constant. Assume that the pressure at the center of the element is p and that the dimensions of the element are δx, δy, and δz.[1] The forces acting on the fluid element in the vertical direction are: (*a*) the **body force**, the action of gravity on the mass within the element, and (*b*) the **surface forces**, transmitted from the surrounding fluid and

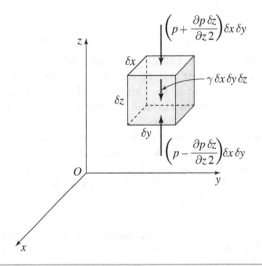

FIGURE 3.2　Forces acting on a static differential fluid element.

acting at right angles against the top, bottom, and sides of the element. Because the fluid is at rest, the element is in equilibrium and the summation of forces acting on the element in any direction must be zero. If forces are summed in the horizontal direction, that is, x or y, the only forces acting are the pressure forces on the vertical faces of the element. To satisfy $\Sigma F_x = 0$ and $\Sigma F_y = 0$, the pressures on the opposite vertical faces must be equal. Thus, $\partial p/\partial x = \partial p/\partial y = 0$ for the case of the fluid at rest.

Summing forces in the vertical direction and setting the sum equal to zero,

$$\sum F_z = \left(p - \frac{\partial p}{\partial z}\frac{\delta z}{2}\right)\delta x\,\delta y - \left(p + \frac{\partial p}{\partial z}\frac{\delta z}{2}\right)\delta x\,\delta y - \gamma\,\delta x\,\delta y\,\delta z = 0$$

This results in $\partial p/\partial z = -\gamma$ which, since p is independent of x and y, we can write as

$$\frac{dp}{dz} = -\gamma \qquad (3.2)$$

This is the general expression that relates variation of pressure in a static fluid to vertical position. The minus sign indicates that as z gets larger (increasing elevation), the pressure gets smaller.

To evaluate the pressure anywhere in a fluid at rest, we must integrate Eq. (3.2) between appropriately chosen limits. For incompressible fluids ($\gamma = $ constant), we can integrate Eq. (3.2) directly. For compressible fluids, however, we must express γ algebraically as a function of z or p if we wish to determine pressure accurately as a function of elevation. The variation of pressure in the earth's atmosphere is an important problem, and different approaches for estimating this are illustrated in the following example.

Sample Problem 3.1

Compute the atmospheric pressure at elevation 20,000 ft, considering the atmosphere as a static fluid. Assume conditions at sea level conform to standard atmospheric conditions. Use two methods: (a) isentropic conditions and (b) air temperature decreasing linearly with elevation at the standard lapse rate of 0.003 56°F/ft.

Solution

From Appendix A, Table A.3, the conditions of the standard atmosphere at sea level are $T_1 = 59.0°F$, $p_1 = 14.70$ psia, $\gamma_1 = 0.076\,48$ lb/ft^3, where subscript 1 indicates conditions at our reference elevation, sea level.

(a) Isentropic

From Sec. 2.7: $\quad pv^{1.4} = \dfrac{p}{\rho^{1.4}} = $ constant; so $\quad \dfrac{p}{\gamma^{1.4}} = $ constant $= \dfrac{p_1}{\gamma_1^{1.4}}$

Eq. (3.2):
$$\frac{dp}{dz} = -\gamma, \text{ where } \gamma = \gamma_1 \left(\frac{p}{p_1}\right)^{1/1.4} = \gamma_1 \left(\frac{p}{p_1}\right)^{0.714}$$

so
$$dp = -\gamma_1 \left(\frac{p}{p_1}\right)^{0.714} dz$$

Integrating:
$$\int_{p_1}^{p} p^{-0.714} \, dp = -\gamma_1 p_1^{-0.714} \int_{z_1}^{z} dz$$

$$p^{0.286} - p_1^{0.286} = -0.286 \gamma_1 p_1^{-0.714}(z - z_1)$$

$$p^{0.286} = (14.70 \times 144)^{0.286} - 0.286(0.07648)(14.70 \times 144)^{-0.714}(20{,}000)$$

$$p = 942 \text{ lb/ft}^2 \text{ abs} = 6.54 \text{ psia} \qquad \textbf{\textit{ANS}}$$

(*b*) Temperature decreasing linearly with elevation

For the standard lapse rate (Fig. 2.2): $T = a + bz$,

where $a = 59.00 + 459.67 = 518.67°R$ and $b = -0.003\,560°R/ft$

Eqs. (3.2) and (2.4):
$$\frac{dp}{dz} = -\rho g; \quad \rho = \frac{p}{RT}$$

Combining to eliminate ρ, which varies, rearranging, and substituting for T,

$$\frac{dp}{p} = -\frac{g \, dz}{R(a + bz)}$$

Integrating:
$$\int_{1}^{2} \frac{dp}{p} = -\frac{g}{R} \int_{1}^{2} \frac{dz}{a + bz}$$

$$\ln\left(\frac{p_2}{p_1}\right) = -\frac{g}{Rb} \ln\left(\frac{a + bz_2}{a + bz_1}\right) = \ln\left(\frac{a + bz_2}{a + bz_1}\right)^{-g/Rb}$$

i.e.,
$$\frac{p_2}{p_1} = \left(\frac{a + bz_2}{a + bz_1}\right)^{-g/Rb}$$

Here
$$\frac{-g}{Rb} = \frac{-32.174}{1716(-0.003\,560)} = 5.27$$

and, from Table A.3: $p_1 = 14.696$ psia when $z_1 = 0$.

Thus
$$\frac{p_2}{14.696} = \left(\frac{518.67 - 0.003\,560 \times 20{,}000}{518.67 + 0}\right)^{5.27} = 0.459$$

$$p_2 = 14.696(0.459) = 6.75 \text{ psia} \qquad \textbf{\textit{ANS}}$$

Sample Problem 3.2

When ascending in the atmosphere, the eardrum will experience a pressure difference on either side, especially if the Eustachian tube is blocked. In this case, the middle ear can remain at initial atmospheric pressure while the pressure in the outer ear will drop as elevation increases. If the risk of eardrum rupture begins when the difference across the eardrum is $\Delta p = 35$ kPa, at what elevation will this occur? Assume initial (middle ear) conditions are standard atmospheric conditions at elevation 0 ($T_1 = 15°C$, $p_1 = 101.3$ kPa, $\gamma_1 = 12.01$ N/m³).

Solution

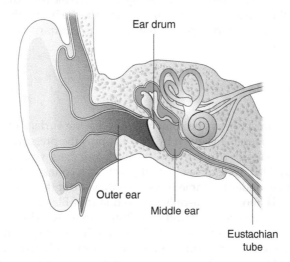

Ear drum

Outer ear

Middle ear

Eustachian tube

In a manner similar to Sample Problem 3.1:

Final (outer ear) pressure: $101,300 - 35,000 = 66,300$ Pa

$$\int_{z_1}^{z} dz = -\frac{1}{\gamma_1 p_1^{-0.714}} \int_{p_1}^{p} p^{-0.714}\, dp$$

$$(z - z_1) = -\frac{1}{0.286\gamma_1 p_1^{-0.714}} (p^{0.286} - p_1^{0.286})$$

$$(z - z_1) = -\frac{1}{0.286(12.01)(101,300)^{-0.714}}[(66,300)^{0.286} - (101,300)^{0.286}]$$

$$(z - z_1) = 3367 \text{ m} \quad \textbf{\textit{ANS}}$$

The latter approach of Sample Problem 3.1 corresponds to the standard atmosphere, described in Sec. 2.9 and in Table A.3 of Appendix A.

For the case of an *incompressible* fluid,

Incompressible: $$p - p_1 = -\gamma(z - z_1)$$ (3.3)

where p is the pressure at an elevation z. This expression is generally applicable to liquids, since they are only very slightly compressible. Only where there are large changes in elevation, as in the ocean, do we need to consider the compressibility of the liquid, to arrive at an accurate determination of pressure variation. For small changes in elevation, Eq. (3.3) will give accurate results when applied to gases.

For the case of a liquid at rest, it is convenient to measure distances vertically downward from the free liquid surface. If h is the distance below the free liquid surface and if the pressure of air and vapor on the surface is arbitrarily taken as zero, we can also write Eq. (3.3) as

Incompressible: $$p = \gamma h$$ (3.4)

In fact, there must always be some pressure on the surface of any liquid, so the total pressure at any depth h is given by Eq. (3.4) plus the pressure on the surface. In many situations this surface pressure may be disregarded, as we point out in Sec. 3.4.

From Eq. (3.4), we can see that all points in a connected body of constant density fluid at rest are under the same pressure if they are at the same depth below the liquid surface. This is known as Pascal's law, in honor of Blaise Pascal (1623–1662), a French mathematician who clarified and contributed to early principles of hydrostatics, and after whom we now name the unit of pressure in the SI system. Pascal's law indicates that a surface of equal pressure for a liquid at rest is a horizontal plane. Strictly speaking, it is a surface everywhere normal to the direction of gravity and is approximately a spherical surface concentric with the earth. For practical purposes, a limited portion of this surface may be considered a plane area.

Sample Problem 3.3
Similar to Sample Problem 3.2, to what depth can a diver descend before risking eardrum rupture, assuming the ear cannot equilibrate the pressures on either side of the eardrum?

Solution
From Sample Problem 3.2: $\Delta p = 35\ kPa$

From Eq. (3.4): $$\Delta h = \frac{\Delta p}{\gamma_w} = \frac{35\ \text{kN/m}^2}{9.81\ \text{kN/m}^3} = 3.57\ \text{m}\quad\textbf{ANS}$$

Note that for Sample Problem 3.2 the pressure in the middle ear exceeds that of the outer ear when ascending in the atmosphere. For a diver, the case is reversed, so the stress on the eardrum is opposite. In both cases, the key to avoid discomfort or injury is to equilibrate the pressure on either side of the eardrum.

3.3   Pressure Expressed in Height of Fluid

Imagine an open tank of liquid with no pressure acting on its surface (Fig. 3.3), though in reality the minimum pressure upon any liquid surface is the pressure of its own vapor. Disregarding this for the moment, by Eq. (3.4), the pressure at any depth h is $p = \gamma h$. If we assume γ to be constant, there is a definite relation between p and h. That is, pressure (i.e., force per unit area) is equivalent to a height h of some fluid of constant specific weight γ. Often we find it more convenient to express pressure in terms of a height of a column of fluid rather than in pressure per unit area.

Even if the surface of the liquid is under some pressure, we only need to convert this pressure into an equivalent height of the fluid in question and add this to the value of h shown in Fig. 3.3, to obtain the total pressure.

For the preceding discussion we considered a liquid, but, providing it is appropriate, it is equally possible to apply it to a gas or vapor by specifying some *constant* specific weight γ for the gas or vapor in question. Thus we may relate pressure p to the height of a column of *any* fluid by the expression

$$h = \frac{p}{\gamma} \tag{3.5}$$

This relationship is true for any consistent system of units. If p is in pounds per square foot, γ must be in pounds per cubic foot, and then h will be in feet. In SI units, we may express p in kilopascals (kilonewtons per square meter), in which case if γ is in kilonewtons per cubic meter, h will be in meters. When we express pressure in this way, in terms of a height of fluid, we commonly refer to it as **pressure head** (see Sec. 5.8). Because we commonly express pressure in pounds per square inch (or kPa in SI units), and since we usually assume the value of γ for water to be 62.4 lb/ft³ (9.81 kN/m³), a convenient relationship is

$$h \text{ (ft of H}_2\text{O)} = \frac{144 \times \text{psi}}{62.4} = 2.308 \times \text{psi}$$

or

$$h \text{ (m of H}_2\text{O)} = \frac{\text{kPa}}{9.81} = 0.1020 \times \text{kPa}$$

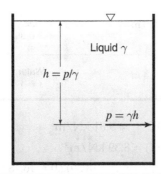

Figure 3.3   Pressure-depth relation for an incompressible liquid in an open tank.

Often we find it more convenient to express pressures occurring in one fluid in terms of the height of another fluid, e.g., barometric pressure in millimeters of mercury.

An important property follows from Eq. (3.3), which we can express as:

Incompressible:
$$\frac{p}{\gamma} + z = \frac{p_1}{\gamma} + z_1 = \text{constant} \qquad (3.6)$$

This shows that for an incompressible fluid at rest, at any point in the fluid the sum of the elevation z and the pressure head p/γ is equal to the sum of these two quantities at any other point. The significance of this statement is that, in a fluid at rest, with an increase in elevation there is a decrease in pressure head, and vice versa. This concept is depicted in Fig. 3.4.

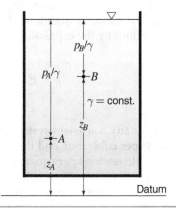

FIGURE 3.4   Variation of pressure with depth in an incompressible fluid.

Sample Problem 3.4
An open tank contains water 1.40 m deep covered by a 2-m-thick layer of oil ($s = 0.855$). What is the pressure head at the bottom of the tank, in terms of a water column?

Solution 1

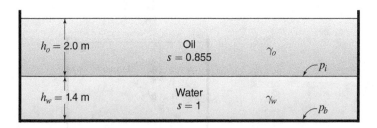

From Appendix A or D: $\gamma_w = 9.81$ kN/m³

Sec. 2.3: $\gamma_o = 0.855(9.81) = 8.39$ kN/m³

Eq. (3.4) for interface: $p_i = \gamma_o h_o = (8.39)2.0 = 16.78$ kN/m² $= 16.78$ kPa

Eq. (3.5) for water equivalent of oil:

$$h_{oe} = \frac{p_i}{\gamma_w} = \frac{16.78 \text{ kN/m}^2}{9.81 \text{ kN/m}^3} = 1.710 \text{ m of water}$$

So $h_{we} = h_w + h_{oe} = 1.40 + 1.710 = 3.11 \text{ m of water}$ ***ANS***

Solution 2

From Eq. (3.4) for bottom of tank:

$$p_b = \gamma_o h_o + \gamma_w h_w = (8.39)2.0 + 9.81(1.4) = 30.51 \text{ kN/m}^2 = 30.51 \text{ kPa}$$

Eq. (3.5) for total water equivalent:

$$h_{we} = \frac{P_b}{\gamma_w} = \frac{30.51 \text{ kN/m}^2}{9.81 \text{ kN/m}^3} = 3.11 \text{ m of water}$$ ***ANS***

3.4 Absolute and Gage Pressures

If we measure pressure relative to absolute zero, we call it ***absolute*** pressure; when we measure it relative to atmospheric pressure as a base, we call it ***gage*** pressure. This is because practically all pressure gages register zero when open to the atmosphere, and so they measure the difference between the pressure of the fluid they are connected to and that of the surrounding air.

If the pressure is below that of the atmosphere, we call it a ***vacuum***, and its gage value is the amount by which it is *below* that of the atmosphere. What we call a "high vacuum" is really a low absolute pressure; a perfect vacuum would correspond to absolute zero pressure.

All values of absolute pressure are positive, since a negative value would indicate tension, which we normally consider impossible in any fluid.[2] Gage pressures are positive if they are above that of the atmosphere and negative if they are vacuum (Fig. 3.5).

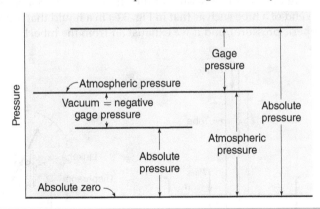

FIGURE 3.5   The relationship between gage and absolute pressures.

[2] For an exception to this statement, see footnote 9 in Chap. 2.

We can see from the preceding discussion that the following relation holds,

Incompressible: $$p_{abs} = p_{atm} + p_{gage} \qquad (3.7)$$

where p_{gage} may be positive or negative (vacuum).

We also call the atmospheric pressure the **barometric** pressure, and it varies with elevation above sea level (Sec. 2.9). Also, at a given place it varies slightly from time to time because of changes in meteorological conditions.

In thermodynamics it is essential to use absolute pressure, because most thermal properties are functions of the actual (absolute) pressure of the fluid, regardless of the atmospheric pressure. For example, the property relation for a perfect gas [Eq. (2.4)] is an equation in which we must use absolute pressure. In fact, we must use absolute pressures in *most* problems involving gases and vapors.

Pressure does not usually much affect the properties of liquids, so we commonly use gage pressures in problems dealing with liquids. Also, we usually find that the atmospheric pressure appears on both sides of an equation, and hence cancels. Thus the value of atmospheric pressure is usually of no significance when dealing with liquids, and, for this reason as well, we almost universally use gage pressures with liquids. About the only situation where we need to consider the absolute pressure of a liquid is where its pressure approaches or equals the saturated vapor pressure (Sec. 2.13). Throughout this text we shall take all numerical pressures to be gage pressures unless they are specifically given as absolute pressures. But whenever confusion is possible, we should specify gage pressures with units like psig or kPa gage.

3.5 Measurement of Pressure

There are many ways to measure pressure in a fluid. Some of these are discussed in this section.

Barometer

We measure the absolute pressure of the atmosphere with a barometer. If we immerse the open end of a tube such as that in Fig. 3.6a in a liquid that is open to the atmosphere (atmospheric pressure), and if we exhaust air from the tube, liquid will rise in the tube.

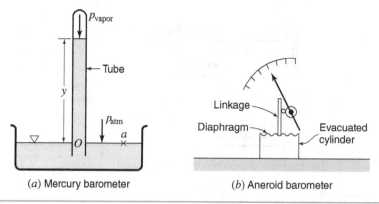

(a) Mercury barometer   (b) Aneroid barometer

FIGURE 3.6 Types of barometers.

If the tube is long enough and if we have removed all the air, the only pressure on the surface of the liquid in the tube will be that of its own vapor pressure, and the liquid will have reached its maximum possible height.

From the concepts developed in Sec. 3.2, we see that the pressure at O within the tube and at a on the surface of the liquid outside the tube must be the same; that is, $p_O = p_a = p_{atm}$. But, from Eq. (3.4) and Sec. 3.2,

$$p_O = \gamma y + p_{vapor}$$

Because of the static equilibrium, we may equate the pressures at O to obtain

$$p_{atm} = \gamma y + p_{vapor} \tag{3.8}$$

If the vapor pressure on the surface of the liquid in the tube were negligible, then we would have

$$p_{atm} = \gamma y$$

The liquid used in barometers of this type is usually mercury, because its density is sufficiently great to enable a reasonably short tube to be used, and also because its vapor pressure is negligibly small at ordinary temperatures. If we used some other liquid, the tube would need to be so high as to be inconvenient and its vapor pressure at ordinary temperatures would be appreciable; so a nearly perfect vacuum at the top of the column would not be attainable. Consequently, the height attained by the liquid would be less than the true barometric height and we would have to make a correction to the reading. When using a mercury barometer, to get as accurate a measurement of atmospheric pressure as possible, we should make corrections for capillarity and vapor pressure to the reading (Secs. 2.12 and 2.13).

An *aneroid* barometer measures the difference in pressure between the atmosphere and an evacuated cylinder by means of a sensitive elastic diaphragm and linkage system as depicted in Fig. 3.6b.

Since we use atmospheric pressure at sea level so widely and often, it is good to keep in mind equivalent forms of expression. By using Eq. (3.5) we find that we can express standard sea-level atmospheric pressure in the following different ways:

14.696 psia (2116.2 psfa)   or   101.325 kPa abs (1013.25 mb abs)

29.92 inHg   or   760 mmHg

33.91 ft of water   or   10.34 m of water.

For most engineering work, we generally round them to three or four significant figures.

Sample Problem 3.5

What would be the reading on a barometer containing carbon tetrachloride at 68°F at a time when the atmospheric pressure was equivalent to 30.26 inHg?

Solution

$$p_{atm} = 30.26 \text{ inHg} \times \frac{14.696 \text{ psia}}{29.92 \text{ inHg}} = 14.86 \text{ psia}$$

Table A.4 for carbon tetrachloride at 68°F:

$$\rho = 3.08 \text{ slugs/ft}^3, \quad p_{\text{vapor}} = 1.90 \text{ psia}$$

From Eq. (3.8):
$$y = \frac{p_{\text{atm}} - p_{\text{vapor}}}{\rho g}$$
$$= \frac{(14.86 - 1.90)\,144}{3.08\,(32.2)}$$
$$= 18.82 \text{ ft of carbon tetrachloride} \quad \textbf{\textit{ANS}}$$

Bourdon Gage

We commonly measure pressures or vacuums with the ***Bourdon gage*** of Fig. 3.7. In this gage, a curved tube of elliptical cross section changes its curvature with changes in pressure inside the tube; higher pressures tend to "straighten" it. The moving end of the tube rotates a hand on a dial through a linkage system. When a pressure and vacuum gage is combined into one we call this a ***compound gage*** (Fig. 3.8). The pressure indicated by such gages is that at their centers. If the connecting piping is filled completely with fluid of the same density as that in A of Fig. 3.7, and if the pressure gage is graduated to read in pounds per square inch, as is customary, then

$$p_A \text{ (psi)} = \text{gage reading (psi)} + \frac{\gamma h}{144}$$

where γ is expressed in pounds per cubic foot and h in feet.

A vacuum gage, or the negative-pressure portion of a compound gage, is traditionally graduated to read in millimeters or inches of mercury. For vacuums,

$$\text{inHg vacuum at } A = \text{gage reading (inHg vacuum)} - \frac{\gamma h}{144}\left(\frac{29.92}{14.70}\right)$$

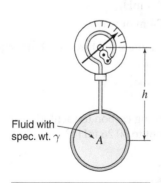

Fluid with spec. wt. γ

A

h

Figure 3.7   Bourdon gage.

Figure 3.8   P3S Series Compound pressure and vacuum gauge. Positive pressure in psi and vacuum inHg, both also in kPa. Many sizes and configurations are available. (Courtesy of Winters Instruments.)

Here, once again, we assume that this fluid completely fills the connecting tube of Fig. 3.7. The elevation-correction terms, i.e., those containing h, may be positive or negative, depending on whether the gage is above or below the point where we want to determine the pressure. The expressions given are for the situation depicted in Fig. 3.7. When measuring liquid pressures, the gage is usually set to measure the pressure at the centerline of the pipe. When measuring gas pressures, the elevation correction terms are generally negligible.

These expressions, when written in SI units, require no conversion factors; however, we must take care in dealing with decimal points when adding terms.

Pressure Transducer

A *transducer* is a device that transfers energy (in any form) from one system to another. A Bourdon gage, for example, is a mechanical transducer in that it has an elastic element that converts energy from the pressure system to a displacement in the mechanical measuring system. An *electrical pressure transducer* converts the displacement of a mechanical system (usually a metal diaphragm) to an electric signal, either actively if it generates its own electrical output or passively if it requires an electrical input that it modifies as a function of the mechanical displacement. In one type of pressure transducer (Fig. 3.9) an electrical strain gage is attached to a diaphragm. As the pressure changes, the deflection of the diaphragm changes. This, in turn, changes the electrical output, which, through proper calibration, can provide pressure. If we connect such a device to a strip-chart recorder we can use it to give a continuous record of pressure. Instead of a strip-chart recorder, we may record the data at fixed time intervals on a tape or disk using a computer data acquisition system and/or we may display it on a panel in digital form.

Piezometer Column

A piezometer column is a simple device for measuring moderate pressures of liquids. It consists of a sufficiently long tube (Fig. 3.10) in which the liquid can freely rise without overflowing. The height of the liquid in the tube will give the value of the pressure head, p/γ (Secs. 3.3 and 5.8), directly. To reduce capillary error (Sec. 2.12) the tube diameter should be at least 0.5 in (12 mm).

If we wish to measure the pressure of a *flowing* fluid, we should take special precautions in making the connection. The hole must be absolutely normal to the interior

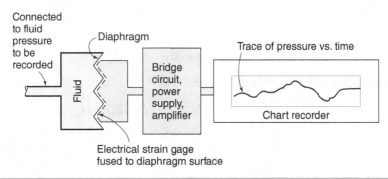

FIGURE 3.9 Schematic of an electrical strain-gage pressure transducer with a strip-chart recorder.

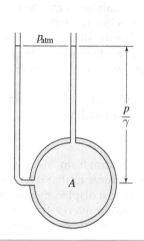

FIGURE 3.10   Piezometer (for measuring p/γ in liquids only).

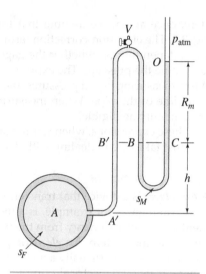

FIGURE 3.11   Open-end manometer (for measuring p/γ in liquids or gases).

surface of the wall, and the piezometer tube or the connection for any other pressure-measuring device must not project beyond the surface. There can be no burrs and surface roughness near the hole, and it is well to round the edge of the hole slightly. Also, the hole should be small, preferably not larger than $\frac{1}{8}$ in (3 mm) diameter.

Simple Manometer

Since the open piezometer tube is too tall and cumbersome for use with liquids under high pressure, and it cannot be used with gases, the simple manometer or mercury U tube of Fig. 3.11 is a convenient device for measuring many pressures. To determine the *gage pressure* or the *gage pressure head* at A, in terms of the liquid at A, we may write a *gage equation* based on the fundamental relations of hydrostatic pressures [Eq. (3.3)]. We can use any units of pressure or pressure head in the gage equation, providing the resulting dimensions of each term are the same. Let us define s_M as the specific gravity of the *manometer* (M) fluid (or gage fluid) and s_F as the specific gravity of the *fluid* (F) whose pressure is being measured. Also, let us identify a manometer reading by R_m; in Fig. 3.11 this is the height OC. If y' is the height of a column of measured fluid (F) that would exert the same pressure at C as does the column of manometer fluid OC, height R_m, then, from Eq. (3.4),

$$\text{gage pressure } p_C = \gamma_M R_m = \gamma_F y'$$

and by rearranging, making use of Sec. 2.3,

$$y' = (\gamma_M/\gamma_F)\,R_m = (\rho_M/\rho_F)\,R_m = (s_M/s_F)R_m$$

Thus the gage pressure at C, in terms of the fluid whose pressure we are measuring, as required, is $\gamma(s_M/s_F)R_m$. This is also the pressure at B because the fluid in BC is in balance. The pressure at A is greater than this by γh, assuming the fluid in the connecting tube $A'B$ is of the same specific weight as that of the fluid at A. For this simple case we can write down the pressure at A directly. But for more complicated gages it is helpful to follow a sequence of steps:

1. Commence the equation at the open end of the manometer with the pressure there (which is known).
2. Proceed through the entire tube to A, adding pressure terms when descending and subtracting them when ascending, all in terms of equivalent pressures of measured fluid (F).
3. For continuous portions of the same fluid with the same end elevations, like BC and $B'B$, following Pascal's law [Eq. (3.4)] they are in balance, so heads are equal and the calculation procedure can directly jump from one portion to the other.
4. Equate the result to the pressure at A.

Thus, for Fig. 3.11,

$$0 + \gamma \left(\frac{s_M}{s_F} \right) R_m + \gamma h = p_A \tag{3.9a}$$

where γ is the specific weight of the liquid at A. If desired, we can perform the same analysis by expressing the terms in units of head rather than pressure. Then, proceeding from O to A in Fig. 3.11,

$$0 + \left(\frac{s_M}{s_F} \right) R_m + h = \frac{p_A}{\gamma} \tag{3.9b}$$

If we multiply this through by γ ($= \gamma_F$), we see it is the same as Eq. (3.9a).

If we want the absolute pressure or the **absolute pressure head** at A then the zero of the first term in Eq. (3.9) must be replaced by the atmospheric pressure (or pressure head) expressed in terms of the fluid whose pressure we are measuring. When measuring the pressure in liquids, an air-relief valve V (Fig. 3.11) will provide a means for the escape of gas should any become trapped in tube $A'B$. If the fluid in A is a gas, the pressure and pressure head contribution from the distance h is generally very small and we can neglect it because of the relatively small specific gravity (or density) of the gas.

When measuring a vacuum, for which we might use the arrangement in Fig. 3.12, the resulting gage equation for pressure head, subject to the same conditions as in the preceding case, is

$$0 - \left(\frac{s_M}{s_F} \right) R_m + h = \frac{p_A}{\gamma} \tag{3.10}$$

Again, it would simplify the equation if we were measuring pressure in a gas, because the h term is then negligible. In measuring vacuums in liquids the arrangement in Fig. 3.13 is advantageous, since gas and vapors cannot become trapped in the tube. For this case,

$$0 - \left(\frac{s_M}{s_F} \right) R_m - h = \frac{p_A}{\gamma}$$

or

$$\frac{p_A}{\gamma} = - \left[h + \left(\frac{s_M}{s_F} \right) R_m \right] \tag{3.11}$$

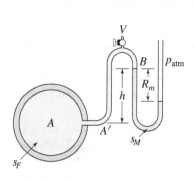

FIGURE 3.12   Negative-pressure manometer.

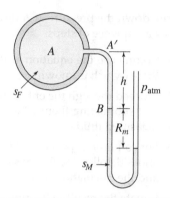

FIGURE 3.13   Negative-pressure manometer.

Although we generally use mercury as the measuring fluid in the simple manometer, we sometimes use other liquids (carbon tetrachloride, for example). As the specific gravity of the measuring fluid approaches that of the fluid whose pressure we are measuring, the reading becomes larger for a given pressure, thus increasing the accuracy of the instrument, providing the two specific gravities are accurately known.

Differential Manometers

In many cases we need to know only the *difference* between two pressures, and for this purpose we can use differential manometers, such as shown in Fig. 3.14. In Fig. 3.14a the

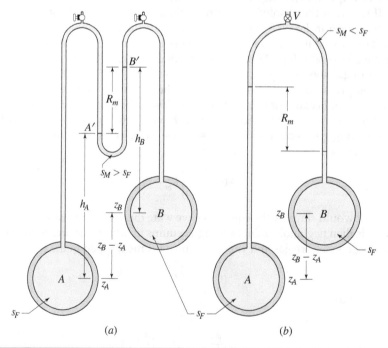

FIGURE 3.14   Differential manometers. (a) For measuring Δp in liquids or gases. (b) For measuring Δp in liquids only.

measuring fluid has a *greater* density than that of the fluid whose pressure difference we seek. If the fluids in A and B (Fig. 3.14a) have the same density, then, proceeding in a similar manner as before, through the manometer tubing from A to B, we obtain

$$\frac{p_A}{\gamma} - h_A - \left(\frac{s_M}{s_F}\right) R_m + h_B = \frac{p_B}{\gamma}$$

So, by rearranging

$$\frac{p_A}{\gamma} - \frac{p_B}{\gamma} = h_A - h_B + \left(\frac{s_M}{s_F}\right) R_m$$

But, from Fig. 3.14a,

$$h_A + R_m = h_B + (z_B - z_A)$$

where z represents elevation, so

$$h_A - h_B = (z_B - z_A) - R_m$$

so that

$$\frac{p_A}{\gamma} - \frac{p_B}{\gamma} = (z_B - z_A) - R_m + \left(\frac{s_M}{s_F}\right) R_m = (z_B - z_A) + \left(\frac{s_M}{s_F} - 1\right) R_m \qquad (3.12a)$$

Later, we will find it convenient to also write this as

$$\left(\frac{p_A}{\gamma} + z_A\right) - \left(\frac{p_B}{\gamma} + z_B\right) = \Delta\left(\frac{p}{\gamma} + z\right) = \left(\frac{s_M}{s_F} - 1\right) R_m \qquad (3.12b)$$

where the left of these equations provides a definition of $\Delta(p/\gamma + z)$

Equation (3.12) is applicable only if the fluids in A and B have the same density. If these densities are different, we can find the pressure head difference by expressing all head components between A and B in terms of one or other of the fluids, as in Sample Problem 3.6. We must emphasize that by far the most common mistakes made in working differential-manometer problems are to omit the factor $(s_M/s_F - 1)$ for the gage difference R_m, or to omit the -1 from this factor. The term $(s_M/s_F - 1)R_m$ accounts for the *difference* in pressure heads due to the two columns of liquids (M) and (F) of height R_m in the U tube.

The differential manometer, when used with a heavy liquid such as mercury, is suitable for measuring large pressure differences. For a small pressure difference, however, a *light* fluid, such as oil, or even air, is preferable, in which case the manometer is arranged as in Fig. 3.14b. Of course, the manometer fluid must be one that will not mix with the fluid in A or B. By the same method of analysis as above, we can show for Fig. 3.14b that, for identical liquids in A and B,

$$\frac{p_A}{\gamma} - \frac{p_B}{\gamma} = (z_B - z_A) + \left(1 - \frac{s_M}{s_F}\right) R_m \qquad (3.13a)$$

or

$$\Delta\left(\frac{p}{\gamma} + z\right) = \left(1 - \frac{s_M}{s_F}\right) R_m \qquad (3.13b)$$

Here (s_M/s_F), the ratio of the specific gravities (or densities or specific weights), has a value less than one. As the density of the manometer fluid approaches that of the fluid being measured, $(1 - s_M/s_F)$ approaches zero, and we will obtain larger values of R_m for small pressure differences, thus increasing the sensitivity of the gage. Once again, we must modify the equation if the densities of fluids A and B are different.

To determine pressure difference between liquids, we often use air or some other gas as the measuring fluid, with the manometer arrangement of Fig. 3.14b. Air can be pumped through valve V until the pressure is sufficient to bring the two liquid columns to suitable levels. Any change in pressure raises or lowers both liquid columns by the same amount, so that the difference between them is constant. In this case the value of (s_M/s_F) can be considered to be zero, since the density of gas is so much less than that of a liquid.

Another way to obtain increased sensitivity is simply to incline the gage tube so that a vertical gage difference R_m is transposed into a reading that is magnified by $1/\sin\alpha$, where α is the angle of inclination with the horizontal.

Sample Problem 3.6

In Fig. S3.6 liquid A weighs 53.5 lb/ft³ (8.4 kN/m³) and liquid B weighs 78.8 lb/ft³ (12.4 kN/m³). Manometer liquid M is mercury. If the pressure at B is 30 psi (207 kPa), find the pressure at A. Express all pressure heads in terms of the liquid in bulb B.

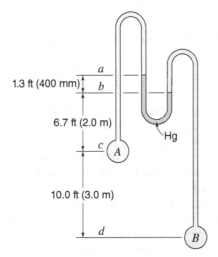

1.3 ft (400 mm)
6.7 ft (2.0 m)
Hg
10.0 ft (3.0 m)

FIGURE S3.6

Solution
Proceeding from A to B:

From Eq. (3.13): $\dfrac{p_A}{\gamma_B} - (z_a - z_c)\dfrac{\gamma_A}{\gamma_B} + (z_a - z_b)\dfrac{\gamma_M}{\gamma_B} + (z_b - z_d)\dfrac{\gamma_B}{\gamma_B} = \dfrac{p_B}{\gamma_B}$

BG units: $\dfrac{p_A}{\gamma_B} - (1.3 + 6.7)\dfrac{53.5}{78.8} + 1.3\dfrac{13.56(62.4)}{78.8} + (6.7 + 10.0) = \dfrac{p_B}{\gamma_B}$

$$\dfrac{p_A}{\gamma_B} - 5.43 + 13.96 + 16.7 = \dfrac{30(144)}{78.8} = 54.8 \text{ ft}$$

$$\dfrac{p_A}{\gamma_B} = 29.6 \text{ ft} \qquad p_A = 29.6\dfrac{78.8}{144} = 16.19 \text{ psi} \qquad ANS$$

SI units: $\dfrac{p_A}{\gamma_B} - 0.4 + 2.0\dfrac{8.4}{12.4} + 0.4\dfrac{13.56(9.81)}{12.4} + 2.0 + 3.0 = \dfrac{p_B}{\gamma_B}$

$\dfrac{p_A}{\gamma_B} - 1.626 + 4.29 + 5.00 = \dfrac{207\ \text{kN/m}^2}{12.4\ \text{kN/m}^3} = 16.69\ \text{m}$

$\dfrac{p_A}{\gamma_B} = 9.03\ \text{m}, \quad p_A = 9.03(12.4) = 112.0\ \text{kN/m}^2 = 112.0\ \text{kPa} \qquad \textbf{\textit{ANS}}$

3.6 Force on a Plane Area

As we noted previously in Sec. 3.1, no tangential force can exist within a fluid at rest. All forces are then normal to the surfaces in question. If the pressure is uniformly distributed over an area, the force is equal to the pressure times the area, and the point of application of the force is at the centroid of the area. For submerged horizontal areas, the pressure is uniform. In the case of compressible fluids (gases), the pressure variation with vertical distance is very small because of the low specific weight; therefore, *when we compute the static fluid force exerted by a gas, we usually treat p as a constant.* Thus, for such cases,

$$F = \int p\,dA = p\int dA = pA \tag{3.14}$$

In the case of liquids the distribution of pressure is generally not uniform, so further analysis is necessary. Let us consider a vertical plane whose upper edge lies in the free surface of a liquid (Fig. 3.15). Let this plane be perpendicular to the plane of the figure, so that MN is merely its trace, or edge. The gage pressure will vary from zero at M to NK at N. The total force on one side of the plane is the sum of the products of the elementary areas and the pressure upon them. From the pressure distribution, we can see that the resultant of this system of parallel forces must act at a point *below* the centroid of the area, since the centroid of an area is the point where the resultant of a system of *uniform* parallel forces would act.

If we lower the plane to position $M'N'$, the *proportionate* change of pressure from M' to N' is less than it was from M to N. Hence the resultant pressure force will act nearer to the centroid of the plane surface. The deeper we submerge the plane, the smaller the proportional pressure variation becomes, and the closer the resultant moves to the centroid.

In Fig. 3.16 let MN be the edge of a plane area making an angle θ with the horizontal. To the right we see the projection of this area onto a vertical plane. The pressure distribution over the sloping area forms a *pressure prism* ($MNKJ$ times width in Fig. 3.16), whose volume is equal to the total force F acting on the area. If the width x is constant then we can easily compute the volume of the pressure prism, using a mean pressure $= 0.5\,(MJ + NK)$, and so obtain F.

If x varies, we must integrate to find F. Let h be the variable depth to any point and let y be the corresponding distance from OX, the intersection of the plane containing the area and the free surface.

Choose an element of area so that the pressure over it is uniform. Such an element is a horizontal strip, of width x, so $dA = x\,dy$. As $p = \gamma h$ and $h = y\sin\theta$, the force dF on the horizontal strip is

$$dF = p\,dA = \gamma h\,dA = \gamma y \sin\theta\,dA$$

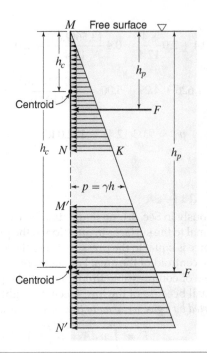

FIGURE 3.15 Pressure distributions on two vertical plane areas (viewed from edges).

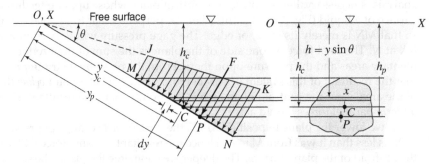

FIGURE 3.16 Pressure distribution on a sloping plane area (viewed from edge). *C* is centroid, *P* is center of pressure. Sloping *y* distances correspond to vertical *h* distances.

Integrating,
$$F = \int dF = \gamma \sin\theta \int y\,dA = \gamma \sin\theta\, y_c A \qquad (3.15)$$

where y_c is, by definition, the distance from OX along the sloping plane to the centroid C of the area A. If h_c is the vertical depth to the centroid, then $h_c = y_c \sin\theta$, and in general we have

$$F = \gamma h_c A \qquad (3.16)$$

Thus we find the total force on any plane area submerged in a liquid by multiplying the specific weight of the liquid by the product of the area and the depth of its centroid.

The value of F is independent of the angle of inclination of the plane so long as the depth of its centroid is unchanged.[3]

Since γh_c is the pressure at the centroid, we can also say that the total force on any plane area submerged in a liquid is the product of the area and the pressure at its centroid.

3.7   Center of Pressure

The point of application of the resultant pressure force on a submerged area is called the *center of pressure*. We need to know its location whenever we wish to work with the moment of this force.

The most general way of looking at the problem of forces on a submerged plane area is through the use of the recently discussed *pressure prism concept* (Sec. 3.6 and Fig. 3.16). The line of action of the resultant pressure force must pass through the centroid of the pressure prism (volume). As noted earlier, this concept is very convenient to apply for simple areas such as rectangles. For example, if the submerged area in Fig. 3.15 is of constant width then we know that the centroid of the pressure prism on area MN is $\frac{2}{3} MN$ below M.

If the shape of the area is not so regular, i.e., if the width x in Fig. 3.16 varies, then we must take moments and integrate. Taking OX in Fig. 3.16 as an axis of moments, the moment of an elementary force $dF = \gamma y \sin \theta dA$ is

$$y \, dF = \gamma y^2 \sin \theta dA$$

and if y_p denotes the distance to the center of pressure, using the basic concept that the moment of the resultant force equals the sum of the moments of the component forces,

$$y_p F = \gamma \sin \theta \int y^2 dA = \gamma \sin \theta I_o$$

where we recognize that I_o is the moment of inertia of the plane area about axis OX, and by definition $I_o = \int y^2 dA$.

If we divide this expression for $y_p F$ by the value of F given by Eq. (3.15), we obtain

$$y_p = \frac{\gamma \sin \theta \, I_o}{\gamma \sin \theta \, y_c A} = \frac{I_o}{y_c A} \tag{3.17}$$

The product $y_c A$ is the static moment of area A about OX. Therefore Eq. (3.17) tells us that we can obtain the distance from the center of pressure to the axis where the plane (extended) intersects the liquid surface by dividing the moment of inertia of the area A about the surface axis by its static moment about the same axis.

---

[3] For a plane submerged as in Fig. 3.16, it is obvious that Eq. (3.16) applies to one side only. As the pressure forces on the two sides are identical but opposite in direction, the net force on the plane is zero. In most practical cases where the thickness of the plane is not negligible, the pressures on the two sides are not the same.

We may also express this in another form, by noting from the parallel axis theorem that

$$I_o = Ay_c^2 + I_c$$

where I_c is the moment of inertia of an area about its centroidal axis. By substituting for I_o into Eq. (3.17),

$$y_p = \frac{Ay_c^2 + I_c}{y_c A}$$

so

$$y_p = y_c + \frac{I_c}{y_c A} \tag{3.18}$$

From this equation, we again see that the location of the center of pressure P is independent of the angle θ; that is, we can rotate the plane area about axis OX without affecting the location of P. Also, we see that P is always *below* the centroid C and that, as the depth of immersion is increased, y_c increases and therefore P approaches C.

For convenient reference, Table A.7 of Appendix A contains values of y_c and I_c for a variety of area shapes.

We can find the lateral position of the center of pressure P by considering that the area is made up of a series of elemental horizontal strips. The center of pressure for each strip is at the midpoint of the strip. Since the moment of the resultant force F must be equal to the moment of the distributed force system about any axis, say, the y axis,

$$X_p F = \int x_p p \, dA \tag{3.19}$$

where X_p is the lateral distance from the selected y axis to the center of pressure P of the resultant force F, and x_p is the lateral distance to the center of any elemental horizontal strip of area dA on which the pressure is p.

Sample Problem 3.7

A rectangular plate 5 ft by 4 ft is at an angle of 30° with the horizontal, and the 5-ft side is horizontal. The top edge is 1 ft below the water surface. It is hinged at point A, and a force at B is applied to counteract the hydrostatic force. Determine the magnitude of the force at B required for the system to be in equilibrium.

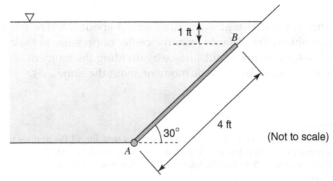

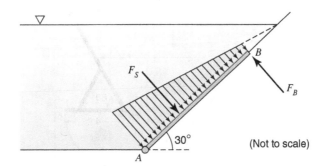

(Not to scale)

Solution

For water, use $\gamma = 62.4$ lb/ft³ (Appendices A and D).

$$h_c = 1 + \frac{1}{2}4 \text{ ft·sin } 30° = 2.0 \text{ ft}$$

Eq. (3.16): $F = \gamma h_c A = \left(62.4\frac{\text{lb}}{\text{ft}^3}\right)(2.0 \text{ ft})(4 \times 5 \text{ ft}^2) = 2496 \text{ lb}$

Determine where the force acts using Eq. (3.18):

$$y_c = \frac{h_c}{\sin 30°} = 4.0 \text{ ft}$$

$$y_p = y_c + \frac{I_c}{y_c A} = 4.0 + \frac{(1/12)\cdot4^3\cdot5}{4.0(4 \times 5)} = 4.333 \text{ ft,}$$

Find the reaction at B by taking moments about A. Assuming clockwise is positive:

$$\sum M_A = 0 = 2498(6 - 4.333) - F_B(4.0)$$

$$F_B = 1040 \text{ lb} \qquad ANS$$

Sample Problem 3.8

Figure S3.8 represents a gate, 2 ft wide perpendicular to the sketch. It is pivoted at hinge H. The gate weighs 500 lb. Its center of gravity is 1.2 ft to the right of and 0.9 ft above H. For what values of water depth x above H will the gate remain closed? Neglect friction at the pivot and neglect the thickness of the gate.

: Programmed computing aids could help solve problems marked with this icon.

64   Chapter Three

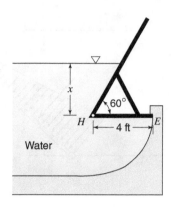

Figure S3.8

Solution
In addition to the reactive forces R_H at the hinge and R_E at end E, there are three forces acting on the gate: its weight W, the vertical hydrostatic force F_v upward on the rectangular bottom of the gate, and the slanting hydrostatic force F_s acting at right angles to the sloping rectangular portion of the gate. The magnitudes of the latter three forces are:

Given: $$W = 500 \text{ lb}$$

Eq. (3.16): $$F_v = \gamma h_c A = \gamma(x)(4 \times 2) = 8\gamma x$$

Eq. (3.16): $$F_S = \gamma h_c A = \gamma\left(\frac{x}{2}\right)\left(\frac{x}{\cos 30°} \times 2\right) = 1.155\gamma x^2$$

A diagram showing these three forces follows:

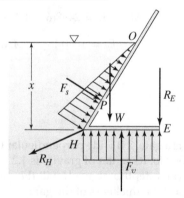

The moment arms of W and F_v with respect to H are 1.2 ft and 2.0 ft, respectively. The moment arm of F_s gets larger as the water depth increases because the location of the center of pressure changes. We can find the location of the center of pressure of F_s from Eq. (3.18):

$$y_p = y_c + \frac{I_c}{y_c A}, \quad \text{where, from Table A.7, } I_c = \frac{bh^3}{12}$$

with $h = x/\cos 30°$ and $y_c = 0.5h$. So

$$y_p = \frac{0.5x}{\cos 30°} + \frac{(1/12)2(x/\cos 30°)^3}{(0.5x/\cos 30°)[2(x/\cos 30°)]}$$

i.e., for F_s: $OP = y_p = 0.577x + \frac{2x}{12\cos 30°} = 0.770x$

Hence the moment arm of F_s with respect to H is $PH = x/\cos 30° - 0.770x = 0.385x$. [Note: In this case we need not use Eq. (3.18) to find the lever arm of F_s because we know the line of action of F_s for the triangular distributed load on the rectangular area is at the one-third point between H and O, i.e., $HP = (1/3)(x/\cos 30°) = 0.385x$.]

When the gate is about to open (incipient rotation), $R_E = 0$ and the sum of the moments of all forces about H is zero, viz.,

$$\sum M = F_s(0.385x) - F_v(2.0) + W(1.2) = 0$$

i.e., $1.155\gamma x^2(0.385x) - 8\gamma x(2) + 500(1.2) = 0$

Substituting $\gamma = 62.4$ lb/ft^3 gives

$$27.73x^3 - 998.4x + 600 = 0$$

This is a cubic, polynomial equation (Appendix B). With a polynomial solver, available on some hand calculators (see Appendix B), we may find the three roots directly. With an equation solver, available on some scientific calculators and in some spreadsheets and mathematics software, we may obtain the root closest to a guessed value we provide.

Without any of these aids, we can solve this equation by trials ("trial and error"), seeking an x value that makes the left side of the equation equal to zero. After two trials, we can use linear interpolation (or extrapolation) to estimate the next, better trial value. We then repeat this until x is sufficiently accurate, e.g., accurate to three significant figures after rounding:

Trial x	Left Side
0.1	500.2
0.5	104.3
0.6	6.95
0.61	−2.73
0.607	0.173

We can find the other two roots by more, similar, trials. We could use a spreadsheet to facilitate such trials. But, more conveniently, dividing the cubic by $(x - 0.607)$ yields a quadratic [Eq. (B.6)] from which we can easily find that the other two roots [Eq. (B.7)] are $x = 5.67$ and -6.28.

Thus $x = 0.607$ ft or 5.67 ft or a negative (meaningless) root. Therefore, from inspection of the moment equation, the gate will remain closed when 0.607 ft $< x <$ 5.67 ft.   *ANS*

Sample Problem 3.9

The cubic tank shown in Fig. S3.9 is half full of water. Find (*a*) the pressure on the bottom of the tank, (*b*) the force exerted by the fluids on a tank wall, and (*c*) the location of the center of pressure on a wall.

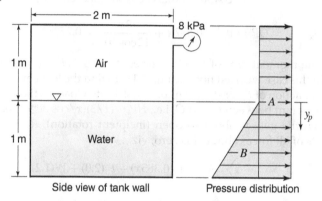

Side view of tank wall   Pressure distribution

FIGURE S3.9

Solution

(*a*)
$$p_{bott} = p_{air} + \gamma_{water} h_{water} = 8 \text{ kN/m}^2 + (9.81 \text{ kN/m}^3)(1 \text{ m})$$
$$= 17.81 \text{ kN/m}^2 = 17.81 \text{ kPa} \quad ANS$$

(*b*) The force acting on the tank end is divided into two components, labeled *A* and *B* on the pressure distribution sketch. Component *A* has a *uniform* pressure distribution, due to the pressure of the confined air, which acts throughout the water:

$$F_A = p_{air} A_{air} = (8 \text{ kN/m}^2)(4 \text{ m}^2) = 32.0 \text{ kN}$$

For component *B*, i.e., the varying water pressure distribution on the lower half of the tank wall, the centroid *C* of the area of application is at

$$h_c = y_c = 0.5(1 \text{ m}) = 0.5 \text{ m below the water top surface,}$$

so, from Eq. (3.16),

$$F_B = \gamma_{water} h_c A_{water} = 9.81 (0.5)2 = 9.81 \text{ kN}$$

So, the total force on the tank wall is

$$F = F_A + F_B = 32.0 + 9.81 = 41.8 \text{ kN} \quad ANS$$

(*c*) The locations of the centers of pressure of the component forces, as distances y_p below the water top surface, are

$$(y_p)_A = 0 \text{ m}$$

below the water top surface, to the centroid of the 2-m-square area for the uniform air pressure.

$$(y_p)_B = \frac{2}{3} h_{water} = \frac{2}{3}(1 \text{ m}) = 0.667 \text{ m}$$

below the water top surface for the varying pressure on the rectangular wetted wall area. We could also find this using Eq. (3.18) with $y_c = 0.5$ m, $I_c = bh^3/12 = 2(1)^3/12 = 0.1667$ m⁴, and $A = bh = 2$ m².

Taking moments; $F(y_p) = F_A(y_p)_A + F_B(y_p)_B$

from which $y_p = 0.1565$ m below the water top surface   *ANS*

Sample Problem 3.10

Water and oil in an open storage tank are in contact with the end wall as shown in Fig. S3.10. (*a*) Find the pressure at the bottom (lowest point) of the tank caused by the liquids. Also find (*b*) the total force exerted on the end wall by the liquids, and (*c*) the depth of its center of pressure.

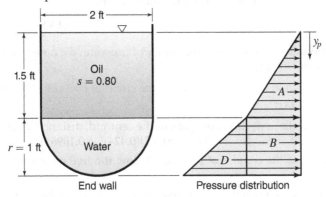

FIGURE S3.10

Solution

(*a*) $p_{bott} = \gamma_{oil}h_{oil} + \gamma_{water}h_{water}$

$= (0.8 \times 62.4 \text{ lb/ft}^3)(1.5 \text{ ft}) + (62.4 \text{ lb/ft}^3)(1.0 \text{ ft})$

$= 137.3 \text{ lb/ft}^3 = 0.953 \text{ psi}$ *ANS*

(*b*) The force acting on the end consists of three components, labeled *A*, *B*, and *D*, on the pressure distribution sketch. Note that component *B* has a *uniform* pressure distribution, due to the oil (*A*) above, which acts throughout the liquid below.

As a preliminary, we note for the semicircular end area ($r = 1$ ft) that

(i) $A = \pi r^2/2 = \pi 1^2/2 = 1.571 \text{ ft}^2$;

(ii) From Appendix A, Table A.7, the centroid is $\dfrac{4r}{3\pi} = 0.424$ ft from the center of the circle, i.e., below the water top surface.

For component *A*, i.e., the varying oil pressure distribution on the 1.5-ft height of the end wall, the centroid *C* of the area of application is at

$h_c = y_c = 0.5(1.5 \text{ ft}) = 0.75 \text{ ft}$ below the free oil surface,

68   Chapter Three

so, from Eq. (3.16),

$$F_A = \gamma_{oil} h_c A_{oil} = (0.8 \times 62.4)0.75(1.5 \times 2) = 112.3 \text{ lb}$$

For component B, the force F_B on the water-wetted area of the end wall due to the uniform pressure produced by the 1.5-ft depth of oil above is

$$F_B = pA = \gamma h A = (0.8 \times 62.4)\, 1.5(\pi 1^2/2) = 117.6 \text{ lb}$$

For component D, i.e., the varying pressure distribution due to the water (only) on the water-wetted area of the end wall, the centroid C is at

$$h_c = y_c = 0.424 \text{ ft below the water top surface,}$$

so:
$$F_D = \gamma h_c A = 62.4(0.424)\pi 1^2/2 = 41.6 \text{ lb}$$

The total force F on the end of the tank is therefore

$$F = F_A + F_B + F_D = 272 \text{ lb} \qquad ANS$$

(c) As a preliminary to locating the center of pressure, we note that for the *semi*circular end area with $D = 2$ ft,

(i) From Table A.7: I about the center of the circle, is $I = \pi D^4/128 = \pi 2^2/128 = 0.393 \text{ ft}^4$, and

(ii) By the parallel axis theorem: I_c about the centroid, distance 0.424 ft below the center of the circle, is $I_c = I - Ad^2 = 0.393 - (\pi 1^2/2)(0.424)^2 = 0.1098 \text{ ft}^4$.

The locations of the centers of pressure, below the free oil surface, of the component forces are:

$(y_p)_A = \frac{2}{3}(1.5 \text{ ft}) = 1.000 \text{ ft}$ for the varying oil pressure on the oil-wetted area, and

$(y_p)_B = y_c = 1.5 + 0.424 = 1.924 \text{ ft}$ to the *centroid* of the water-wetted semi-circular area, for the uniform pressure on this area due to 1.5 ft of oil above the water; and

Eq. (3.18): $(y_p)_D = y_c + \dfrac{I_c}{y_c A} = 0.424 + \dfrac{0.1098}{0.424(\pi 1^2/2)} = 0.589 \text{ ft}$

below the water top surface, for the varying water pressure on the water-wetted semi-circular area,

$$= 1.5 + 0.589 = 2.09 \text{ ft below the free oil surface.}$$

Finally,
$$Fy_p = F_A(y_p)_A + F_B(y_p)_B + F_D(y_p)_D$$
$$= 1.567 \text{ ft} \qquad ANS$$

3.8   Force on a Curved Surface

On any curved or warped surface such as MN in Fig. 3.17a, the force on the various elementary areas that make up the curved surface are different in direction and magnitude, so an algebraic summation is impossible. Hence we can apply Eq. (3.16) only to a plane area. But for nonplanar areas, we can find **component forces** in certain directions,

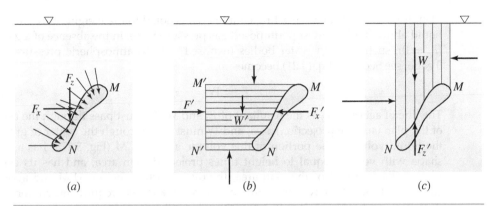

Hydrostatic forces on curved surfaces.

and often without integration. While often less efficient, the method of component forces can be applied to plane areas as well.

Horizontal Force on Curved Surface

We may project any irregular curved area MN (Fig. 3.17a) onto a vertical plane whose trace is $M'N'$ (Fig. 3.17b). The projecting elements, which are all horizontal, enclose a volume whose ends are the vertical plane $M'N'$ and the irregular area MN. This volume of liquid is in static equilibrium. A force F' acts on the projected vertical area $M'N'$. The horizontal force component F_x' acts on the irregular end area MN and is equal and opposite to the F_x of Fig. 3.17a. Gravity force W' is vertical, and the lateral forces on all the horizontal projection elements are normal to these elements and hence normal to F'. Thus the only horizontal forces on $MNN'M'$ are F' and F_x', and therefore

$$F' - F_x' = 0$$

and
$$F_x = F_x' = F' \tag{3.20}$$

Hence the horizontal force in any given direction on any area is equal to the force on the *projection* of that area onto a vertical plane normal to the given direction. The line of action of F_x must be the same as that of F'. Equation (3.20) applies to gases as well as liquids. In the case of a gas the horizontal force on a curved surface is equal to the pressure multiplied by the projection of that area onto a vertical plane normal to the force.

Vertical Force on a Curved Surface

We can find the vertical force F_z on a curved or warped area, such as MN in Fig. 3.17a, by considering the volume of liquid enclosed by the area and vertical elements extending to the free surface (Fig. 3.17c). This volume of liquid is in static equilibrium. The only vertical forces on this volume of liquid are the force $F_G = p_G A$ due to any gas (at pressure p_G) above the liquid, the gravity force W downward, and F_z' the upward vertical force on the irregular area MN. The force F_z' (Fig. 3.17c) is equal and opposite to the force F_z (Fig. 3.17a). Any other forces on the vertical elements are normal to the elements, and so are horizontal. Therefore,

$$F_z' - W - F_G = 0$$

and
$$F_z = F_z' = W + F_G \tag{3.21}$$

Therefore, the vertical force acting on any area is equal to the weight of the volume of liquid above it, plus any superimposed gas pressure force. In the absence of a gas force ($F_G = 0$), such as open water bodies (exposed only to atmospheric pressure, $p_{atm} = 0$ gage, see Sec. 3.4), Eq. (3.21) becomes:

$$F_z = W = \gamma \mathcal{V} \tag{3.22}$$

The line of action of F_z is the resultant of F_G and W. F_G must pass through the centroid of the plan (surface projection) area, and W must pass through the center of gravity of the liquid volume. The portion of this volume above $M'M$ (Fig. 3.17b) has a regular shape with volume equal to height times projected plan area, and has its centroid beneath the centroid of the plan area; the other portion, below $M'M$ and above the curved surface MN, may have a difficult shape and so may require integration to find its volume and centroid. If only a gas is involved, the procedure is similar, but is much simpler because W is negligible.

For the case where a force acts on the lower side of the surface but not on the upper side, the vertical force component is the same in magnitude as that given by Eq. (3.21) but opposite in sense.

Resultant Force on a Curved Surface

In general, there is no single resultant force on an irregular area, because the horizontal and vertical forces, as found in the above discussion, may not be in the same plane. But in certain cases these two forces will lie in the same plane and then we can combine them into a single force.

Sample Problem 3.11
Repeat Sample Problem 3.7 using the method of components.

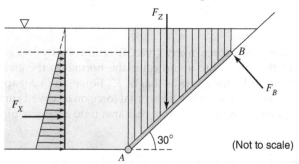

Solution
Horizontal component (on a vertical projection of the gate):

$$h_c = 1 + \tfrac{1}{2}4\sin 30° = 2.0 \text{ ft}$$

$$A = 4\sin 30° \times 5 = 10 \text{ ft}^2$$

Eq. (3.16): $F_x = \gamma h_c A = 62.4(2.0)(10) = 1248 \text{ lb}$

Adapting Eq. (3.18) for a vertical surface (refer to Fig. 3.15):

$$h_p = h_c + \frac{I_c}{h_c A} = 2 + \frac{(1/12)(5)(2)^3}{2(10)} = 2.1667 \text{ ft or } 2.17 \text{ ft}$$

Vertical component:

$$\text{Eq. (3.22)} \quad F_z = 62.4(2.0)(4\cos 30°)(5) = 2161.6 \text{ lb}$$

F_z acts through the center of volume, which for a rectangular plate is identical to the centroid of the shaded trapezoid. The horizontal distance from point A to the centroid, x_p, can be calculated by breaking the trapezoid into a rectangle and triangle:

$$x_p = \frac{\sum x_i A_i}{\sum A_i}$$

$$= \frac{[(4\cos 30°)/2]1(4\cos 30°) + [(4\cos 30°)/3](4\cos 30°)2/2}{1(4\cos 30°) + (2/2)4\cos 30°}$$

$$= 1.443 \text{ ft}$$

Find the reaction at B by taking moments about A:

$$\sum M_A = 0 = 1248(3 - 2.167) + 2161.6(1.443) - F_B(4.0)$$

$$F_B = 1040 \text{ lb} \quad ANS$$

Sample Problem 3.12

Assessing the hydrostatic forces on a dam face is critical for many reasons, one of which is that the resultant force, F_R, can tend to cause an overturning moment (rotating the structure clockwise about point A in the diagram). The dam in this example has a curved upstream face with a circular arc of $r = 6$ m, and the depth of water is 3 m, so $\theta = 30°$. Consider a 1-m section of the dam perpendicular to the diagram. (a) Find the magnitude of the resultant force, F_R on the face of the dam and the angle, α, at which it acts. (b) With the width of the top of the dam $w = 1$ m, determine whether the hydrostatic force would cause an overturning moment.

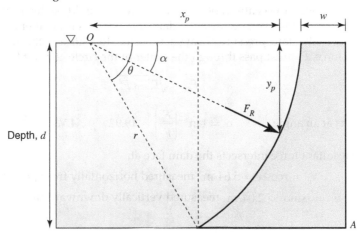

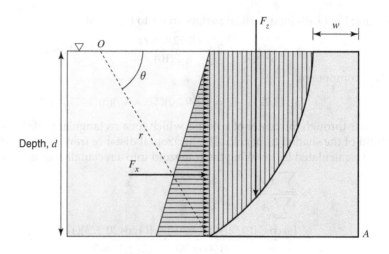

Solution

(*a*) The horizontal force is that on a vertical projection of the wetted dam face, which is a rectangle of height equal to the water depth.

$$h_c = \frac{1}{2}3 \text{ m} = 1.5 \text{ m}$$

Eq. (3.16): $F_x = \gamma h_c A = 9.81 \text{ kN/m}^3 (1.5 \text{ m})(3 \times 1 \text{ m}^2) = 44.145 \text{ kN}$

This acts through the centroid of the triangular prism, at one-third of the water depth measured up from the bottom.

The vertical force is the volume represented by the inclined shaded area, which is a circular segment (see Table A.7) less a triangular region.

$$\text{Volume} = 1 \text{ m}\left(\frac{\theta r^2}{2} - \frac{bh}{2}\right) = 1 \cdot \left(\frac{30°(\pi/180°)6^2}{2} - \frac{(6\cos 30°)3}{2}\right) = 1.631 \text{ m}^3$$

Eq. (3.22): $F_z = (9.81 \text{ kN/m}^3)(1.631 \text{ m}^3) = 16.00 \text{ kN}$ *ANS*

F_z acts through the centroid of the volume, which could be determined using composite volumes. However, because the dam face is circular another approach may be used. Since hydrostatic pressure always acts normal to a surface, the line of action of the resultant force, F_R must pass through the center of the circle at point O.

$$F_R = \sqrt{F_x^2 + F_z^2}$$

which acts at an angle $\alpha = \tan^{-1}\left(\frac{F_z}{F_x}\right) = 19.92°$ *ANS*

(*b*) The resultant force intersects the dam face at:

$$x_p = r\cos\alpha = 5.64 \text{ m} \quad \text{measured horizontally from point O}$$
$$y_p = r\sin\alpha = 2.04 \text{ m} \quad \text{measured vertically downward from point O}$$

Taking moments about point A, assuming clockwise is positive:

$$\sum M_A = F_x(3 - y_p) - F_z(r + w - x_p) = 44.145(3 - 2.04) - 16.00(6 + 1 - 5.64)$$
$$= 20.6 \text{ kN·m}$$

A positive result indicates the hydrostatic pressure produces an overturning moment. The weight of the dam would need to provide a righting moment to counteract this with an adequate safety factor. *ANS*

Sample Problem 3.13
Find the horizontal and vertical components of the force exerted by the fluids on the horizontal cylinder in Fig. S3.13 if (*a*) the fluid to the left of the cylinder is a gas confined in a closed tank at a pressure of 35.0 kPa; (*b*) the fluid to the left of the cylinder is water with a free surface at an elevation coincident with the uppermost part of the cylinder. Assume in both cases that atmospheric pressure occurs to the right of the cylinder.

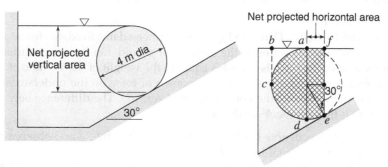

FIGURE S3.13

Solution
The net projection on a vertical plane of the portion of the cylindrical surface under consideration (see left-hand diagram) is, from the right-hand diagram, $ef = 2 + 2\cos 30°$ = 3.73 m.

(*a*) For the gas,

$$F_x = pA_z = (35.0 \text{ kN/m}^2)(3.73 \text{ m}) = 130.5 \text{ kN/m to the right} \quad ANS$$

The vertical force of the gas on the surface *ac* is equal and opposite to that on the surface *cd*. Hence the net projection on a horizontal plane for the gas is $af = 2\sin 30° = 1$ m. Thus

$$F_z = pA_x = (35.0 \text{ kN/m}^2)(1 \text{ m}) = 35.0 \text{ kN/m upward} \quad ANS$$

(b) For the fluid,

Eq. (3.16): $F_x = \gamma h_c A = (9.81 \text{ kN/m}^3)\left(\frac{1}{2} \times 3.73 \text{ m}\right)(3.73 \text{ m})$

$= 68.3 \text{ kN/m to the right}$ *ANS*

Net $F_Z =$ upward force on surface *cde* − downward force on surface *ca*

$=$ weight of volume *abcdefa* − weight of volume *abca*

$=$ weight of cross-hatched volume of liquid

$= (9.81 \text{ kN/m}^3)\left[\dfrac{210}{360}\pi 2^2 + \dfrac{1}{2}(1 \times 2\cos 30°) + (1 \times 2)\right] \text{m}^2$

$= 100.0 \text{ kN/m upward}$ *ANS*

3.9 Buoyancy and Stability of Submerged and Floating Bodies

Submerged Body

When a body such as *DHCK* in Fig. 3.18 is immersed in a fluid, the forces acting on it are gravity and the pressures of the surrounding fluid. On its upper surface the vertical component of the force is F_z and is equal to the weight of the volume of fluid *ABCHD*. In a similar manner, the vertical component of force on the undersurface is F_z' and is equal to the weight of the volume of fluid *ABCKD*. The difference between these two volumes is the volume ∀ of the body *DHCK*.

Buoyancy. Let us denote the buoyant force of a fluid by F_B, and observe that it is vertically upward and equal to $F_z' - F_z$, which is equal to the weight of the volume of fluid *DHCK*. That is, *the **buoyant force** on any body is equal to the weight of fluid displaced*, or in equation form,

$$F_B = \gamma_{\text{fluid}}\,∀$$

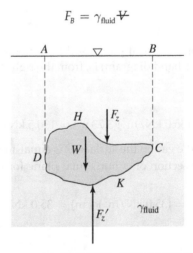

FIGURE 3.18   Forces acting on an immersed body.

This is probably the best-known discovery of Archimedes (287–212 B.C.E.), a Greek philosopher acclaimed as the father of hydrostatics, and one of the earliest known pioneers of fluid mechanics.

If the body in Fig. 3.18 is in equilibrium, W is equal and opposite to F_B, which means that the densities of the body and the fluid are equal. If W is greater than F_B, the body will sink. If W is less than F_B, the body will rise until its density and that of the fluid are equal, as in the case of a balloon in the air or, in the case of a liquid with a free surface, the body will rise to the surface until the weight of the displaced liquid equals the weight of the body. If the body is less compressible than the fluid, there is a definite level at which it will reach equilibrium. If it is more compressible than the fluid, it will rise indefinitely, provided the fluid has no definite upper limit.

Stability. When we give a body in equilibrium a slight angular displacement (tilt or *list*), a horizontal distance a then separates W and F_B, which in combination create moments that tend to rotate the body, as we can see in Fig. 3.19. If the moments tend to restore the body to its original position, the *lesser* of the two moments is called the **righting moment** (Fig. 3.19), and we say the body is in **stable equilibrium**. The stability of submerged or floating bodies depends on the relative positions of the buoyant force and the weight of the body. The buoyant force acts through the **center of buoyancy** B, which corresponds to the center of gravity of the displaced fluid. *The criterion for stability of a fully submerged body* (balloon or submarine, etc.) *is that the center of buoyancy is above the center of gravity of the body.* From Fig. 3.19 we can see that if B were initially below G, the center of gravity, then the moment created by a tilt would tend to increase the displacement.

Floating Body

For a body in a liquid with a free surface, if its weight W is less than that of the same volume of liquid, it will rise and float on the surface as in Fig. 3.20, so that $W = F_B$. The forces then acting on body $AHBK$ are gravity and the pressures of the fluids in contact with it. The vertical component of force on the undersurface is F_z' and this is equal to the weight of the volume of liquid AKB. This volume is the volume of liquid displaced by the body.

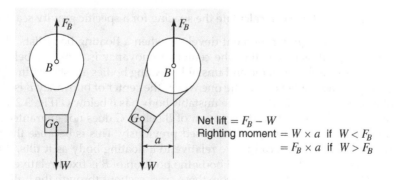

Net lift $= F_B - W$
Righting moment $= W \times a$ if $W < F_B$
$\qquad\qquad\qquad = F_B \times a$ if $W > F_B$

FIGURE 3.19 Submerged body (balloon).

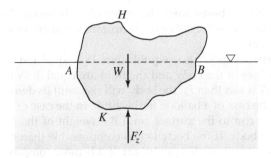

FIGURE 3.20   Forces acting on a floating body.

Buoyancy. The buoyant force F_B is vertically upward and equal to F_z'. So, just as for a fully submerged body, the buoyant force acting on a floating body is equal to the weight of liquid displaced. Thus a *floating body displaces a volume of liquid equivalent to its weight*. For equilibrium, the two forces W and F_B must be equal and opposite, and must lie in the same vertical line.

The atmospheric pressure is transmitted through the liquid to act equally on all surfaces of the body. As a result, it has zero net effect. Any buoyancy due to the weight of air displaced by the portion of the body above the liquid surface is usually negligible in comparison with the weight of liquid displaced.

A practical application of the buoyancy principle is the **hydrometer**, an instrument we use to measure the specific gravity of liquids. It has a thin, uniform stem of constant cross-sectional area, say A. Weights make it float upright as in Fig. 3.21*a*, with a reference mark that is at the water surface when floating in pure water ($s = 1.0$). When floating in a denser liquid of specific gravity s (Fig. 3.21*b*), the volume of liquid displaced is smaller, so less is submerged and the reference mark is some height Δh above the water surface. If the submerged volume in pure water is $\forall$ then in the denser liquid it is $\forall - A\Delta h$, and the hydrometer's weight

$$W = \gamma_w \forall = (s\gamma_w)(\forall - A\Delta h)$$

from which

$$\Delta h = \frac{\forall}{A}\left(\frac{s-1}{s}\right) \tag{3.23}$$

Using Eq. (3.23) we can calculate the spacing for a specific gravity scale on the stem.

Stability. If a righting moment develops when a floating body lists, the body will be stable regardless of whether the center of buoyancy is above or below the center of gravity. Examples of stable and unstable floating bodies are shown in Fig. 3.22. In these examples the stable body is the one where the center of buoyancy B is above the center of gravity G (Fig. 3.22*a*), and the unstable body has B below G (Fig. 3.22*b*). However, for floating bodies note that the location of B below G does not guarantee instability as it does for submerged bodies, discussed previously. This is because the position of the center of buoyancy B can move relative to a floating body as it tilts, due to its shape, whereas for a fully submerged body the position of B is fixed relative to the body. Figure 3.23 illustrates this point; from these cross sections through the hull of a ship we can see that it is stable even though B is below G. Because of the cross-sectional shape, as the

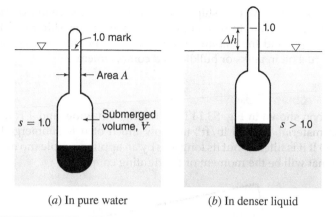

(a) In pure water   (b) In denser liquid

FIGURE 3.21 Hydrometer floating in two different liquids.

ship tilts to the right (Fig. 3.23b) the center of gravity of the displaced water (i.e., B) moves to the right further than the line of action of the body weight W, and so the buoyancy provides a righting moment $F_B \times a$. Clearly, therefore, the stabilities of many floating bodies (those with B below G) depend upon their shapes.

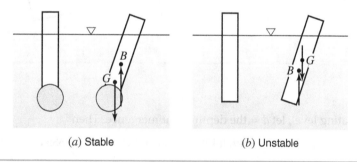

(a) Stable   (b) Unstable

FIGURE 3.22 Stable and unstable floating bodies.

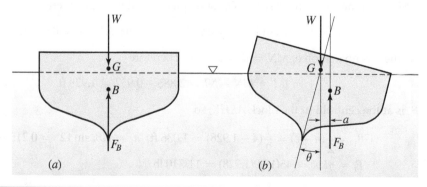

FIGURE 3.23 Movement of center of buoyancy as a floating body tilts.

If liquid in the hull of a ship is not constrained, the center of mass of the floating body will move toward the center of buoyancy when the ship rolls, thus decreasing the righting couple and the stability. For this reason, floating vessels usually store liquid ballast or fuel oil in tanks or bulkheaded compartments.

Sample Problem 3.14

The pontoon shown in Fig. S3.14 is 15 ft long, 9 ft wide, and 4 ft high, and is built of uniform material, $\gamma = 45$ lb/ft³. (a) How much of it is submerged when floating in water? (b) If it is tilted about its long axis by an applied couple (no net force), to an angle of 12°, what will be the moment of the righting couple?

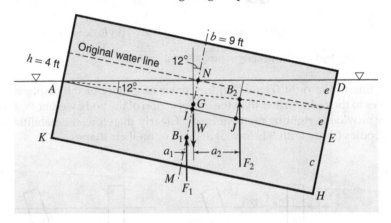

FIGURE S3.14

Solution

(a) Floating level, let d = the depth of submergence. Then

$$W = F_B; \quad 15(9)4(45) = 15(9)d(62.4); \quad d = 2.885 \text{ ft} \quad \textbf{ANS}$$

(b) At 12° tilt, let AD be the water line (see Fig. S3.14).

Divide the buoyancy force into two components B_1 and B_2, due to the rectangular block $AEHK$ and the triangular prism ADE of displaced water, respectively.

$$DE = 2e = b \tan 12° = 9 \tan 12° = 1.913 \text{ ft}; \quad NI = e = 0.957 \text{ ft}$$

As there is no net force, $MN = d = 2.885$ ft. Therefore

$$c = IM = MN - NI = 2.885 - 0.957 = 1.928 \text{ ft}$$

B_1 is at the centroid of the block $AEHK$, so

$$GB_1 = \frac{1}{2}(h - c) = \frac{1}{2}(4 - 1.928) = 1.036 \text{ ft}; \quad a_1 = GB_1 \sin 12° = 0.215 \text{ ft}$$

$$F_1 = \gamma Lbc = 45(15)9(1.928) = 11,710 \text{ lb}$$

B_2 is at the centroid of the triangle ADE, so

$$JE = b/3, \ IJ = b/6 = 1.5 \text{ ft}, \ B_2J = \frac{2}{3}e = 0.638 \text{ ft}$$

G is at the centroid of the major rectangle, so $MG = h/2 = 2$ ft,

$$GI = MG - MI = MG - c = 2 - 1.928 = 0.0719 \text{ ft}$$

$$a_2 = IJ \cos 12° + (B_2J - GI) \sin 12° = 1.585 \text{ ft}$$

$$F_2 = \gamma Lbe = 45 \,(15)9\,(0.957) = 5810 \text{ lb}$$

Counterclockwise moments about G:

$$\text{Righting moment} = F_2 a_2 - F_1 a_1 = 5810(1.585) - 11{,}710\,(0.215)$$
$$= 6690 \text{ lb·ft} \qquad ANS$$

3.10   Liquid Masses Subjected to Acceleration

Under certain conditions there may be no relative motion between the particles of a liquid mass yet the mass itself may be in motion. If a body of liquid in a tank is transported at a uniform velocity, the conditions are those of ordinary fluid statics. But if it is subjected to acceleration, special treatment is required. Consider the case of a liquid mass in an open tank moving horizontally with a linear acceleration a_x, as in Fig. 3.24a. A free-body diagram (Fig. 3.24b) of a small particle (mass m) of liquid on the surface indicates that the forces exerted by the surrounding liquid on the particle are such that $F_z = F_B = -W$ and $F_X = ma_x$. F_z counterbalances W, so there is no acceleration in the

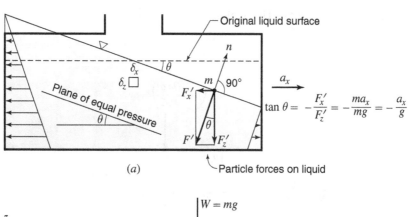

(a)

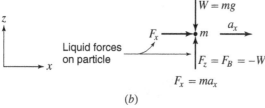

(b)

FIGURE 3.24   Liquid mass subjected to horizontal acceleration.

z direction. F_x is the force required to produce acceleration a_x of the particle. Equal and opposite to these forces are F_x' and F_z' of Fig. 3.24a, the forces exerted by the particle on the surrounding fluid. The resultant of these forces is F'. The liquid surface must be at right angles to F', for if it were not, the particle would not maintain its fixed relative position in the liquid. Hence, (Fig. 3.24a) $\tan\theta = -a_x/g$. The liquid surface and all other planes of equal hydrostatic pressure must be inclined at angle θ with the horizontal as in Fig. 3.24a.

Next let us consider the more general case where a fluid mass is accelerating in both the x and z directions. Figure 3.25 is a free-body diagram of an elemental cube of fluid, volume $\delta x\,\delta y\,\delta z$ with pressure p at its center. Applying the equation of motion in the x direction,

$$\sum F_x = ma_x$$

$$\left(p - \frac{\partial p}{\partial x}\frac{\delta x}{2}\right)\delta y\,\delta z - \left(p + \frac{\partial p}{\partial x}\frac{\delta x}{2}\right)\delta y\,\delta z = \rho\delta x\,\delta y\,\delta z\,a_x$$

which reduces to

$$\frac{\partial p}{\partial x} = -\rho a_x \qquad (3.24)$$

Thus, as seems intuitive, fluid pressure reduces in the direction of acceleration. In the vertical direction

$$\sum F_z = ma_z$$

or

$$\left(p - \frac{\partial p}{\partial z}\frac{\delta z}{2}\right)\delta x\,\delta y - \left(p + \frac{\partial p}{\partial z}\frac{\delta z}{2}\right)\delta x\,\delta y - \gamma\delta x\,\delta y\,\delta z = \rho\delta x\,\delta y\,\delta z\,a_z$$

where $\gamma = \rho g$ [Eq. (2.1)]. This yields

$$\frac{\partial p}{\partial z} = -\rho(a_z + g) \qquad (3.25)$$

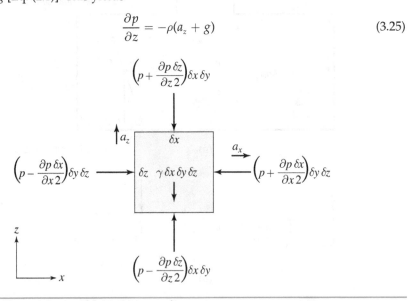

FIGURE 3.25   Elemental cube of fluid, thickness δy.

Therefore, the pressure decreases with elevation z in a static fluid (Sec. 3.3), and it does so more rapidly if the fluid is being accelerated upward.

We can use Eqs. (3.24) and (3.25) to obtain a general result for a liquid mass that is accelerating in both the x and z directions. The chain rule for the total differential of dp in terms of its partial derivatives is

$$dp = \frac{\partial p}{\partial x}dx + \frac{\partial p}{\partial z}dz$$

So, substituting the expressions for $\partial p/\partial x$ and $\partial p/\partial z$ from Eqs. (3.24) and (3.25), we get

$$dp = -\rho(a_x)dx - \rho(a_z + g)dz \qquad (3.26)$$

Along a line of constant pressure, $dp = 0$. From Eq. (3.26), if $dp = 0$,

For $p = $ constant: $\qquad \dfrac{dz}{dx} = -\dfrac{a_x}{a_z + g} \qquad (3.27)$

This defines the slope $dz/dx = \tan\theta$ of a line of constant pressure within the accelerated liquid mass; the liquid surface is one such line.

In obtaining Eq. (3.27) from Eq. (3.26) we divided out the mass, represented by ρ. If we consider a liquid particle of mass m within the liquid, then using Newton's second law ($F = ma$):

$$\tan\theta = \frac{dz}{dx} = -\frac{ma_x}{ma_z + mg} = -\frac{F_x}{F_z + F_B}$$

So we see that the three accelerations in the right side of Eq. (3.27) represent the three forces exerted by the liquid on the liquid particle, as depicted in Fig. 3.26. These forces together produce the net force F, which is normal to lines of constant pressure. Note that g is upward-acting, in the positive z direction, because here it represents buoyancy $F_B (= -W)$.

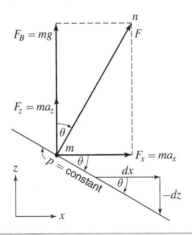

FIGURE 3.26 Liquid forces on an accelerating particle.

From Eqs. (3.24) and (3.25), we may obtain the resultant of $\partial p/\partial x$ and $\partial p/\partial z$, namely

$$\frac{\partial p}{\partial n} = -\rho \sqrt{a_x^2 + (a_z + g)^2} \tag{3.28}$$

where n is at right angles to the lines of equal pressure and in the direction of the most rapidly decreasing pressure (Fig. 3.26). When $a_x = a_z = 0$, this equation reduces to $\partial p/\partial n = -\rho g = -\gamma$, which is essentially the same as the basic hydrostatic equation given by Eq. (3.2). Equation (3.28) indicates that, if liquid in a container experiences an upward acceleration, this increases pressures within the liquid; downward acceleration decreases them.

Sample Problem 3.15

At a particular instant an airplane is traveling upward at a velocity of 180 m/s in a direction that makes an angle of 40° with the horizontal. At this instant the airplane is losing speed at the rate of 4 m/s². Also, it is moving on a concave-upward circular path having a radius of 2600 m. Determine for the given conditions the slope of the free liquid surface in the fuel tank of this vehicle.

Solution

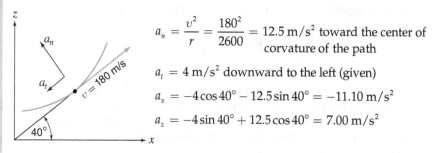

$$a_n = \frac{v^2}{r} = \frac{180^2}{2600} = 12.5 \text{ m/s}^2 \text{ toward the center of corvature of the path}$$

$$a_t = 4 \text{ m/s}^2 \text{ downward to the left (given)}$$

$$a_x = -4\cos 40° - 12.5\sin 40° = -11.10 \text{ m/s}^2$$

$$a_z = -4\sin 40° + 12.5\cos 40° = 7.00 \text{ m/s}^2$$

Eq. (3.27): Slope of the free surface $= \dfrac{dz}{dx} = -\left(\dfrac{-11.10}{7.00 + 9.81}\right) = +0.660$ **ANS**

$$\theta = \tan^{-1}(0.660) = 33.4° \quad \textit{ANS}$$

CHAPTER 4

Basics of Fluid Flow

In this chapter we shall deal with fluid velocities and accelerations and their variations in space without considering any forces involved. As we mentioned in Sec. 1.1, this subject, that deals with velocities and flow paths without considering forces or energy, is known as *kinematics*.

Because only certain types of flow can be treated by the methods of kinematics, and because there are many different types of flow, we summarize these first to provide perspective. We shall also introduce some related concepts, most notably the control volume and the flow net.

4.1 Types of Flow

When speaking of fluid flow, we often refer to the flow of an *ideal fluid* (Sec. 2.10). We presume that such a fluid has no viscosity. This is an idealized situation that does not exist; however, there are instances in engineering problems where the assumption of an ideal fluid is helpful. When we refer to the flow of a *real fluid*, the effects of viscosity are introduced into the problem. This results in the development of shear stresses between neighboring fluid particles when they are moving at different velocities. In the case of an ideal fluid flowing in a straight conduit, all particles move in parallel lines with equal velocity (Fig. 4.1a). In the flow of a real fluid the velocity adjacent to the wall will be zero; it will increase rapidly within a short distance from the wall and produce a velocity profile such as shown in Fig. 4.1b.

Flow can also be classified as that of an *incompressible* or *compressible* fluid. Since liquids are relatively incompressible, we generally treat them as wholly incompressible fluids. Under particular conditions where there is little pressure variation, we may also consider the flow of gases to be incompressible, though generally we should consider the effects of the compressibility of the gas.

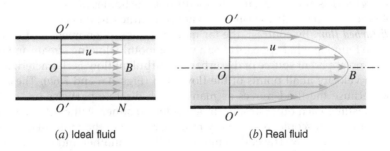

(a) Ideal fluid   (b) Real fluid

FIGURE 4.1 Typical velocity profiles.

In addition to the flow of different types of fluids, i.e., real, ideal, incompressible, and compressible, there are various classifications of flow. Flow may be *steady* or *unsteady* with respect to time (see Sec. 4.3). It may be *laminar* or *turbulent*, as discussed in the following section. Other classifications of flow include *rotational* or *irrotational*, *supercritical* or *subcritical* (Chap. 8), etc. These and other common ways in which we can classify flow are listed in Table 4.1, in many cases with definitions.

4.2   Laminar and Turbulent Flow

Whether laminar or turbulent flow occurs in a given situation, or how much of each occurs, is very important because of the strongly different effects these two different types of flow have on a variety of flow features, including on energy losses, velocity profiles, and mixing of transported materials.

Osborne Reynolds demonstrated in 1883 that there are two distinctly different types of fluid flow. He injected a fine, threadlike stream of colored liquid having the same density as water at the entrance to a large glass tube through which water was flowing from a tank. A valve at the discharge end permitted him to vary the flow. When the velocity in the tube was small, he saw this colored liquid as a straight line throughout the length of the tube, showing that the particles of water moved in parallel straight lines. As he gradually increased the velocity of the water by opening the valve further, at a certain velocity the flow changed. The line first became wavy, and then at a short distance from the entrance it broke into numerous vortices beyond which the color became uniformly diffused so that no streamlines could be distinguished. Later observations have shown that in this latter type of flow the velocities are continuously subject to irregular fluctuations.

The first type is known as *laminar, streamline,* or *viscous* flow. The significance of these terms is that the fluid appears to move by the sliding of laminations of infinitesimal thickness over adjacent layers, with relative motion of fluid particles occurring at a molecular scale; that the particles move in definite and observable paths or streamlines, as in Fig. 4.2; and also that the flow is characteristic of a viscous fluid or is one in which viscosity plays a significant part (Fig. 2.4 and Sec. 2.11).

The second type is known as *turbulent* flow, and is illustrated in Fig. 4.3, where (a) represents the irregular motion of a large number of particles during a very brief time interval, while (b) shows the erratic path followed by a single particle during a longer time interval. A distinguishing characteristic of turbulence is its irregularity, there being no definite frequency as in wave action, and no observable pattern as in the case of large swirls.

Large swirls and irregular movements of large bodies of fluid, which can be traced to obvious sources of disturbances, do not constitute turbulence, but may be described as *disturbed flow*. By contrast, the far more common phenomenon of turbulence may often be found in what appears to be a very smoothly flowing stream and one in which there is no apparent source of disturbance. Turbulent flow is characterized by fluctuations in velocity at all points of the flow field (Figs. 4.6 and 7.6b). These fluctuations arise because the fluid moves as many small, discrete particles or "packets" called *eddies*, jostling each other around in a random manner. Although small, the smallest eddies are macroscopic in size, very much larger than the molecular sizes of the particles in laminar flow. The eddies interact with one another and with the general flow.

One-dimensional, two-dimensional, or ***three-dimensional flow***

See Sec. 4.8 for discussion.

Real fluid flow or ***ideal fluid flow*** (also referred to as ***viscid*** and ***inviscid flow***)

Real fluid flow implies frictional (viscous) effects. Ideal fluid flow is hypothetical; it assumes no friction (i.e., viscosity of fluid = 0).

Incompressible fluid flow or ***compressible fluid flow***

Incompressible fluid flow assumes that the fluid has constant density (ρ constant). Although liquids are slightly compressible, we usually assume them to be incompressible. Gases are compressible; their density is a function of absolute pressure and absolute temperature [$\rho = f(p, T)$].

Steady or ***unsteady flow***

Steady flow means steady with respect to time. Thus, all properties of the flow at every point remain constant with respect to time. In unsteady flow, the flow properties at a point change with time.

Pressure flow or ***gravity flow***

Pressure flow implies that flow occurs under pressure. Gases always flow in this manner. When a liquid flows with a free surface (e.g., a partly full pipe), we refer to the flow as gravity flow, because gravity is the primary moving force. Liquids also flow under pressure (e.g., a pipe flowing full).

Spatially constant or ***spatially variable flow***

Spatially constant flow occurs when the fluid density and the local average flow velocity are identical at all points in a flow field. If these quantities change along or across the flow lines, the flow is spatially variable. Examples of different types of spatially varied flow include the local flow field around an object, flow through a gradual contraction in a pipeline, and the flow of water in a uniform gutter of constant slope receiving inflow over the length of the gutter.

Laminar or ***turbulent flow***

See Sec. 4.2 for a discussion of the difference between these two types of flow.

Established or ***unestablished flow***

We discuss these in Sec. 7.9.

Uniform or ***varied flow***

We ordinarily use these classifications when dealing with open-channel (gravity) flow (Chap. 8). In uniform flow the cross section (shape and area) through which the flow occurs remains constant.

Subcritical or ***supercritical flow***

We use these classifications with open-channel flow (Chap. 8).

Subsonic or ***supersonic flow***

We use these classifications with compressible flow.

Rotational or ***irrotational flow***

We use these in mathematical hydrodynamics.

Other classifications of flow include ***converging*** or ***diverging, disturbed, isothermal*** (constant temperature), ***adiabatic*** (no heat transfer), and ***isentropic*** (frictionless adiabatic).

TABLE 4.1 Classification of types of flow[a]

[a] Note that in a given situation these different types of flow may occur in combination. For example, we usually consider flow of a liquid in a pipe to be one-dimensional, incompressible, real fluid flow that may be steady or unsteady, and laminar or turbulent. Such flow is commonly spatially constant and established.

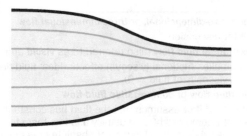

FIGURE 4.2   Laminar (streamline or viscous) flow.

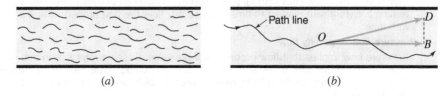

(a)   (b)

FIGURE 4.3   Turbulent flow.

They are the cause of the effective mixing action experienced with turbulent flow. They are often caused by rotation, particularly near boundaries, and so the eddies themselves often rotate. They change shape and size with time as they move along with the flow. Each eddy dissipates its energy through viscous shear with its surroundings and eventually disappears. New eddies are continuously forming. Large eddies (large-scale turbulence) have smaller eddies within them giving rise to small-scale turbulence. The resulting fluctuations in velocity are rapid and irregular, and often we can only detect them by a fast-acting probe such as a hot-wire or hot-film anemometer (Sec. 9.4).

At a certain instant the flow passing point O in Fig. 4.3b may be moving with the velocity OD. In turbulent flow OD will vary continuously both in direction and in magnitude. Fluctuations of velocity are accompanied by fluctuations in pressure, which is the reason why manometers or pressure gages attached to a pipe containing flowing fluid usually show pulsations. In this type of flow an individual particle will follow a very irregular and erratic path, and no two particles may have identical or even similar motions. Thus a rigid mathematical treatment of turbulent flow is impossible, and instead we must use statistical methods of evaluation.

Criteria governing the conditions under which the flow will be laminar and those under which it will be turbulent are discussed in Sec. 7.3.

4.3   Steady Flow and Uniform Flow

A *steady flow* is one in which all conditions at any point in a stream remain constant with respect to *time*, but the conditions may be different at different points. A truly *uniform flow* is one in which the velocity is the same in both magnitude and direction at a given instant at every point in the fluid. Both of these definitions must be modified somewhat, since true steady flow is found only in laminar flow. In turbulent flow there are continual fluctuations in velocity and pressure at every point, as was just explained.

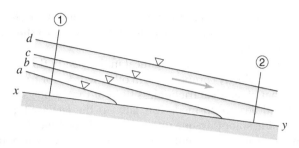

FIGURE 4.4   Unsteady flow in a canal.

But if the values fluctuate equally on both sides of a constant average value, we call the flow steady flow. However, a more precise term for this condition would be *mean steady flow*.

Likewise, this strict definition of uniform flow can have little meaning for the flow of a real fluid where the velocity varies across a section, as in Fig. 4.1*b*. But when the size and shape of cross section are constant along the length of channel under consideration, we say the flow is *uniform*.

Steady (or unsteady) and uniform (or nonuniform) flow can exist independently of each other, so that any of four combinations is possible. Thus the flow of liquid at a constant rate in a long straight pipe of constant diameter is *steady uniform* flow, the flow of liquid at a constant rate through a conical pipe is *steady nonuniform* flow, while at a changing rate of flow these cases become *unsteady uniform* and *unsteady nonuniform flow*, respectively.

Unsteady flow is a transient phenomenon, which may in time become either steady flow or zero flow. An example is given in Fig. 4.4, where *a* denotes the surface of a stream that has just been admitted to the bed of a canal by the sudden opening of a gate. After a time the water surface will be at *b*, later at *c*, and finally it reaches equilibrium at *d*. The unsteady flow has then become mean steady flow. Another example of transient phenomenon is when a valve is closed at the discharge end of a pipeline (Sec. 10.6), thus causing the velocity in the pipe to decrease to zero. In the meantime, there will be fluctuations in both velocity and pressure within the pipe.

Unsteady flow may also include periodic motion such as that of waves on beaches, tidal motion in estuaries, and other oscillations. The difference between such cases and that of mean steady flow is that the deviations from the mean are very much greater and the time scale is also much longer.

4.4   Path Lines, Streamlines, and Streak Lines

A *path line* (Fig. 4.3*b*) is the trace made by a *single* particle over a *period* of time. If a camera were to take a time exposure of a flow in which a fluid particle was colored so it would register on the negative, the picture would show the course followed by the particle. This would be its path line. The path line shows the direction of the velocity of the particle at successive instants of time.

Streamlines show the mean direction of a *number* of particles at the *same* instant of time. If a camera were to take a very short time exposure of a flow in which there were a large number of particles, each particle would trace a short path, which would indicate its velocity during that brief interval. A series of curves drawn tangent to the means of the velocity vectors are streamlines. It follows that there can be no net velocity normal to a streamline.

Path lines and streamlines are identical in the steady flow of a fluid in which there are no fluctuating velocity components, in other words, for truly steady flow. Such flow may be either that of an ideal frictionless fluid or that of one so viscous and moving so slowly that no eddies are formed. This latter is the *laminar* type of flow, wherein the layers of fluid slide smoothly, one upon another. In turbulent flow, however, path lines and streamlines are not coincident, the path lines being very irregular while the stream-lines are everywhere tangent to the local mean temporal velocity. The lines in Fig. 4.2 represent both path lines and streamlines if the flow is laminar; they represent only streamlines if the flow is turbulent.

In experimental fluid mechanics, a dye or other tracer is frequently injected into the flow to trace the motion of the fluid particles. If the flow is laminar, a ribbon of color results. This is called a *streak line*, or *filament line*. It is an instantaneous picture of the positions of all particles in the flow that have passed through a given point (namely, the point of injection). When using fluid-tracer techniques it is important to choose a tracer with physical characteristics (especially density) the same as those of the fluid being observed. Thus the smoke rising from an incense stick, while giving the appearance of a streak line, does not properly represent the movement of the ambient air in the room because it is less dense (warmer) than the air and therefore rises more rapidly.

4.5   Flow Rate and Mean Velocity

We call the quantity of fluid flowing per unit time across any section the *flow rate*. We may express it (1) in terms of *volume flow rate (discharge)* using BG units such as cubic feet per second (cfs), gallons per minute (gpm), million gallons per day (mgd), or (2) in terms of *mass flow rate* (slugs per second), or (3) in terms of *weight flow rate* (pounds per second). In SI units, cubic meters per second (m³/s), kilograms per second (kg/s), and kilonewtons per second (kN/s) are fairly standard for expressing volume, mass, and weight flow rates, respectively. When dealing with incompressible fluids, we commonly use volume flow rate, whereas weight flow rate or mass flow rate is more convenient with compressible fluids.

Figure 4.5 depicts a streamline in steady flow lying in the xz plane. The element of area dA lies in the yz plane. The mean velocity at point P is u. The volume flow rate passing through the element of area dA is

$$dQ = \mathbf{u} \cdot d\mathbf{A} = (u \cos \theta)dA = u(\cos \theta \, dA) = u \, dA' \tag{4.1}$$

where dA' is the projection of dA on the plane normal to the direction of u. This indicates that the *volume flow rate is equal to the magnitude of the mean velocity multiplied by the flow area at right angles to the direction of the mean velocity.* We can compute the mass flow rate and the weight flow rate by multiplying the volume flow rate by the density and specific weight of the fluid respectively.

If the flow is turbulent, the *instantaneous velocity component* u_t along the streamline will fluctuate with time, even though the flow is nominally steady. A plot of u_t as a function of time is shown in Fig. 4.6. The average value of u_t over a period of time determines the time (temporal) mean value of velocity u at point P.

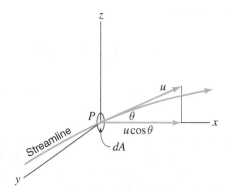

FIGURE 4.5   A streamline in steady flow.

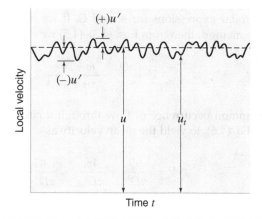

FIGURE 4.6   Fluctuating velocity at a point due to turbulence.

The difference between u_t and u, which we shall denote by u', is called the ***turbulent fluctuation*** of this component; it may be either positive or negative, but the time mean value of u' must be zero. Likewise, the time means of all velocity components perpendicular to the streamline must also be zero. At any instant, then,

$$u_t = u + u' \tag{4.2}$$

and we can evaluate u for any finite time t from $u = (1/t)\int_0^t u_t dt$.

In a real fluid the local time mean velocity u will vary across the section in some manner, such as that shown in Fig. 4.1b, and so we can express the flow rate as

$$Q = \int_A u\, dA = AV \tag{4.3}$$

or, for constant-density flow,

$$\dot{m} = \rho \int_A u \, dA = \rho AV = \rho Q \tag{4.4}$$

or

$$G = g\dot{m} = \gamma \int_A u \, dA = \gamma AV = \gamma Q \tag{4.5}$$

where u is the time mean velocity through an infinitesimal area dA, while V is the **mean, or average, velocity** over the entire sectional area A;[1] Q is the volume flow rate (cfs or m³/s), $\dot{m}$ is the mass flow rate (slugs/sec or kg/s),[2] and G is the weight flow rate (lb/sec or kN/s).[3] If u is known as a function of A, we can integrate Eqs. (4.3), (4.4), and (4.5). If only an average value of V is known for each finite subarea of the total sectional area, then

$$Q = A_a V_a + A_b V_b + \cdots + A_n V_n = AV$$

We can write similar expressions for $\dot{m}$ and G. If we have determined the flow rate directly by some method, then from Eqs. (4.3)–(4.5) we can find the mean velocity,

$$V = \frac{Q}{A} = \frac{\dot{m}}{\rho A} = \frac{G}{\gamma A} \tag{4.6}$$

For the very common occurrence of flow through a circular pipe, we can substitute $A = \pi D^2/4$ into Eq. (4.6), to yield the mean velocity as

Circular pipe: $$V = \frac{4Q}{\pi D^2} = \frac{4\dot{m}}{\pi D^2 \rho} = \frac{4G}{\pi D^2 \gamma} \tag{4.7}$$

Sample Problem 4.1

Air at 100°F and under a pressure of 40 psia flows in a 10-in-diameter ventilation duct at a mean velocity of 30 fps. Find the mass flow rate.

Solution

Table A.5 for air: $R = 1715$ ft·lb/(slug·°R)

From Eq. (2.4): $\rho = \dfrac{p}{RT} = \dfrac{40(144)}{1715(460 + 100)} = 0.006\,00$ slug/ft³

Eq. (4.4): $\dot{m} = \rho AV = (0.006\,00)\dfrac{\pi}{4}\left(\dfrac{10}{12}\right)^2 (30) = 0.0981$ slug/sec   ***ANS***

---

[1] Note that we define area A by the surface at right-angles to the velocity vectors.
[2] Here, as used on m, and subsequently, the overdot represents the time derivative, as is standard practice.
[3] In Eqs. (4.4) and (4.5) the ρ and γ should be to the right of the integral sign if the density of the fluid varies across the flow.

4.6 Fluid System and Control Volume

The concept of a free body diagram, as used in the statics of rigid bodies and in fluid statics (e.g., Fig. 3.1), is usually inadequate for the analysis of moving fluids. Instead, we frequently find the concepts of a *fluid system* and a *control volume* to be useful in the analysis of fluid mechanics.

A *fluid system* refers to a *specific mass of fluid* within the boundaries defined by a closed surface. The shape of the system, and so the boundaries, may change with time, as when liquid flows through a constriction or when gas is compressed; as a fluid moves and deforms, so the system containing it moves and deforms. The size and shape of a system is entirely optional.

In contrast, a *control volume* (CV) refers to a *fixed region in space, which does not move or change shape* (Fig. 4.7). We usually choose it as a region that the fluid flows into and out of. We call its closed boundaries the *control surface*. Again, the size and shape of a control volume is entirely optional, although we often choose the boundaries to coincide with some solid or other natural flow boundaries. Actually, the control surface may be in motion through space relative to an absolute frame of reference; this is acceptable provided the motion is limited to constant-velocity translation.

We shall now derive a general relationship between a system and a control volume that provides an important basis for the equations of continuity, energy, and momentum for moving fluids. This relationship is derived from what we commonly call the *control volume approach*, more formally known as the **Reynolds transport theorem**. Addressing the motion of fluid as it moves through a given region, the control volume approach is also called the **Eulerian approach**, in contrast to the **Lagrangian approach** in which we describe the motion of each particle by its position as a function of time.

Let X represent the total amount of some fluid property (scalar or vector), such as mass, energy, or momentum, contained within specified boundaries at a specified time. It will probably help to think of X as mass for most of this section. The specified boundaries will be either those of a system, indicated by a subscript S, or those of a control volume, indicated by a subscript CV. Consider the general flow situation of Fig. 4.7. At time t, the boundaries of the system and the control volume were chosen to coincide, so $(X_S)_t = (X_{CV})_t$. At instant Δt later, the system has moved a little through the control volume and possibly slightly changed its shape; a small amount of new fluid ΔV_{CV}^{in} has entered the control volume, and another small amount of system fluid ΔV_{CV}^{out} has left the control volume, where V represents volume. These small volumes carry small

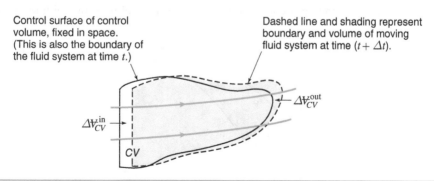

Control surface of control volume, fixed in space. (This is also the boundary of the fluid system at time t.)

Dashed line and shading represent boundary and volume of moving fluid system at time $(t + \Delta t)$.

ΔV_{CV}^{out}

ΔV_{CV}^{in}

CV

FIGURE 4.7   Fluid system, control volume, and differences.

amounts of property X (mass, etc.) with them, so that ΔX_{CV}^{in} enters and ΔX_{CV}^{out} leaves the control volume. Comparing X in the various volumes, we see that

$$(X_S)_{t+\Delta t} = (X_{CV})_{t+\Delta t} + \Delta X_{CV}^{out} - \Delta X_{CV}^{in}$$

Subtracting the equation for t from that for $t + \Delta t$, we obtain

$$(X_S)_{t+\Delta t} - (X_S)_t = (X_{CV})_{t+\Delta t} - (X_{CV})_t + \Delta X_{CV}^{out} - \Delta X_{CV}^{in}$$

or
$$\Delta X_S = \Delta X_{CV} + \Delta X_{CV}^{out} - \Delta X_{CV}^{in} \qquad (4.8)$$

and dividing by Δt and letting $\Delta t \to 0$, we get

$$\frac{dX_S}{dt} = \frac{dX_{CV}}{dt} + \frac{dX_{CV}^{out}}{dt} - \frac{dX_{CV}^{in}}{dt} \qquad (4.9)$$

These equations will be used in subsequent studies of continuity, energy, and momentum. The left-hand side of Eq. (4.9) is the rate of change of the total amount of any extensive property X within the moving system. The next term, dX_{CV}/dt, is the rate of change of that same property, but contained within the fixed control volume. The last two terms are the net rate of outflow of X passing through the control surface. So Eq. (4.9) states that *the difference between the rate of change of X (e.g., mass) within the system and that within the control volume is equal to the net rate of outflow from the control volume.*

4.7   Equation of Continuity

Although continuity is a strongly intuitive concept, it took early investigators a long time to formalize it. Pioneers in this effort include the Greek scientist Hero (or Heron) of Alexandria (*circa* 100 A.D.), the prodigious Italian artist and scientist Leonardo da Vinci (1452–1519), and Benedetto Castelli (1577–1643), a pupil of Galileo.

Let Fig. 4.8 represent a short length of a **stream tube**, which may be assumed, for practical purposes, to be a bundle of streamlines. Since the stream tube is bounded on all sides by streamlines and since there can be no net velocity normal to a streamline (Sec. 4.4), no fluid can leave or enter the stream tube except at the ends. The fixed volume between the two end sections is a control volume, of volume $\mathcal{V}$ let us say.

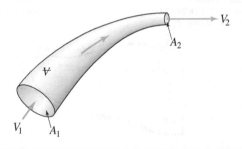

FIGURE 4.8   Portion of stream tube as control volume.

Using the relation of a system to a control volume developed in Sec. 4.6, and letting the general property X now be the mass m, Eq. (4.9) becomes

$$\frac{dm_S}{dt} = \frac{dm_{CV}}{dt} + \frac{dm_{CV}^{out}}{dt} - \frac{dm_{CV}^{in}}{dt} \qquad (4.10)$$

But according to Newtonian physics (i.e., disregarding the possibility of converting mass to energy), the mass of a system must be conserved, so

$$\frac{dm_S}{dt} = 0 \qquad (4.11)$$

In addition, because the volume $\mathcal{V}$ of the control volume is fixed, $m_{CV} = \mathcal{V}\bar{\rho}_{CV}$ where $\bar{\rho}_{CV}$ is the mean density within the control volume, so

$$\frac{dm_{CV}}{dt} = \mathcal{V}\frac{d\bar{\rho}_{CV}}{dt} = \mathcal{V}\frac{\partial\bar{\rho}_{CV}}{\partial t} \qquad (4.12)$$

since $\bar{\rho}_{CV}$ can vary only with time within the control volume. Also, from Fig. 4.8, $\Delta m_{CV}^{out} = \rho_2 \Delta \mathcal{V}_2 = \rho_2 A_2 V_2 \Delta t$, so that

$$\frac{dm_{CV}^{out}}{dt} = \rho_2 A_2 V_2 \qquad (4.13)$$

and similarly,

$$\frac{dm_{CV}^{in}}{dt} = \rho_1 A_1 V_1 \qquad (4.14)$$

Substituting Eqs. (4.11)–(4.14) into (4.10) and rearranging slightly, we obtain

$$\rho_1 A_1 V_1 - \rho_2 A_2 V_2 = \mathcal{V}\frac{\partial\bar{\rho}_{CV}}{\partial t} \qquad (4.15)$$

This is the general equation of continuity for flow through regions with fixed boundaries, in which $\partial\bar{\rho}_{CV}/\partial t$ is the time rate of change of the mean density of the fluid in $\mathcal{V}$. The equation states that the net rate of mass inflow to the control volume is equal to the rate of increase of mass within the control volume.

For steady flow (Sec. 4.3), $\partial\bar{\rho}_{CV}/\partial t = 0$ in Eq. (4.15) and

Steady flow:
$$\rho_1 A_1 V_1 = \rho_2 A_2 V_2 = \dot{m} \qquad (4.16a)$$
$$\gamma_1 A_1 V_1 = \gamma_2 A_2 V_2 = g\dot{m} = G \qquad (4.16b)$$

These are the continuity equations that apply to steady, compressible, or incompressible flow within fixed boundaries.

If the fluid is incompressible, $\rho = $ constant; hence $\rho_1 = \rho_2$ and $\partial\rho/\partial t = 0$ in Eq. (4.15), and thus

Incompressible flow:
$$A_1 V_1 = A_2 V_2 = Q \qquad (4.17)$$

This is the continuity equation that applies to incompressible fluids for both steady and unsteady flow within fixed boundaries.[4]

---

[4] The continuity equations (4.16) and (4.17) apply to any stream tube in a flow system. Most commonly the continuity equation is applied to the stream tube that coincides with the boundaries of the flow.

Equations (4.16) and (4.17) are generally adequate for the analysis of flows in conduits with solid boundaries, but they are not suitable for the compressible and possibly unsteady flow of air around an airplane, for example. For the case of unsteady flow of a liquid in an open channel (Fig. 4.4), the principle of conservation of mass indicates that the rate of flow past section 1 minus the rate of flow past section 2 is equal to the time rate of change of the volume of liquid $\mathcal{V}$ contained in the channel between the two sections. Thus

$$Q_1 - Q_2 = \frac{d\mathcal{V}}{dt} \tag{4.18}$$

4.8   One-, Two-, and Three-Dimensional Flow

In true one-dimensional flow the velocity at all points has the same direction and (for an incompressible fluid) the same magnitude. Such a case is rarely of practical interest. However, we apply the term *one-dimensional method of analysis* to the flow between boundaries that are really three-dimensional, with the understanding that the "one dimension" is taken *along the central streamline* of the flow. We consider average values of velocity, pressure, and elevation across a section normal to this streamline to be typical of the flow as a whole. Thus we call the equation of continuity in Sec. 4.7 the one-dimensional equation of continuity, even though we can apply it to flow in conduits that curve in space and in which the velocity varies across sections normal to the flow. When we need high accuracy, in the following chapters, we will need to remember to refine the equations derived by the one-dimensional method of analysis in order to account for the variation in conditions across the flow section.

If all streamlines in the flow are plane curves and are identical in a series of parallel planes, we call the flow *two-dimensional*. In Fig. 4.9a the channel has a constant dimension perpendicular to the plane of the figure. Thus every cross section normal to the flow must be a rectangle of this constant width. The flow depicted in Fig. 4.9b is *three-dimensional*, although in this particular case the flow is also axially symmetric, which simplifies the analysis. A generalized three-dimensional flow, such as the flow of cool air from an air conditioning outlet into a room, is quite difficult to analyze. We often approximate such flows as two-dimensional or as axially symmetric flow. This offers the advantages that we can more easily draw diagrams describing the flow, and the mathematical treatment is much simpler.

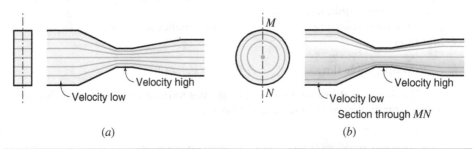

(a)   (b)

FIGURE 4.9   (a) Two-dimensional and (b) three-dimensional (axially symmetric) flow of an ideal fluid.

4.9   The Flow Net

In the case of steady two-dimensional flow of an *ideal* fluid within any boundary configuration, we can represent and determine the streamlines and velocity distribution by a *flow net*, such as that shown in Fig. 4.10. This is a network of *streamlines* and lines normal (perpendicular) to them called *equipotential lines*. Flow nets help us visualize flow patterns, the streamlines indicate the mean flow directions, and the spacing between both sets of lines at any point is inversely proportional to the local flow velocity. Furthermore, by sketching and adjusting streamlines and equipotential lines until they are approximately orthogonal at all points, we can graphically solve for flow patterns, and thus local flow directions and velocities. Thus, for example, the maximum velocity around the inside bend in Fig. 4.10 (at the smallest "square") is, from the ratio of the square sizes, about $2.7U_0$.

A fundamental property of the flow net is that it provides the one and only representation of the ideal flow within the given boundaries. It is also independent of the actual magnitude of the flow; and, for the *ideal* fluid, it is the same whether the flow is in one direction or the reverse.

In a number of simple cases we can obtain mathematical expressions, known as *stream functions*, from which we can plot streamlines. But even the most complex cases we can solve by plotting a flow net by a trial-and-error method. Although it is possible to construct nets for three-dimensional flow, we will restrict treatment here to the simpler two-dimensional net, which will more clearly illustrate the method. Consider the two-dimensional stream tube of Fig. 4.11. Assuming a constant unit thickness perpendicular to the paper, the continuity equation gives $V_1 \Delta n_1 = V_2 \Delta n_2$.

Consider next a region of uniform flow divided into a number of strips of equal width, separated by streamlines, as in Fig. 4.9a. Each strip represents a stream tube, and the flow is equally divided among the tubes. As the flow approaches a bend or obstruction, the streamline must curve so as to conform to the boundaries, but each stream tube

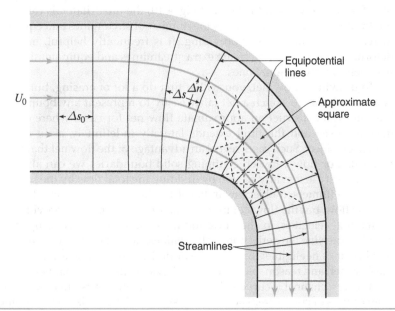

FIGURE 4.10   Flow net (two-dimensional flow).

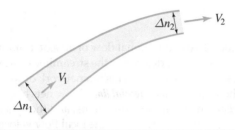

FIGURE 4.11   A two-dimensional stream tube.

still carries the same flow. Thus the spacing between all streamlines in the entire field is everywhere inversely proportional to the local velocities so that, for any section normal to the velocity,

$$V \Delta n = \text{constant} \tag{4.19}$$

When drawing the streamlines, we need to start by estimating not only the spacing between them but also their directions at all points. As an aid in the latter, we also draw the normal, or equipotential lines. As an analogy consider heat flow through a homogeneous material enclosed between perfectly insulated boundaries. We might consider the heat to flow along the equivalent of streamlines. As no heat can flow along a line of constant temperature, it follows that everywhere the heat must flow perpendicularly to isothermal lines. Likewise, streamlines must be everywhere perpendicular to equipotential lines. Because solid boundaries, across which there can be no flow, also represent streamlines, it follows that *equipotential lines must meet the boundaries everywhere at right-angles.*

If, as is usually most convenient, the equipotential lines are spaced the same distance apart as the streamlines in the region of uniform two-dimensional flow (as at the ends of Fig. 4.10), the flow net for that region is composed of perfect squares. In a region of deformed flow (as in the bend of Fig. 4.10) the quadrilaterals cannot remain square, but they will approach squares as the number of streamlines and equipotential lines are increased indefinitely by subdividing. It is frequently helpful, in regions where the deformation is marked, to create extra streamlines and equipotential lines spaced midway between the original ones.

In drawing a flow net, you will at first do a lot of erasing, but with some practice you will be able to sketch a net fairly easily to represent any boundary configuration. We can even construct an approximate flow net for cases where one solid boundary does not exist and the fluid extends laterally indefinitely, as in the flow around an immersed object. Such cases reveal an advantage of the flow net that is not evident from Fig. 4.10. For flow between confining solid boundaries we can always determine the mean velocity across any section by dividing the total flow by the section area. For flow around an immersed object, as in Fig. 4.12, there is no fixed area by which to divide a definite flow, but the flow net in combination with Eq. (4.19) provides a good means of estimating velocities in the surrounding region. With increasing distance from the body's centerline the deflection of streamlines around the body reduces, until the deflection becomes negligible; this distance or deflection must be estimated in order to draw the flow net, and reasonable estimates will yield closely similar velocities near the body.

Where a channel is curved, the equipotential lines must diverge because they radiate from centers of curvature. The distance between the associated streamlines must vary in the same way as that between the equipotential lines. Therefore, as in Fig. 4.10, the areas are smallest along the inner radius of the bend and increase toward the outside.

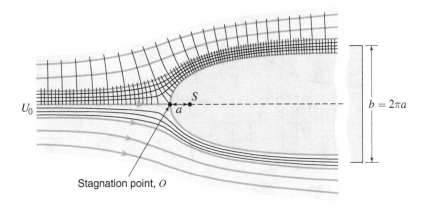

U_0 S a $b = 2\pi a$

Stagnation point, O

Figure **4.12** Two-dimensional flow of a frictionless fluid past a solid[5] whose surface is perpendicular to the plane of the paper. Streamlines or path lines for steady flow.

We can check the accuracy of the final flow net by drawing diagonals, as indicated by a few dashed lines in Fig. 4.10. If the net is correct, these dashed lines will also form a network of lines that cross each other at right angles and produce areas that approach squares in shape.

4.10 Use and Limitations of the Flow Net

Although the flow net is based on an ideal frictionless fluid, we can apply it to the flow of a real fluid within certain limits. Such limits are dictated by the extent that factors which the ideal-fluid theory neglects affect the real fluid. The principal factor of this type is fluid friction.

The viscosity effects of a real fluid are most pronounced at or near a solid boundary and diminish rapidly with distance from the boundary. Hence, for an airplane or a submerged submarine, we can consider the fluid as frictionless, except when very close to the object. The flow net always indicates a velocity next to a solid boundary, whereas a real fluid must have zero velocity adjacent to a wall due to the no-slip condition (Sec. 2.11). The region in which the velocity is so distorted, however, is confined to a relatively thin layer called the **boundary layer** (Secs. 7.8–7.11), outside of which the real fluid behaves very much like the ideal fluid.

The effect of the boundary friction is minimized when the streamlines are converging, but in a diverging flow there is a tendency for the streamlines not to follow the boundaries if the rate of divergence is too great. In a sharply diverging flow, such as is shown schematically in Fig. 4.13, there may be a **separation** of the boundary layer from the wall, resulting in **eddies** (Sec. 4.2) and even reverse flow in that region. The flow is badly disturbed in such a case, and the flow net may then be of limited value.

A practical application of the flow net is to the flow around a body, as shown in Fig. 4.12. An example of this is the upstream portion of a bridge pier below the surface where surface wave action is not a factor. Except for a thin layer adjacent to the body, this diagram represents the flow in front of and around the sides of the body. The central streamline branches at the forward tip of the body to form two streamlines along the walls.

---

[5] This surface shape is the boundary between the given flow field and that issuing from a source of strength $Q = bdU_0$ located at S, where d is the source length perpendicular to the figure.

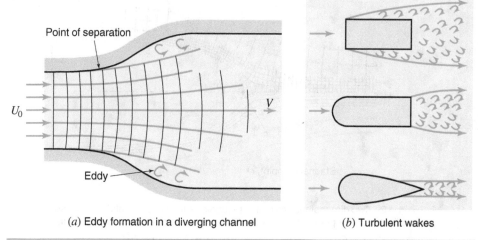

Point of separation

U_0

V

Eddy

(*a*) Eddy formation in a diverging channel

(*b*) Turbulent wakes

FIGURE 4.13   Separation in diverging flow.

At the forward tip the velocity must be zero, so we call this point a ***stagnation point***. Other common applications are to flows over spillways, and to seepage flows through earth dams and through the ground under a concrete dam. In the first two of these cases, the flow has a *free surface* at atmospheric pressure. To draw flow nets for free surface flows, we must make use of more advanced principles that are not covered in this text.

Considering the limitations of the flow net in diverging flow, we can see that, while the flow net gives a fairly accurate picture of the velocity distribution in the region near the upstream part of any solid body, it may give little information concerning the flow conditions near the rear because of the possibility of separation and eddies. We call the disturbed flow to the rear of a body a ***turbulent wake*** (Fig. 4.13*b*). We can greatly reduce the space occupied by the wake by streamlining the body, i.e., by giving the body a long slender tail, which tapers to a sharp edge for two-dimensional flow or to a point for three-dimensional flow.

Sample Problem 4.2

Figure 4.12 represents flow toward and around a bridge pier where $b = 5$ ft and $U_0 = 10$ fps. (*a*) Make a plot of the velocity along the flow centerline to the left of the solid, and along the boundary of the solid. (*b*) By what percentage does the maximum velocity along the boundary exceed the uniform velocity? (*c*) How far from the stagnation point does a velocity of 7.5 fps occur?

Solution

(*a*)

Eq. (4.19): $V\Delta n = \text{const.} = U_0 \Delta n_0$

So $V = (\Delta n_0/\Delta n)10$ fps.

Use $b = 5$ ft to scale 1 ft distances along the centerline and around the boundary of the solid. On Fig. 4.12 measure the net "square" sizes, in both the flow (ΔL) and perpendicular (ΔW) directions, using three or four squares where appropriate and taking the average. Calculate Δn and V as shown in the table:

Distance from stagnation point, ft	-6	-5	-4	-3	-2	-1	0	1	2	3	4	5	6	7	8
Average ΔL, mm	0.98	1.02	1.06	1.06	1.17	1.25	—	0.88	0.70	0.67	0.66	0.73	0.75	0.78	0.84
Average ΔW, mm	0.88	0.91	0.94	1.00	1.30	1.80	—	0.95	0.80	0.74	0.75	0.76	0.79	0.90	0.93
$\Delta n = \frac{1}{2}(\Delta L + \Delta W)$, mm	0.93	0.97	1.00	1.03	1.24	1.53	—	0.92	0.75	0.71	0.71	0.74	0.77	0.84	0.88
$\Delta n_o/\Delta n = 0.93/\Delta n$	1.00	0.96	0.93	0.90	0.75	0.61	—	1.02	1.24	1.32	1.32	1.25	1.21	1.11	1.05
$V = 10(\Delta n_o/\Delta n)$, fps	10.0	9.6	9.3	9.0	7.5	6.1	0	10.2	12.4	13.2	13.2	12.5	12.1	11.1	10.5

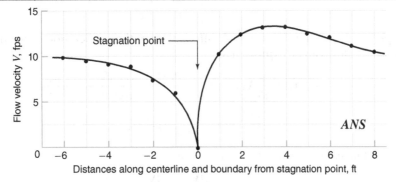

(b) $V_{max}/U_0 = 13.3/10.0 = 1.33$

Therefore V_{max} is 33% greater than U_0 *ANS*

(c) From the plot shown, a velocity of 7.5 fps occurs at about −1.9 ft and +0.4 ft from the stagnation point   *ANS*

Note: If the flow net had been constructed perfectly, the respective average ΔL and ΔW values would have been identical. Even so, the results obtained here are quite accurate, because the respective values of ΔL and ΔW were averaged. This problem can also be solved analytically using principles of hydrodynamics, which yield $V_{max}/U_0 = 1.260$.

4.11   Frame of Reference in Flow Problems

In flow problems we are really concerned with the *relative* velocity between the fluid and the body. It makes no difference whether the body is at rest and the fluid flows past it or whether the fluid is at rest and the body moves through the fluid. There are thus two frames of reference. In one the observer (or the camera) is at rest with respect to the solid body. If the observer at rest with respect to a bridge pier views a steady flow past it or is on a ship moving at constant velocity through still water, the streamlines appear to him to be unchanging and therefore the flow is steady. But if he floats with the current past the pier or views a ship going by while he stands on the bank, the flow pattern that he observes is changing with time. Then the flow is unsteady.

The same flow may therefore be either steady or unsteady according to the frame of reference. The case that is usually of more practical importance is steady, ideal flow, for which the streamlines and path lines are identical. In unsteady flow streamlines and path lines are entirely different from each other, and they also bear no resemblance to those of steady flow.

4.12   Velocity and Acceleration in Steady Flow

In a typical three-dimensional flow field, the velocities may be everywhere different in magnitude and direction. Also, the velocity at any point in the field may change with time. Let us first consider the case where the flow is steady and thus independent of time. If the velocity of a fluid particle has components u, v,[6] and w parallel to the x, y, and z axes, then, for steady flow,

$$u_{st} = u(x, y, z) \tag{4.20a}$$
$$v_{st} = v(x, y, z) \tag{4.20b}$$
$$w_{st} = w(x, y, z) \tag{4.20c}$$

Applying the chain rule of partial differentiation, the acceleration of the fluid particle for steady flow can be expressed as

$$\mathbf{a}_{st} = \frac{d}{dt}\mathbf{V}(x, y, z) = \frac{\partial \mathbf{V}}{\partial x}\frac{dx}{dt} + \frac{\partial \mathbf{V}}{\partial y}\frac{dy}{dt} + \frac{\partial \mathbf{V}}{\partial z}\frac{dz}{dt} \tag{4.21}$$

where

$$|\mathbf{V}| = \sqrt{u^2 + v^2 + w^2}$$

Noting that $dx/dt = u$, $dy/dt = v$, and $dz/dt = w$,

$$\mathbf{a}_{st} = u\frac{\partial \mathbf{V}}{\partial x} + v\frac{\partial \mathbf{V}}{\partial y} + w\frac{\partial \mathbf{V}}{\partial z} \tag{4.22}$$

This vector equation can be written as three scalar equations:

$$(a_x)_{st} = u\frac{\partial u}{\partial x} + v\frac{\partial u}{\partial y} + w\frac{\partial u}{\partial z} \tag{4.23a}$$

$$(a_y)_{st} = u\frac{\partial v}{\partial x} + v\frac{\partial v}{\partial y} + w\frac{\partial v}{\partial z} \tag{4.23b}$$

$$(a_z)_{st} = u\frac{\partial w}{\partial x} + v\frac{\partial w}{\partial y} + w\frac{\partial w}{\partial z} \tag{4.23c}$$

These equations show that even though the flow is steady, the fluid may possess an acceleration by virtue of a change in velocity with change in position. This type of acceleration we commonly refer to as *convective acceleration*. With incompressible fluid flow, there is a convective acceleration wherever the effective flow area changes along the flow path. This is also true for compressible fluid flow, but, in addition, convective acceleration of a compressible fluid occurs wherever the density varies along the flow path regardless of any changes in the effective flow area.

At times we find it convenient to superimpose the coordinate system on the streamline pattern in such a way that the x axis is tangential to the streamline at a particular

---

[6] This text uses a rounded lower case v (vee) to help distinguish it from the capital V and from the Greek ν (nu) used for kinematic viscosity.

point of interest. In such a case we shall let s indicate distance along the streamline. Thus $\mathbf{V} = \mathbf{V}(s)$, and, since the perpendicular velocity components in Eq. (4.22) are zero, we can conveniently express the acceleration of the fluid particle along the streamline at this point as

$$\mathbf{a}_{st} = V\frac{\partial \mathbf{V}}{\partial s} \qquad (4.24)$$

In the terminology of curvilinear motion, we refer to this as the **tangential acceleration**. In uniform flow with $\rho = $ constant this acceleration is zero.

At this point in our discussion we should recall that a particle moving steadily along a curved path has a **normal acceleration** a_n toward the center of curvature of the path. From mechanics,

$$a_n = \frac{V^2}{r} \qquad (4.25)$$

where r is the radius of the path. A particle moving on a curved path will always have a normal acceleration, regardless of its behavior in the tangential direction.

Sample Problem 4.3
A two-dimensional flow field is given by $u = 2y$, $v = x$. (a) Sketch the flow field. (b) Derive a general expression for the velocity and acceleration (x and y are in units of length L; u and v are in units of L/T). (c) Find the acceleration in the flow field at point A ($x = 3.5$, $y = 1.2$).

Solution
(a) Velocity components u and v are plotted to scale, and streamlines are sketched tangentially to the resultant velocity vectors. This yields the following general picture of the flow field:

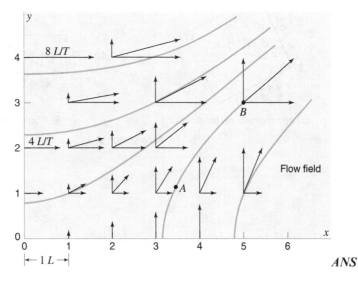

ANS

(b)
$$V = (u^2 + v^2)^{1/2} = (4y^2 + x^2)^{1/2} \qquad \textit{ANS}$$

Eq. (4.23a):
$$a_x = u\frac{\partial u}{\partial x} + v\frac{\partial u}{\partial y} = 2y(0) + x(2) = 2x$$

Eq. (4.23b):
$$a_y = u\frac{\partial v}{\partial x} + v\frac{\partial v}{\partial y} = 2y(1) - x(0) = 2y$$

$$a = (a_x^2 + a_y^2)^{1/2} = (4x^2 + 4y^2)^{1/2} = 2(x^2 + y^2)^{1/2} \qquad \textit{ANS}$$

(c) At A (3.5, 1.2),

$$(a_A)_x = 2x = 2(3.5) = 7.00 \ L/T^2; \quad (a_A)_y = 2y = 2(1.2) = 2.40 \ L/T^2$$

and
$$a_A = [(a_A)_x^2 + (a_A)_y^2]^{1/2} = [(7.00)^2 + (2.40)^2]^{1/2} = 7.40 \ L/T^2 \qquad \textit{ANS}$$

Rough check. Imagine a velocity vector at point A. This vector would have a magnitude approximately midway between that of the adjoining vectors, or $V_A \approx 4 \ L/T$. The radius of curvature of the sketched streamline at A is roughly 3L. Thus $(a_A)_n \approx 4^2/3 \approx 5.3 \ L/T^2$. The tangential acceleration of the particle at A may be approximated by noting that the velocity along the streamline increases from about $3.2 \ L/T$, where it crosses the x axis, to about $7.8 \ L/T$ at B. The distance along the streamline between these two points is roughly $4 \ L$. Hence a very approximate value of the tangential acceleration at A is

$$(a_A)_t = V\frac{\partial V}{\partial s} \approx 4\left(\frac{7.8 - 3.2}{4}\right) \approx 4.6 \ L/T^2$$

Vector diagrams of these roughly computed normal and tangential acceleration components plotted below allow us to compare them with the true acceleration as given by the analytic expressions.

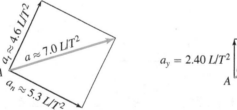

(a) Approximate vector diagram   (b) True vector diagram

Acceleration at A

4.13   Velocity and Acceleration in Unsteady Flow

For unsteady flow, Eqs. (4.20a–c) take the form

$$u = u(x, y, z, t) \ldots \tag{4.26}$$

Following a similar procedure to that of the preceding section, we obtain the vector equation

$$\mathbf{a} = \left(u \frac{\partial \mathbf{V}}{\partial x} + v \frac{\partial \mathbf{V}}{\partial y} + w \frac{\partial \mathbf{V}}{\partial z} \right) + \frac{\partial \mathbf{V}}{\partial t} \tag{4.27}$$

and the following set of scalar equations

$$a_x = \left(u \frac{\partial u}{\partial x} + v \frac{\partial u}{\partial y} + w \frac{\partial u}{\partial z} \right) + \frac{\partial u}{\partial t} \tag{4.28a}$$

$$a_y = \left(u \frac{\partial v}{\partial x} + v \frac{\partial v}{\partial y} + w \frac{\partial v}{\partial z} \right) + \frac{\partial v}{\partial t} \tag{4.28b}$$

$$a_z = \left(u \frac{\partial w}{\partial x} + v \frac{\partial w}{\partial y} + w \frac{\partial w}{\partial z} \right) + \frac{\partial w}{\partial t} \tag{4.28c}$$

In the set of equations, given by Eqs. (4.28a)–(4.28c) we recognize the three terms in parentheses as the convective accelerations (Sec. 4.12). The $\partial \mathbf{V}/\partial t$, $\partial u/\partial t$, $\partial v/\partial t$ and $\partial w/\partial t$ terms, however, represent the acceleration caused by the unsteadiness of the flow; we commonly refer to this latter type of acceleration as the *local acceleration*.

If we let s represent distance along an instantaneous streamline, in the same manner as the previous section, we now have $\mathbf{V} = \mathbf{V}(s, t)$, and the tangential acceleration of a fluid particle along the streamline is

$$\mathbf{a} = V \frac{\partial \mathbf{V}}{\partial s} + \frac{\partial \mathbf{V}}{\partial t} \tag{4.29}$$

The first term on the right-hand side of this equation is the convective acceleration, which becomes zero in uniform flow (straight and parallel streamlines) with $\rho = $ constant.

Sample Problem 4.4

Figure S4.4 is of a cross section along the centerline of a circular pipe with a conically converging nozzle. An incompressible ideal fluid flows through at $Q = (0.1 + 0.05t)$ cfs, where t is in sec. Find the average velocity and acceleration of the flow at points D and B when $t = 5$ sec.

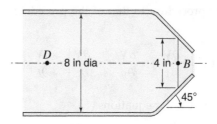

FIGURE S4.4

Solution

As a first step we sketch an approximate flow net to provide a general picture of the flow. We note that the flow is symmetric about the pipe axis (axisymmetric flow), so *the net is not a true two-dimensional flow net* (see Fig. 4.9).

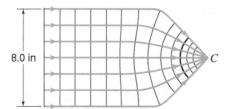 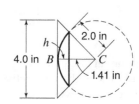

Since D and B are both on the pipe axis, $v = 0$ and $w = 0$ due to symmetry, so Eqs. (4.28) for these points reduce to

$$a_x = u\frac{\partial u}{\partial x} + \frac{\partial u}{\partial t}, \qquad a_y = 0, \qquad a_z = 0$$

At section D the streamlines are parallel and hence the area at right angles to the velocity vectors is a plane circle,

$$A_D = \frac{\pi}{4}\left(\frac{8}{12}\right)^2 = 0.349 \text{ ft}^2$$

So

$$u = \frac{Q}{A_D} = \frac{0.1 + 0.05t}{0.349} = \frac{2+t}{6.98}$$

and

$$\frac{\partial u}{\partial x} = 0, \quad \frac{\partial u}{\partial t} = \frac{1}{6.98}$$

Thus at $t = 5$ sec: $\quad V_D = u = \dfrac{2+5}{6.98} = 1.003$ fps $\qquad$ *ANS* and $$a_D = a_x = 1.003(0) + \frac{1}{6.98} = 0.1432 \text{ ft/sec}^2 \quad \textit{ANS}$$

At section B, however, the perpendicular flow area is the partial spherical surface through B, with center C and radius $r = 2$ in (see sketch). By table lookup, or by integration, this area is $2\pi rh$, where $h = r - r\cos 45° = 0.293r$. Thus $A_B = 2\pi r(0.293r) = 1.840r^2$.

On the centerline near B, $u = \dfrac{Q}{A_B} = \dfrac{0.1 + 0.05t}{1.840r^2} = \dfrac{2 + t}{36.8r^2}$

and since $x = \text{constant} - r$,

$$\frac{\partial u}{\partial x} = -\frac{\partial u}{\partial r} = -\left[\frac{-2(2 + t)}{36.8r^3}\right] = \frac{2 + t}{18.40r^3}$$

and $$\frac{\partial u}{\partial t} = \frac{1}{36.8r^2}$$

Thus at $r = 2$ in and $t = 5$ sec:

$$V_B = u = \frac{2 + 5}{36.8(2/12)^2} = 6.85 \text{ fps} \quad \textit{ANS}$$

and $$a_B = a_X = 6.85\left[\frac{2 + 5}{18.40(2/12)^3}\right] + \frac{1}{36.8(2/12)^2}$$

$$= 563 \text{ (convective)} + 0.978 \text{ (local)}$$

$$= 564 \text{ ft/sec}^2 \quad \textit{ANS}$$

Note: For the flow net shown in the sketch, the velocity at C is infinite because the flow area at that point is zero. This, of course, cannot occur; in the real case a jet somewhat similar to that of Fig. 9.13 will form downstream of the nozzle opening.

CHAPTER 5

Energy in Steady Flow

I n this chapter we shall approach flow from the viewpoint of energy considerations. The first law of thermodynamics tells us that energy can be neither created nor destroyed. But it can, of course, be changed in form. It follows that all forms of energy are equivalent.

In Secs. 5.5–5.7 we derive flow equations based on such energy considerations, but before this, in Secs. 5.2 and 5.3 we shall see how we can derive some of these equations from Newton's second law.

First we introduce the various forms of energy present in fluid flow.

5.1 Energies of a Flowing Fluid

Kinetic Energy

A body of mass m when moving at a velocity V possesses a kinetic energy, $KE = \frac{1}{2}mV^2$. Thus, if a fluid were flowing with all particles moving at the same velocity, its kinetic energy would also be $\frac{1}{2}mV^2$; for a unit *weight* of the fluid we can write this as

$$\frac{KE}{\text{Weight}} = \frac{\frac{1}{2}mV^2}{\gamma \forall} = \frac{\frac{1}{2}(\rho \forall)V^2}{\rho g \forall} = \frac{V^2}{2g} \qquad (5.1a)$$

where $\forall$ represents the volume of the fluid mass. In BG units we express $V^2/2g$ in ft·lb/lb = ft and in SI units as N·m/N = m. Similarly,

$$\frac{KE}{\text{Mass}} = \frac{\frac{1}{2}mV^2}{m} = \frac{V^2}{2} \qquad (5.1b)$$

and

$$\frac{KE}{\text{Volume}} = \frac{\frac{1}{2}mV^2}{\forall} = \frac{\frac{1}{2}(\rho \forall)V^2}{\forall} = \frac{\rho V^2}{2} \qquad (5.1c)$$

The units of $V^2/2$ of course are ft²/sec² in BG units or m²/s² in SI units. The units of $\rho V^2/2$ are lb/ft² or N/m², which are units of pressure.

In most situations the velocities of the different fluid particles crossing a section are not the same, so it is necessary to integrate all portions of the stream to obtain the true value of the kinetic energy. It is convenient to express the true value in terms of the

107

mean velocity V and a factor α (alpha), known as the *kinetic-energy correction factor*. Then

$$\frac{\text{True KE}}{\text{Weight}} = \alpha \frac{V^2}{2g} \tag{5.2}$$

In order to obtain an expression for α, consider the case where the axial components of the velocity vary across a section, as in Fig. 4.1b. If u is the local axial velocity component at a point, the mass flow per unit of time through an elementary area dA is $\rho dQ = \rho u \, dA$. Thus the true flow of kinetic energy per unit of time across area dA is $\frac{1}{2}(\rho u \, dA)u^2 = \frac{1}{2}\rho u^3 dA$. The weight rate of flow through dA is $\gamma dQ = \rho g u \, dA$. Thus, for the entire section,

$$\frac{\text{True KE/time}}{\text{Weight/time}} = \frac{\text{true KE}}{\text{weight}} = \frac{\frac{1}{2}\rho \int u^3 dA}{\rho g \int u \, dA} = \frac{\int u^3 dA}{2g \int u \, dA} \tag{5.3}$$

Comparing Eq. (5.3) with Eqs. (5.2) and (4.3), we get

$$\alpha = \frac{1}{V^2} \frac{\int u^3 dA}{\int u \, dA} = \frac{1}{AV^3} \int u^3 dA \tag{5.4}$$

and we get the same result if we use True KE/Mass, where mass flow rate is, $\rho \int dQ = \rho \int u \, dA$, or if we use True KE/Volume, where the volume flow rate is $\int dQ = \int u \, dA$.

As the average of cubes is always greater than the cube of the average, the value of α will always be more than 1. The greater the variation in velocity across the section, the larger will be the value of α. For laminar flow in a circular pipe, $\alpha = 2$ (see Sample Problem 5.1); for turbulent flow in pipes, α ranges from 1.01 to 1.15, but it is usually between 1.03 and 1.06.

In some instances it is very desirable to use the proper value of α, but in most cases the error made in neglecting its divergence from 1.0 is negligible. As precise values of α are seldom known, it is customary in the case of turbulent flow to assume that $\alpha = 1$, i.e., that the kinetic energy is $V^2/2g$ per unit weight of fluid, measured in units of ft·lb/lb = ft or N·m/N = m. In laminar flow the velocity is usually so small that the kinetic energy per unit weight of fluid is negligible.

Potential Energy

The potential energy of a particle of fluid depends on its elevation above an arbitrary datum plane. We are usually interested only in *differences* of elevation, and therefore the location of the datum plane used is determined solely by convenience. A fluid particle of weight W situated a distance z above datum possesses a potential energy of Wz. Thus its potential energy per unit *weight* is z, again measured in units of ft·lb/lb = ft or N·m/N = m.

The particle's potential energy per unit mass is gz, again measured in units of ft^2/sec^2 or m^2/s^2; its potential energy per unit volume is ρgz, again measured in units of lb/ft^2 or N/m^2.

Pressure Head

A particle of fluid has energy due to its pressure above datum, most usually its pressure above atmospheric, although we normally do not refer to this as pressure energy. From Eq. (3.4) this pressure is $p = \gamma h$, and so the depth of liquid that would produce this pressure, or the "pressure head" (Sec. 5.8), is $h = p/\gamma$. We see that the units of p/γ are ft = ft·lb/lb or m = N·m/N, or once again energy per unit *weight*.

Internal Energy

Internal energy is stored energy that is associated with the molecular or internal state of matter; it may be stored in many forms, including thermal, nuclear, chemical, and electrostatic. Here we shall only consider internal *thermal* energy (heat), which is due to the motion of molecules and forces of attraction between them. Texts on thermodynamics describe this more fully. Experiments indicate that the internal thermal energy is primarily a function of temperature. For liquids and solids, the only exception occurs as they approach the vapor phase, when the internal thermal energy also depends on specific volume, or pressure. When a gas behaves as a perfect gas (Sec. 2.7), this also implies that the internal thermal energy is a function of temperature only. We can express internal thermal energy in terms of energy i per unit of mass[1] or in terms of energy I per unit of weight. Note therefore that $i = gI$.

We can take the zero of internal energy at any arbitrary temperature, since we are usually concerned only with differences. For a unit mass of substance at a constant volume, $\Delta i = c_v \Delta T$, where c_v is the specific heat at constant volume, whose units are ft·lb/(slug·°R) in the BG system or N·m/(kg·K) in SI units. Thus we express Δi in ft·lb/slug (N·m/kg in SI units). We usually express internal energy I per unit of weight in ft·lb/lb = ft (N·m/N = m in SI units).[2]

Sample Problem 5.1

In laminar flow through a circular pipe the velocity profile is a parabola (Fig. S5.1), the equation of which is $u = u_m[1 - (r/r_0)^2]$, where u is the velocity at any radius r, u_m is the maximum velocity in the center of the pipe where $r = 0$, and r_0 is the radius to the wall of the pipe. Find α.

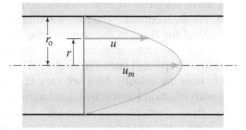

FIGURE S5.1

---

[1] The technical literature commonly represents internal energy per unit mass by the symbol u. In this text, however, we use i for internal energy per unit mass since we use u in several situations for velocity.
[2] In the BG system of measurement, scientists sometimes express internal energy I in Btu/lb; however, we rarely use those units today. Nevertheless, it is important that we be familiar with such units when reading technical papers that were written a number of years ago. 1 Btu $\approx$ 778 ft·lb.

Solution

$$u = u_m\left[1 - \left(\frac{r}{r_0}\right)^2\right], \quad dA = 2\pi r\,dr$$

For Eq. (5.4):
$$\int u^3\,dA = 2\pi u_m^3 \int_0^{r_0}\left[1 - \left(\frac{r}{r_0}\right)^2\right]^3 r\,dr$$

$$= 2\pi u_m^3 \int \left(r - 3\frac{r^3}{r_0^2} + 3\frac{r^5}{r_0^4} - \frac{r^7}{r_0^6}\right)dr$$

$$= 2\pi u_m^3 \left[\frac{r^2}{2} - \frac{3}{4}\frac{r^4}{r_0^2} + \frac{3}{6}\frac{r^6}{r_0^4} - \frac{1}{8}\frac{r^8}{r_0^6}\right]_0^{r_0}$$

$$= 0.25\pi r_0^2 u_m^3$$

and
$$Q = AV = \int u\,dA = 2\pi u_m \int_0^{r_0}\left[1 - \left(\frac{r}{r_0}\right)^2\right]r\,dr$$

$$= 2\pi u_m \int_0^{r_0}\left(r - \frac{r^3}{r_0^2}\right)dr = 2\pi u_m\left[\frac{r^2}{2} - \frac{r^4}{4r_0^2}\right]_0^{r_0} = 0.5\pi u_m r_0^2$$

So
$$V = \frac{Q}{A} = \frac{0.5\pi u_m r_0^2}{\pi r_0^2} = 0.5u_m$$

Eq. (5.4):
$$\alpha = \frac{1}{AV^3}\int u^3\,dA = \frac{0.25\pi r_0^2 u_m^3}{(\pi r_0^2)(0.5u_m)^3} = 2 \quad \textit{ANS}$$

5.2 Equation for Steady Motion of an Ideal Fluid Along a Streamline, and Bernoulli's Theorem

Referring to Fig. 5.1, let us consider frictionless steady flow of an ideal fluid (Sec. 2.10) along the streamline. We shall consider the forces acting in the direction of the streamline on a small element of the fluid in the stream tube, and we shall apply Newton's second law, that is $F = ma$. The cross-sectional area of the element at right angles to the streamline may have any shape and varies from A to $A + dA$. We recall from Secs. 4.12 and 4.13 that in steady flow the velocity does not vary at a point (local acceleration $= 0$), but that it may vary with position (convective acceleration $\neq 0$).

The mass of the fluid element is $m = \rho ds(A + \frac{1}{2}dA) = \rho dsA$ when we neglect second-order terms. The forces tending to accelerate or decelerate this mass along s are the pressure and weight forces. These are calculated as:

1. The pressure forces:

$$pA + \left(p + \frac{1}{2}dp\right)dA - (p + dp)(A + dA) = -dpA$$

The second term is the contribution of a pressure prism reflecting the linearly increasing pressure along the side of the element, the resultant of which acts on

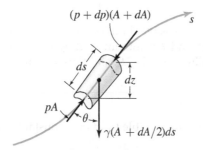

Figure 5.1   Element moving along a streamline (ideal fluid).

the increase in area dA in the direction of movement. Higher order terms ($dpdA$) are neglected.

2. The weight component in the direction of motion (again, ignoring higher order terms):

$$-\gamma ds\left(A + \frac{1}{2}dA\right)\cos\theta = -\rho g ds A\frac{dz}{ds} = -\rho g A dz$$

Applying $\Sigma F = ma$ along the streamline, we get

$$-dpA - \rho g A dz = (\rho ds A)a$$

Dividing by the volume dsA,

$$-\frac{dp}{ds} - \rho g\frac{dz}{ds} = \rho a$$

This states that the pressure gradient along the streamline combined with the weight component in that direction causes the acceleration a of the element. Recalling from Eq. (4.24) that $a = V(dV/ds)$ for steady flow, we get

$$-\frac{dp}{ds} - \rho g\frac{dz}{ds} = \rho V\frac{dV}{ds}$$

Multiplying by ds/ρ and rearranging,

$$\frac{dp}{\rho} + gdz + VdV = 0 \tag{5.5}$$

We commonly refer to this equation as the ***one-dimensional Euler*[3] *equation,*** because Leonhard Euler (1707–1783), a Swiss mathematician, first derived it in about 1750. It applies to both compressible and incompressible steady flow, since the variation of ρ over the elemental length ds is small. Dividing through by g, we can also express Eq. (5.5) as

$$\frac{dp}{\gamma} + dz + d\frac{V^2}{2g} = 0 \tag{5.6}$$

---

[3] Euler is pronounced (oi'lər), to rhyme with boiler.

Compressible Fluid

For the case of a compressible fluid, since $\gamma \neq$ constant, we must introduce an equation relating γ (or ρ) to p and T before integrating Eq. (5.5) or (5.6). We discussed stationary compressible fluid in Secs. 2.7–2.9, and flowing compressible fluid is treated further in Sec. 5.7.

Incompressible Fluid

For the case of an incompressible fluid ($\gamma =$ constant), we can integrate Eq. (5.6) to give

Energy per unit weight:
$$\frac{p}{\gamma} + z + \frac{V^2}{2g} = \text{constant (along a streamline)} \tag{5.7}$$

We know this famous equation as ***Bernoulli's theorem,*** in honor of Daniel Bernoulli (1700–1782), the Swiss physicist who presented this theorem in 1738. If we multiply each term first by g and then by ρ, we obtain the following alternate forms:

Energy per unit mass:
$$\frac{p}{\rho} + gz + \frac{V^2}{2} = \text{constant (along a streamline)} \tag{5.8}$$

and

Energy per unit volume:
$$p + \gamma z + \frac{1}{2}\rho V^2 = \text{constant (along a streamline)} \tag{5.9}$$

Terms in these three equations represent various energies of the flow, as discussed in Secs. 5.1 and 5.8. As noted, in Eq. (5.7) they are in units of energy per unit weight, in Eq. (5.8) they are in units of energy per unit mass, and in Eq. (5.9) they are in units of energy per unit volume. The constant (of integration) is known as the ***Bernoulli constant.***

Because there are so many basic assumptions involved in the derivation of Bernoulli's equation, it is important to remember them all when applying it. They are:

1. It assumes viscous (friction) effects are negligible.
2. It assumes the flow is steady.
3. The equation applies along a streamline.
4. It assumes the fluid to be incompressible.
5. It assumes no energy is added to or removed from the fluid along the streamline.

If we do not comply with any of these restrictions, serious errors can result. However, we do sometimes apply the Bernoulli equation to *real* fluids with good results in situations where the frictional effects are very small. The streamline, shown as two dimensional in Fig. 5.1, may also be three dimensional. If enough is known about the flow at some point on the streamline, we can find the Bernoulli constant; for Eq. (5.7) this constant is known as the total head, which we will discuss further in Sec. 5.8. Note that certain special flows do occur for which Bernoulli's equation holds *throughout* the flow field, not just along a streamline.

If the fluid is not moving, then we see that Eq. (5.7) reduces to Eq. (3.6).

Sample Problem 5.2

Glycerin (specific gravity 1.260) in a processing plant flows in a pipe at a rate of 700 L/s. At a point where the pipe diameter is 600 mm, the pressure is 300 kPa. Find the pressure at a second point where the pipe diameter is 300 mm if the second point is 1.0 m lower than the first point. Neglect head loss.

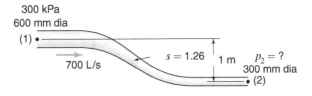

Solution

Appendix A or D:
$$\gamma_{water} = 9810 \text{ N/m}^3 = 9.81 \text{ kN/m}^3$$

Eq. (4.6):
$$V_1 = \frac{0.70 \text{ m}^3/\text{s}}{\pi(0.3)^2 \text{ m}^2} = 2.48 \text{ m/s}, \qquad V_2 = 4V_1 = 9.90 \text{ m/s}$$

Eq. (5.7):
$$\frac{300}{1.260(9.81)} + 0 + \frac{(2.48)^2}{2(9.81)} = \frac{p_2}{1.260(9.81)} - 1.0 + \frac{(9.90)^2}{2(9.81)}$$

from which
$$p_2 = 254 \text{ kN/m}^2 = 254 \text{ kPa} \qquad ANS$$

5.3 Equation for Steady Motion of a Real Fluid Along a Streamline

Let us follow the same procedure as in the previous section, except that now we shall consider a real fluid. The real fluid element in a stream tube depicted in Fig. 5.2 is similar to that of Fig. 5.1, except that now with the real fluid there is an additional force acting because of fluid friction, namely $\tau(P + \frac{1}{2}dP)ds$, where τ (tau) is the shear stress at the boundary of the element and $(P + \frac{1}{2}dP)ds$ is the area over which the shear stress acts, P being the perimeter of the end area A, which may have any shape. Writing $\Sigma F = ma$ along the streamline and neglecting second-order terms, for steady flow we now get

$$-dpA - \rho gAdz - \tau Pds = (\rho dsA)V\frac{dV}{ds}$$

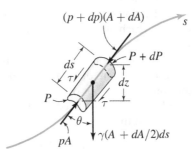

FIGURE 5.2   Element moving along a streamline (real fluid).

Dividing through by ρA and rearranging gives

$$\frac{dp}{\rho} + g\,dz + V\,dV = -\frac{\tau P}{\rho A}\,ds \qquad (5.10)$$

This equation is similar to Eq. (5.5), except that it has an extra term. The extra term $-\tau P\,ds/(\rho A)$ accounts for fluid friction.

As before, we can also express Eq. (5.10) as

$$\frac{dp}{\gamma} + dz + d\frac{V^2}{2g} = -\frac{\tau P}{\gamma A}\,ds \qquad (5.11)$$

These equations apply to steady flow of both compressible and incompressible real fluids.

Compressible Fluid

Once again, when we are dealing with a compressible fluid we must introduce an equation of state relating γ or ρ to p and T before integrating. Energy equations for the flow of compressible real fluid are further developed in Sec. 5.7.

Incompressible Fluid

For an incompressible fluid (γ = constant), we can integrate Eq. (5.11) directly. Integrating from some point 1 to another point 2 on the same streamline, where the distance between them is L, we get for an incompressible real fluid

$$\frac{p_2}{\gamma} - \frac{p_1}{\gamma} + z_2 - z_1 + \frac{V_2^2}{2g} - \frac{V_1^2}{2g} = -\frac{\tau PL}{\gamma A}$$

or

Energy per unit weight:
$$\left(\frac{p_1}{\gamma} + z_1 + \frac{V_1^2}{2g}\right) - \frac{\tau PL}{\gamma A} = \left(\frac{p_2}{\gamma} + z_2 + \frac{V_2^2}{2g}\right) \qquad (5.12)$$

As we did in Sec. 5.2, we may easily convert Eq. (5.12) to represent energy per unit mass, or energy per unit volume. The basic assumptions involved in the derivation of this equation, that we need to bear in mind, are (1) steady flow, (2) of incompressible fluid, (3) along a streamline, (4) with no energy added or removed.

If we compare Eq. (5.12) with Bernoulli Eq. (5.7) for ideal flow we see again the only difference is the additional term $-\tau PL/(\gamma A)$, which represents the loss of energy per unit weight due to fluid friction between points 1 and 2. The dimensions of this energy loss term are length only, which agrees with all the other terms in Eq. (5.12), and so this term is a form of head (Sec. 5.8).

As we noted at the outset, the friction causing this loss of energy occurs over the boundary or surface of the element, of area PL. When, as occurs often, we consider the stream tube to fill the conduit, pipe, or duct conveying the fluid, PL becomes the inside surface area of the conduit wall, and τ becomes the shear stress at the wall, τ_0. Then we can call this energy loss term the

Wall friction head loss:
$$h_f = \frac{\tau_0 PL}{\gamma A} \qquad (5.13)$$

Inserting this into Eq. (5.12), we obtain

Energy per unit weight:
$$\left(\frac{p_1}{\gamma} + z_1 + \frac{V_1^2}{2g} \right) - h_f = \left(\frac{p_2}{\gamma} + z_2 + \frac{V_2^2}{2g} \right) \tag{5.14}$$

If, as is most common, the conduit is a circular pipe of diameter D, then $P/A = \pi D/(\pi D^2/4) = 4/D$, and Eq. (5.13) becomes the

Pipe friction head loss:
$$h_f = \frac{4\tau_0 L}{\gamma D} \tag{5.15}$$

Later we shall see how fluid friction can dissipate energy in many other ways besides through shear stress over the stream tube surface; examples are in the extra turbulence caused when discharging into still water (Sec. 5.12) and by flow through valves and orifices and the like (Chaps. 7 and 9). Fluid friction loss from *any* such cause, *including* wall or pipe friction, we commonly refer to as **head loss,** denoted by h_L. So wall friction head loss is usually a part of, but it may be all of, the total head loss. In a given conduit, then, $h_L \geq h_f$. We shall discuss head loss further in Sec. 5.6.

Sample Problem 5.3
Water flows through a 150-ft-long, 9-in-diameter pipe at 3.8 cfs. At the entry point, the pressure is 30 psi; at the exit point, 15 ft higher than the entry point, the pressure is 20 psi. Between these two points, find (a) the pipe friction head loss, (b) the wall shear stress, and (c) the friction force on the pipe.

Solution

(a) From Eq. (5.14): $h_f = \left(\frac{30(144)}{62.4} + 0 + \frac{V^2}{2g} \right) - \left(\frac{20(144)}{62.4} + 15 + \frac{V^2}{2g} \right)$

$V_1 = V_2$, so terms in V cancel, and

$$h_f = 8.08 \text{ ft} \qquad ANS$$

(b) From Eq. (5.15): $\tau_0 = \frac{h_f \gamma D}{4L} = \frac{8.08(62.4)0.75}{4(150)} = 0.630 \text{ lb/ft}^2 \qquad ANS$

(c) Friction force $= \tau_0 PL = \tau_0 (\pi D)L = 0.630\pi(0.75)150 = 223 \text{ lb} \qquad ANS$

5.4   Pressure in Fluid Flow

Pressure in Conduits of Uniform Cross Section

Let us now consider how pressure varies over a cross section of flow in a uniform conduit. Figure 5.3 shows a small prism of a flowing fluid. Perpendicular to the motion and in the plane of the sketch, the forces acting on the faces of the prism are $p_1 A$ and $p_2 A$ as shown. Forces parallel to the direction of motion, namely the pressure and friction forces and the weight component, must balance out if the flow is steady and parallel. Summing the perpendicular forces, we get $p_1 A + \gamma A y \cos \alpha - p_2 A = 0$, where y is the dimension of the prism as shown, and A is its cross-sectional area. From this, we get

$$p_2 - p_1 = \gamma y \cos \alpha = \gamma h = \gamma(z_1 - z_2) = -\gamma(\Delta z) \tag{5.16}$$

which is similar to Eq. (3.3). Therefore *in any plane perpendicular to the direction of a parallel and steady flow the pressure varies according to the hydrostatic law.* The average pressure is then the pressure at the centroid of such an area. The pressure is lowest near the top of the conduit, and cavitation (Sec. 5.10), if it were to occur, would appear there first. Equation (5.16) tells us that on a horizontal axis through the conduit and perpendicular to its centerline the pressure is everywhere the same. Since the velocity is higher near the center than near the walls, it follows that the local energy head is also higher near the center. This emphasizes the fact noted earlier that a flow equation such as Eq. (5.7) or (5.12) applies along the same streamline, but not between two different streamlines, just as they do not apply between two streams in two separate channels.

Static Pressure

In a flowing fluid, we call the fluid pressure p the *static pressure* because it is the pressure that an instrument would measure if it were *static with respect to the fluid,* i.e., moving *with* the fluid. We measure it with piezometer tubes (Sec. 3.5) and other devices that attempt to minimize disturbance to the flow (see Sec. 9.2).

Stagnation Pressure

The center streamline in Fig. 4.12 shows that the velocity becomes zero at the *stagnation point.* If p/γ denotes the static-pressure head at some distance away where the velocity

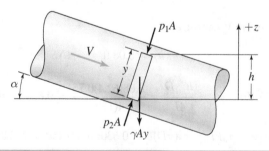

FIGURE 5.3   A prism of flowing fluid in a uniform conduit.

is V, while p_0/γ denotes the pressure head at the stagnation point, then, applying Eq. (5.7) to these two points, $p/\gamma + 0 + V^2/2g = p_0/\gamma + 0 + 0$, or the **stagnation pressure** is

$$p_0 = p + \gamma \frac{V^2}{2g} = p + \rho \frac{V^2}{2} \tag{5.17}$$

Some scientists call the quantity $\gamma V^2/2g$, or $\rho V^2/2$, the **dynamic pressure.**

Equation (5.17) applies to a fluid where we may disregard compressibility. We can show that for a compressible fluid,

$$p_0 = p + \rho \frac{V^2}{2}\left(1 + \frac{V^2}{4c^2} + \cdots\right) \tag{5.18}$$

where c is the sonic (acoustic) velocity. For air at 68°F (20°C), $c \approx 1130$ fps (345 m/s). If $V = 226$ fps (69 m/s) the error in using 1.00 for the compressibility factor, which is the quantity in parentheses, is only one percent. But for higher values of V, the effect becomes much more important. Equation (5.18) is, however, restricted to values of V/c less than one.

5.5 General Energy Equation for Steady Flow of Any Fluid

Let us now use the principles of Sec. 4.6 to consider the energy of the fluid system and control volume defined within the stream tube of Fig. 5.4. The fixed control volume lies between sections 1 and 2, and the (colored) moving fluid system consists of the fluid mass that was contained in the control volume at time t. During a short time interval Δt we shall assume that the fluid moves a short distance Δs_1 at section 1 and Δs_2 at section 2. As we are restricting ourselves in these discussions to steady flow, from Eq. (4.16b), $\gamma_1 A_1 \Delta s_1 = \gamma_2 A_2 \Delta s_2 = g\,\Delta m$, the weight of fluid entering and leaving the control volume during Δt. Recalling the analysis of Sec. 4.6, and letting the general property X now be the energy E, Eq. (4.8) becomes

$$\Delta E_S = \Delta E_{CV} + \Delta E_{CV}^{out} - \Delta E_{CV}^{in}$$

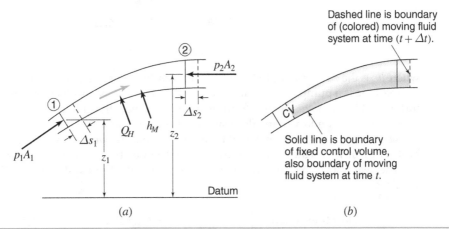

FIGURE 5.4 A stream tube with a defined control volume between sections 1 and 2.

where, as before, subscript S denotes the moving fluid system and subscript CV denotes the fixed control volume. Because the flow is steady, conditions within the control volume do not change so $\Delta E_{CV} = 0$ and

$$\Delta E_S = \Delta E_{CV}^{out} - \Delta E_{CV}^{in} \tag{5.19}$$

Let us now apply the *first law of thermodynamics* to the fluid system. This law states that for steady flow, the external work done on any system plus the thermal energy transferred into or out of the system is equal to the change of energy of the system. In other words, for steady flow during time Δt,

External work done + heat transferred = ΔE_S

Note that work, heat, and energy all have the same units, and thus are interchangeable under certain conditions.

External work can be done on the moving fluid system in various ways. One way is when a force moves through a distance. So here, when the pressure forces acting on the boundaries move, in our case when $p_1 A_1$ and $p_2 A_2$ at the end sections move through Δs_1 and Δs_2, respectively, external work is done. This work is referred to as *flow work.* It can be expressed as

$$\text{Flow work} = p_1 A_1 \Delta s_1 - p_2 A_2 \Delta s_2$$
$$= \frac{p_1}{\gamma_1}(\gamma_1 A_1 \Delta s_1) - \frac{p_2}{\gamma_2}(\gamma_2 A_2 \Delta s_2)$$
$$= \left(\frac{p_1}{\gamma_1} - \frac{p_2}{\gamma_2}\right) g \Delta m$$

The minus signs in the second terms indicate that the force and displacement are in opposite directions.

In addition to flow work, if there is a machine between sections 1 and 2 then there will be *shaft work.* During the short time interval Δt, we can write

$$\text{Shaft work} = \frac{\text{weight}}{\text{time}} \times \frac{\text{energy}}{\text{weight}} \times \text{time}$$
$$= \left(\gamma_1 A_1 \frac{ds_1}{dt}\right) h_M \Delta t = (\gamma_1 A_1 \Delta s_1) h_M$$
$$= (\gamma_1 \Delta \forall_1) h_M = (g \Delta m) h_M$$

where h_M is the energy *added* to the flow by the machine per unit *weight* of flowing fluid. If the machine is a pump, which adds energy to the fluid, h_M is positive; if the machine is a turbine, which removes energy from the fluid, h_M is negative. Note that frictional shear stresses at the boundary of the fluid system also do work on the fluid within the system. These shear stresses are not external to the system and the work they do transforms into heat, which increases the temperature of the fluid within the system (usually very slightly).

The heat transferred from an external source into the fluid system over a time interval Δt is

$$\text{Heat transferred} = \left(\gamma_1 A_1 \frac{ds_1}{dt}\right) Q_H \Delta t = (\gamma_1 A_1 \Delta s_1) Q_H = (g\,\Delta m) Q_H$$

where Q_H is the amount of energy put *into* the flow by the external heat source per unit *weight* of flowing fluid. If the heat flow is out of the fluid, the value of Q_H is negative. Note that because the fluid is flowing through the control volume at some rate (weight/sec), and Q_H is added to each unit weight of fluid, Q_H here does correspond to a *rate* of flow of heat.

So the total energy added to (or removed from) the fluid system during time Δt [the left side of Eq. (5.19)], is

$$\Delta E_S = \text{external work done} + \text{heat transferred}$$
$$= \text{flow work} + \text{shaft work} + \text{heat transferred}$$
$$= \left(\frac{p_1}{\gamma_1} - \frac{p_2}{\gamma_2} + h_M + Q_H \right) g \Delta m \tag{5.20}$$

To evaluate the right-hand side of Eq. (5.19), we first recall, as we noted initially, that for steady flow during time interval Δt, the weights of fluid entering the control volume at section 1 and leaving at section 2 are both equal to $g\,\Delta m$. From Sec. 5.1, we see that the energy (kinetic + potential + internal) carried across the boundary by $g\,\Delta m$ is

$$\Delta E = g\,\Delta m \left(z + \alpha \frac{V^2}{2g} + I \right)$$

Thus the change in energy of the control volume during Δt is

$$\Delta E_{CV}^{\text{out}} - \Delta E_{CV}^{\text{in}} = g\Delta m \left(z_2 + \alpha \frac{V_2^2}{2g} + I_2 \right) - g\Delta m \left(z_1 + \alpha \frac{V_1^2}{2g} + I_1 \right) \tag{5.21}$$

Substituting Eqs. (5.20) and (5.21) into Eq. (5.19), at the same time factoring out $g\,\Delta m$, we get

$$\frac{p_1}{\gamma_1} - \frac{p_2}{\gamma_2} + h_M + Q_H = \left(z_2 + \alpha_2 \frac{V_2^2}{2g} + I_2 \right) - \left(z_1 + \alpha_1 \frac{V_1^2}{2g} + I_1 \right)$$

or

$$\left(\frac{p_1}{\gamma_1} + z_1 + \alpha_1 \frac{V_1^2}{2g} + I_1 \right) + h_M + Q_H = \left(\frac{p_2}{\gamma_2} + z_2 + \alpha_2 \frac{V_2^2}{2g} + I_2 \right) \tag{5.22}$$

This energy equation applies to liquids, gases, and vapors, and to ideal fluids as well as to real fluids with friction, both compressible and incompressible. The only restriction is that it is for steady flow. The new features of this general equation are that it takes into account density changes (via γ), energy changes due to machines (h_M) and due to heat transfer to or from outside the fluid (Q_H), and it accounts for the conversion of other forms of fluid energy into internal heat (I).

The p/γ terms (pressure head, see Sec. 5.1) represent energy possessed by the fluid per unit weight of fluid by virtue of the pressure under which the fluid exists. Under proper circumstances, this pressure can be released and will transform into other forms of energy, i.e., kinetic, potential, or internal energy. Likewise, it is possible for these other forms of energy to transform into pressure head.

In turbulent flow there are other forms of kinetic energy besides that of translation described in Sec. 5.1. These other forms are the rotational kinetic energy of eddies initiated by fluid friction (Sec. 4.2) and the kinetic energy of the turbulent fluctuations of velocity (Sec. 4.5). No specific terms in Eq. (5.22) represent them because their effect appears indirectly. While the kinetic energy of translation can transform into increases in p/γ or z, the kinetic energy due to eddies and turbulent fluctuations can never transform into anything but thermal energy. Thus they appear as an increase in the numerical value of I_2 over the value it would have if there were no friction.

The general energy equation given by Eq. (5.22) and the continuity equation given by Eq. (4.16) are two important keys to the solution of many problems in fluid mechanics. For compressible fluids, we need a third equation, which is the equation of state, Eq. (2.4), which provides a relationship between density (or specific volume) and the absolute values of the pressure and temperature.

In many cases Eq. (5.22) simplifies greatly because certain quantities are equal and thus cancel each other, or are zero. Thus, if two points are at the same elevation, $z_1 - z_2 = 0$. If the conduit is well insulated or if the temperature of the fluid and that of its surroundings are practically the same, Q_H may effectively be zero. On the other hand, Q_H may be very large, as in the case of flow of water through a boiler tube. If there is no machine between sections 1 and 2 then the term h_M drops out. If there is a machine present, we can determine the rate of shaft work it does or is done on it by first solving Eq. (5.22) for h_M.

5.6 Energy Equations for Steady Flow of Incompressible Fluids, Bernoulli's Theorem

For liquids, and even for gases and vapors, where the change in pressure is very small, we can consider the fluid as incompressible for all practical purposes, and thus we can take $\gamma_1 = \gamma_2 = \gamma = $ constant. In turbulent flow the value of α is only a little more than 1.0 (Sec. 5.1), and, as a simplifying assumption, we will assume it equal to 1.0. If the flow is laminar, $V^2/2g$ is usually very small compared with the other terms in Eq. (5.22); hence we introduce little error if we set α equal to 1.0 rather than 2.0, its true value for laminar flow in circular pipes. Thus, for an incompressible fluid, Eq. (5.22) with $\gamma = $ constant and $\alpha = 1.0$ becomes

$$\left(\frac{p_1}{\gamma} + z_1 + \frac{V_1^2}{2g}\right) + h_M + Q_H = \left(\frac{p_2}{\gamma} + z_2 + \frac{V_2^2}{2g}\right) + (I_2 - I_1) \qquad (5.23)$$

Fluid friction produces eddies and turbulence (Sec. 4.2) and these forms of kinetic energy eventually transform into thermal energy. If there is no heat transfer, friction results in an increase in temperature, so that I_2 becomes greater than I_1. Or if the flow is isothermal (T and I both constant), there must be a loss of heat Q_H from the system at a rate equal to the rate at which friction is converting mechanical energy into thermal energy.

A change in the internal energy of a fluid coincides with a change in temperature. If c is the specific heat[4] of the incompressible fluid then, on a *mass* basis,

$$\frac{\Delta(\text{internal energy})}{\text{Unit of mass}} = \Delta i = i_2 - i_1 = c(T_2 - T_1) \tag{5.24}$$

On a unit *weight* basis, the change of internal energy is equal to the heat added to or removed from the fluid plus the heat generated by fluid friction, i.e.,

$$\frac{\Delta(\text{internal energy})}{\text{Unit of weight}} = \Delta I = \frac{\Delta i}{g} = I_2 - I_1 = \frac{c}{g}(T_2 - T_1) = Q_H + h_L \tag{5.25}$$

where h_L is the fluid-friction energy loss from all causes (Sec. 5.3) per unit weight of fluid (ft·lb/lb = ft or N·m/N = J/N = m); we commonly refer to h_L as **head loss.**

The occurrence of head loss follows directly from the **second law of thermodynamics** (the law of degradation of energy). This states that some forms of energy, such as kinetic and potential energies, which will completely convert to other forms, are "superior" to other "inferior" forms, such as heat and internal energy, which will only partially convert to the superior forms. Thus, while it is possible for a given amount of mechanical energy to completely transform into heat, the opposite is only possible in part, resulting in the mechanical energy (head) loss that always occurs with viscous flow.

We see in Eq. (5.25) that if the loss of heat (Q_H negative) is greater than h_L then T_2 will be less than T_1. On the other hand, if there is any absorption of heat (Q_H positive), T_2 will be greater than the value which would have resulted from friction alone. Note that, because the specific heat c of water and other liquids is numerically very large (footnote 4 and Table A.4), Eq. (5.25) dictates that changes in heat energy ΔI, from heat exchange or head loss, cause relatively *small* temperature changes.

We can rewrite Eq. (5.25) as

$$h_L = (I_2 - I_1) - Q_H = \frac{c}{g}(T_2 - T_1) - Q_H \tag{5.26}$$

This states that the head loss is equal to the total internal heat gain minus any heat added from external sources, per unit weight of fluid; in other words, it is the gain in thermal energy *from internal sources only.*

---

[4] For water, $c = 1$ Btu/(mass of standard lb·°R) = 1 Btu(32.2 ft/sec²)/(lb·°R) = 32.2 Btu/(slug·°R). We define the Btu (British thermal unit) in Appendix D, and the slug in Sec. 1.5. In SI units, c for water = 1 cal/(g·K). We can also express these values as 25,000 ft·lb/(slug·°R) and 4187 N·m/(kg·K), equivalent to 25,000 ft²/(sec²·°R) and 4187 m²/(s²·K), respectively. For the specific heats of various liquids, refer to Appendix A, Table A.4.

If we use Eq. (5.26) to substitute h_L for $(I_2 - I_1) - Q_H$, the general steady flow equation given by Eq. (5.23) for an incompressible fluid becomes

$$\left(\frac{p_1}{\gamma} + z_1 + \frac{V_1^2}{2g}\right) + h_M - h_L = \left(\frac{p_2}{\gamma} + z_2 + \frac{V_2^2}{2g}\right) \qquad (5.27)$$

This states that the increase in the total mechanical energy of the fluid, between sections 1 and 2, is equal to that added by a machine minus that dissipated in head loss. If there is no machine between sections 1 and 2, the energy equation for an incompressible fluid becomes

$$\left(\frac{p_1}{\gamma} + z_1 + \frac{V_1^2}{2g}\right) - h_L = \left(\frac{p_2}{\gamma} + z_2 + \frac{V_2^2}{2g}\right) \qquad (5.28)$$

When the head loss is caused only by wall or pipe friction, h_L becomes h_f (Sec. 5.3), and then we see that Eq. (5.28), derived from the energy of the fluid system and control volume, is the same as Eq. (5.14), derived from Newton's second law.

The head loss h_L may be very large in some cases, such as in very long pipelines or almost-closed valves. Although for any real fluid the head loss can never be zero, there are cases when it is so small that it may be neglected with small error.[5] In such special cases

$$\frac{p_1}{\gamma} + z_1 + \frac{V_1^2}{2g} = \frac{p_2}{\gamma} + z_2 + \frac{V_2^2}{2g} \qquad (5.29)$$

from which it follows that

$$\frac{p}{\gamma} + z + \frac{V^2}{2g} = \text{constant} \qquad (5.30)$$

Readers will recognize the energy equation in either of these two last forms as **Bernoulli's theorem,** Eq. (5.7), previously discussed in Sec. 5.2. The constant will be discussed further in Sec. 5.8. We recall that Bernoulli's equation strictly holds for steady flow of an ideal (frictionless) incompressible fluid along a streamline, and along which no energy is added to or removed from the fluid. However, we can apply it to *real* incompressible fluids with good results if the friction effects are very small.

With this, we have now developed the Bernoulli equation from two standpoints: previously from Newton's second law, and now from energy considerations. In Sec. 5.2 we presented the Bernoulli equation in three alternate forms, depending on whether the terms represent energy per unit weight as in Eqs. (5.29) and (5.30), energy per unit mass, or energy per unit volume.

---

[5] It is important to recognize that we can assume frictionless flow when frictional effects are very small. For example, we can determine the pressure around the nose of a streamlined body (Fig. 4.12) quite accurately by assuming frictionless flow; however, we must consider frictional effects if we wish to determine the shear stresses at the boundary.

Sample Problem 5.4
Water flows at 10 m³/s in a 1.5-m-diameter aqueduct; the head loss in a 1000 m length of this pipe is 20 m. Find the increase in water temperature assuming no heat enters or leaves the pipe.

Solution

Using Eq. (5.26): $20 \text{ m} = h_L = \dfrac{c}{g}(T_2 - T_1)$

From Table A.4: c for water $= 4187$ N·m/(kg·K), and, rearranging,

$$\Delta T = T_2 - T_1 = \frac{gh_L}{c} = \frac{(9.81\text{m/s}^2)(20\text{ m})}{4187[(\text{kg·m/s}^2)\cdot\text{m}]/(\text{kg·K})} = 0.0469 \text{ K} = 0.0469°\text{C} \quad ANS$$

Note that temperature *differences* in K are equivalent to differences in °C.

Sample Problem 5.5
Water is pumped through a pipeline to a treatment plant at a rate of 130 cfs. The 5-ft-diameter suction line is 3000 ft long, and the 3-ft-diameter discharge line is 1500 ft long. The pump adds energy to the water at a rate of 40 ft·lb/lb, and the total head loss is 56 ft. If the water pressure at the pipeline entrance is 50 psi, what is the pressure at the exit, which is 15 ft higher than the entrance?

Solution

Eq. (4.6): $V_1 = \dfrac{130}{(\pi/4)(5)^2} = 6.62$ fps, $V_2 = \dfrac{130}{(\pi/4)(3)^2} = 18.39$ fps

Eq. (5.27): $\left(\dfrac{50(144)}{62.4} + 0 + \dfrac{(6.62)^2}{2(32.2)}\right) + 40 - 56 = \left(\dfrac{p_2(144)}{62.4} + 15.0 + \dfrac{(18.39)^2}{2(32.2)}\right)$

from which $p_2 = 34.6$ psi *ANS*

5.7 Energy Equation for Steady Flow of Compressible Fluids
If we choose sections 1 and 2 so that there is no machine between them, and if we assume α as 1.0, Eq. (5.22) becomes

$$\left(\frac{p_1}{\gamma_1} + I_1 + z_1 + \frac{V_1^2}{2g}\right) + Q_H = \left(\frac{p_2}{\gamma_2} + I_2 + z_2 + \frac{V_2^2}{2g}\right) \tag{5.31}$$

For most compressible fluids, i.e., gases or vapors, the quantity p/γ is usually very large compared with $z_1 - z_2$ because of the small value of γ, and therefore we usually omit the z terms. But we should not ignore $z_1 - z_2$ unless we know it is negligible compared with the other quantities.

For gases and vapors, we usually combine the p/γ and the I terms into a single term called *enthalpy*. Thus enthalpy represents a composite energy property possessed by a given mass (or weight) of gas or vapor. In thermodynamics we usually express enthalpy in terms of energy per unit mass (h) rather than energy per unit weight. Thus[6]

$$h = i + \frac{p}{\rho} = gI + \frac{p}{\rho} \tag{5.32}$$

and so

$$I + \frac{p}{\gamma} = \frac{h}{g} \tag{5.33}$$

With these changes, Eq. (5.31) becomes

$$\frac{h_1}{g} + \frac{V_1^2}{2g} + Q_H = \frac{h_2}{g} + \frac{V_2^2}{2g} \tag{5.34}$$

This equation is valid for any gas or vapor and for any process. We will need some knowledge of thermodynamics to evaluate the enthalpies, and in the case of vapors we will need to use vapor tables or charts, because we cannot express their properties by any simple equations.

Sample Problem 5.6
In an air conditioning system, air flows without heat gain or loss through a horizontal pipe of uniform diameter. At section 1 the pressure is 150 psia, the velocity is 80 fps, and the temperature is 70°F; at section 2 the pressure is 120 psia and the temperature is 50°F. Find (a) the change in kinetic energy of the air; (b) the head (mechanical energy) loss in Btu/lb; (c) the change in enthalpy; all between sections 1 and 2. Assume the air to be a perfect gas.

Solution

Eq. (2.4):
$$\frac{p}{\rho T} = R = \text{constant}, \quad \text{so} \quad \frac{p_1}{\rho_1 T_1} = \frac{p_2}{\rho_2 T_2} \tag{1}$$

From Eq. (4.16a):
$$\frac{\dot{m}}{A} = \rho_1 V_1 = \rho_2 V_2 \tag{2}$$

Multiplying (1) by (2) to eliminate ρ:

$$\frac{p_1 V_1}{T_1} = \frac{p_2 V_2}{T_2}$$

---

[6] Values of enthalpy h for vapors commonly used in engineering, such as steam, ammonia, freon, and others, are given in vapor tables or charts. For a perfect gas and practically for real gases, $\Delta h = c_p \Delta T$, where c_p is specific heat at constant pressure. For air at usual pressures, c_p has a value of 6000 ft·lb/(slug·°R) [or 1003 N·m/(kg·K)]. These are equivalent to 6000 ft²/(sec²·°R) [or 1003 m²/(s²·K)].

So
$$V_2 = V_1\left(\frac{p_1 T_2}{p_2 T_1}\right) = 80\left(\frac{150}{120}\right)\frac{(460+50)}{(460+70)} = 96.2 \text{ fps}$$

(a) $\quad \Delta KE = \dfrac{V_2^2}{2g} - \dfrac{V_1^2}{2g} = \dfrac{96.2^2 - 80^2}{2(32.2)} = 44.4 \text{ ft·lb/lb increase} \qquad ANS$

(b) From Eq. (5.28) with $z_1 = z_2$:
$$h_L = \frac{p_1}{\rho_1 g} - \frac{p_2}{\rho_2 g} + \frac{V_1^2 - V_2^2}{2g}$$

Eq. (2.4)
$$\frac{p}{\rho} = RT$$

Table A.5 for air:
$$R = 1715 \text{ ft}^2/(\text{sec}^2\cdot{}^\circ R)$$

So
$$h_L = \frac{R}{g}(T_1 - T_2) + \frac{V_1^2 - V_2^2}{2g}$$

$$= \frac{1715}{32.2}(460 + 70 - 460 - 50) - 44.4 = 1021 \text{ ft}$$

$$h_L = \frac{1021 \text{ ft·lb/lb}}{778 \text{ ft·lb/Btu}} = 1.312 \text{ Btu/lb} \qquad ANS$$

(c) From Eq. (5.34) with $Q_H = 0$:
$$\frac{h_1 - h_2}{g} = \frac{V_2^2 - V_1^2}{2g} = \Delta KE = 44.4 \text{ ft·lb/lb increase}$$

$$\Delta h = h_2 - h_1 = -g(\Delta KE) = -32.2(44.4) = 1493 \text{ ft·lb/slug decrease} \qquad ANS$$

5.8 Head

In Eq. (5.28) each term has the dimensions of *length*. Thus p/γ, called the **pressure head,** represents the energy per unit weight stored in the fluid by virtue of the pressure under which the fluid exists; z, called the **elevation head** or **potential head,** represents the potential energy per pound of fluid; and $V^2/2g$, called the **velocity head,** represents the kinetic energy per pound of fluid. We call the sum of these three terms the **total head,** usually denoted by H, so that

$$H = \frac{p}{\gamma} + z + \frac{V^2}{2g} \qquad (5.35)$$

Although we usually express each term in this equation in feet (or meters), it actually represents *foot pounds of energy per pound of fluid flowing* (newton meters of energy per newton of fluid flowing in SI units). Note also that we call the sum of the middle two terms above, $(p/\gamma + z)$, the **piezometric head** or the **static (pressure) head** (see Sec. 5.11).

For an ideal (frictionless) incompressible fluid with no machine between sections 1 and 2, $H_1 = H_2$, but for a real fluid,

$$H_1 - h_L = H_2 \qquad (5.36)$$

This is merely a brief way of writing Eq. (5.28), in which the total head loss h_L (Sec. 5.3) includes the pipe or wall friction head loss h_f and possibly other losses, to be discussed later. For a real fluid, it is obvious that if there is no input of energy head h_M by a machine between sections 1 and 2, the total head must decrease in the direction of flow.

If there *is* a machine between sections 1 and 2, then

$$H_1 + h_M - h_L = H_2 \qquad (5.37)$$

If the machine is a pump, $h_M = h_p$, where h_p is the energy head put *into* the flow by the pump. If the machine is a turbine, $h_M = -h_t$, where h_t is the energy head *extracted* from the flow by the turbine.

5.9   Power Considerations in Fluid Flow

We recall from mechanics that the power P developed when a force F acts on a translating body, or when a torque T acts on a rotating body, is given by

$$\text{Rate of energy transfer} = \text{Power} = P = FV = T\omega \qquad (5.38)$$

where V is linear velocity in feet per second (or meters per second) and ω is angular velocity in radians per second. The force F represents the component force in the direction of the velocity V.

Substituting ΔpA for F and γh for Δp we can write

$$FV = (\Delta pA)V = (\gamma h)AV$$

and noting from Eq. (4.3) that $AV = Q$, we get

$$P = Q\Delta p = \gamma Qh \qquad (5.39)$$

where γ = the unit weight of fluid, lb/ft³ (N/m³ in SI units)
Q = the rate of flow, ft³/sec (m³/s in SI units)
h = the energy head, ft (m in SI units)
p = the pressure, lb/ft² (Pa in SI units)

We will refer to these equations in Chap. 6, where we discuss the dynamic forces exerted by moving fluids, and again in Chap. 11, in our discussion of turbomachinery.

In an alternative approach, we recall that every term of Eq. (5.28) represents energy per unit weight (i.e., energy head). If we multiply the energy head by the weight rate of flow, the resulting product represents rate of energy transfer, or power, since

$$\text{Power} = \frac{\text{energy}}{\text{time}} = \frac{\text{energy}}{\text{weight}} \times \frac{\text{weight}}{\text{time}} = h \times G = h \times g\dot{m}$$

from which, again, $P = \gamma Qh$

Noting that (from Appendix D) 1 hp = 550 ft·lb/sec (exactly) = 0.745700 kW, we can obtain the convenient conversions:

In BG units,
$$\text{Horsepower} = P = \frac{\gamma Q h}{550} = \frac{Q \Delta p}{550} \qquad (5.40)$$

while in SI units,
$$\text{Kilowatts} = P = \frac{\gamma Q h}{1000} = \frac{Q \Delta p}{1000} \qquad (5.41)$$

In these equations h may be any head (difference) and Δp any pressure difference for which we desire the corresponding power. For example, to find the power extracted from the flow by a turbine (i.e., the rate at which shaft work is done on a turbine, see Sec. 5.5), substitute h_t for h; to find the power of a jet, substitute $V_j^2/2g$ for h, where V_j is the jet velocity; and to find the power lost because of fluid friction, substitute h_L (or h_f) for h.

When power is transmitted through a process or machine, some power is lost in the process due to friction. The *efficiency* η (eta) of the transmission is the fraction of the power input that appears in the output, i.e.,

$$\text{Efficiency } \eta = \frac{\text{power output}}{\text{power input}} \qquad (5.42)$$

We discuss the efficiency of pumps and turbines in more detail in Chap. 11.

Sample Problem 5.7
Find the rate of energy loss due to pipe friction for the pipe of Sample Problem 5.4.

Solution

Eq. (5.41):
$$\text{Rate of energy loss} = \frac{\gamma Q h}{1000}, \quad \text{where } h = h_L$$

$$= \frac{(9810\text{ N/m}^3)(10\text{ m}^3/\text{s})(20\text{ m})}{1000}$$

$$= 1962\text{ kW} \qquad ANS$$

Sample Problem 5.8
A liquid with a specific gravity of 1.26 is being pumped from A to B through the pipeline of Fig. S5.8. At A the pipe diameter is 24 in (600 mm) and the pressure is 45 psi (300 kPa). At B the pipe diameter is 12 in (300 mm) and the pressure is 50 psi (330 kPa). Point B is 3 ft (1.0 m) lower than A. Find the flow rate if the pump puts 22 hp (16 kW) into the flow. Neglect head loss. Note that while the discharge at B is lower than point A, so liquid would flow without a pump in the system, the pump can be used to increase the pressure at B and increase the flowrate.

: Programmed computing aids could help solve problems marked with this icon.

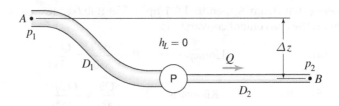

FIGURE S5.8

Solution   (BG units)

Eq. (5.40):

$$\text{Horsepower} = 22 = \frac{(1.26 \times 62.4)Qh_p}{550}$$

Rearranging:

$$h_p = \frac{153.9}{Q}$$

Eq. (5.27) with elevation A as datum, and with $h_L = 0$ (given), using $V = Q/A$, gives

$$\frac{45(144)}{1.26(62.4)} + 0 + \frac{(Q/\pi)^2}{2(32.2)} + \frac{153.9}{Q} = \frac{50(144)}{1.26(62.4)} - 3 + \frac{[Q/(0.25\pi)]^2}{2(32.2)}$$

i.e.,

$$\frac{Q^2}{42.37} + 6.158 - \frac{153.9}{Q} = 0$$

Without a polynomial solver or an equation solver, we can solve this cubic equation by trials (see Sample Problem 3.5) as follows:

Trial Q:	10.0	20.0	17.0	14.0	14.15
Left side:	−22.26	7.903	3.925	−0.210	0.006

Thus $Q = 14.15$ cfs   *ANS*

Note: The other two roots of this equation in Q involve imaginary numbers. We could "automate" the trial calculations by using a spreadsheet.

Solution   (SI Units)

Eq. (5.41):

$$kW = 16 = \frac{(1.26 \times 9810)Qh_p}{1000}$$

Rearranging:

$$h_p = \frac{1.294}{Q}$$

Eq. (5.27) with elevation A as datum, and with $h_L = 0$ (given), using $V = Q/A$, gives

$$\frac{300 \times 10^3}{1.26(9810)} + 0 + \left[\frac{Q}{\pi(0.3)^2}\right]^2 \frac{1}{2(9.81)} + \frac{1.294}{Q} = \frac{330 \times 10^3}{1.26(9810)} - 1.0 + \left[\frac{Q}{\pi(0.15)^2}\right]^2 \frac{1}{2(9.81)}$$

By trials, $Q = 0.418$ m^3/s   *ANS*

Note: The other two roots of this cubic equation in Q involve imaginary numbers.

5.10 Cavitation

The rapid vaporization and recondensation of liquid as it briefly flows through a region of low absolute pressure we call *cavitation,* as we first noted in Sec. 2.13. This phenomenon is not possible in gas flow, because a gas does not change state at low pressure, whereas a liquid will change to a gas (vapor) if the pressure is low enough. We must investigate the possibility of cavitation occurring in liquid flows because it can cause serious damage.

The dangerous, temporary low-pressure conditions associated with cavitation result from temporary high velocities, in accordance with Bernoulli's theorem, Eq. (5.7). In view of that theorem, at a given location (elevation $z =$ constant) in a liquid flow where no energy is added or removed, if the velocity head increases, there must be a corresponding decrease in the pressure head. However, so long as there is some liquid present to evaporate, there is a minimum absolute pressure possible, namely, the vapor pressure of the liquid. The (absolute) vapor pressure depends on the liquid and its temperature (see, e.g., Table A.1 for water), and it is usually less than atmospheric (Sec. 2.13). If conditions are such that calculations indicate the absolute pressure of a liquid is lower than its vapor pressure, this simply means that the assumptions upon which the calculations are based no longer apply. Thus the critical condition for cavitation is

$$(p_{crit})_{abs} = p_v$$

But

$$(p_{crit})_{abs} = p_{atm} + (p_{crit})_{gage}$$

So that

$$(p_{crit})_{gage} = -(p_{atm} - p_v) \tag{5.43}$$

where p_{atm}, p_v, and p_{crit} represent the (absolute) atmospheric pressure, the (absolute) vapor pressure, and the critical (or minimum) possible pressure, respectively, in liquid flow. Equation (5.43) states that the gage pressure head in a flowing liquid can be negative, but no more negative than $p_{atm} - p_v$. Note that the same equations can of course be expressed in terms of pressure *head,* by dividing all pressures by γ.

If at any point in a liquid the local velocity is so high that the pressure falls to its vapor pressure, the liquid will then vaporize (or boil) at that point, and bubbles of vapor will form. As the fluid moves on into a region of higher pressure, the bubbles of vapor will suddenly condense; in other words, they *collapse* or *implode.* When this occurs adjacent to solid walls, the collapse begins as a jet of liquid entering the bubble from the side opposite the wall. Figure 5.5 is a microphotograph of such a jet. Investigators have estimated that jet velocities reach 360 ft/sec (110 m/s), and that they cause pressures of 500 atmospheres (7350 psi, or 50700 kPa) or greater when the jet strikes the wall,[7] with reports of pressures as high as 7800 atmospheres.[8] They also estimated that the implosion heats the liquid immediately surrounding the cavity to about 3800°F (2100°C) for less than a microsecond. Although the jets are very small, they occur continuously with a high frequency; combined with the high temperatures and the shock waves caused by bubble collapse, they may damage the wall material.

Such action often severely and quickly damages turbine runners, pump impellers, and ship screw propellers, because it rapidly makes holes in the metal. Similar damage can occur immediately downstream of partly open valves. Overflow spillways

---

[7] K.S. Suslick, The Chemical Effects of Ultrasound, *Scientific American,* Vol. 260, No. 2, pp. 80–86, 1989.
[8] Crowe, C.T., *et al., Engineering fluid Mechanics, 9th Edition,* Wiley, 2009.

Figure 5.5   Photomicrograph of imploding bubble, with liquid jet moving downward through the center. (Bubble diameter is about 0.006 in, or 0.15 mm.)(Courtesy of Dr. Larry Crum)

(Sec. 9.13), stilling basins (Sec. 8.18), and other types of hydraulic structures built of concrete also can experience damage by cavitation. The damaging action is known as *pitting*. Not only is cavitation destructive, but it can cause a drop in efficiency of the machine or propeller or other device, and it can produce undesirable cavitation noise and vibration.

In order to avoid cavitation, we need to keep the absolute pressure at every point above the vapor pressure. There are various ways we can ensure this. In one way, we can raise the general pressure level, by placing the device below the intake level so that the liquid flows to it by gravity rather than being drawn up by suction. In another way, we can design the machine so that there are no local velocities high enough to produce such a low pressure. In a third way, we can admit atmospheric air into the low-pressure zone; we often do this downstream of partly open valves and on overflow spillways (see Sec. 9.13).

Figure 5.6 shows photographs of blades for an axial-flow pump set up in a transparent-lucite working section where the pressure level was varied. For *a*, *b*, and *c*, the water

(*a*)

(*b*)

(*c*)

(*b*)

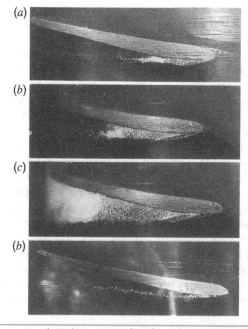

Figure 5.6   Cavitation vapor pocket phenomena: flow from right to left around a blade of an axial-flow pump, illustrating the effect of reducing absolute pressure in (*a*), (*b*), and (*c*), and the effect of a slight change of shape in (*d*). (Courtesy of the Archives, California Institute of Technology)

velocity was the same around the same vane but with decreasing absolute pressures. We see that the vapor pocket under the vane became larger at lower pressures. For d, the stream flow and the pressure were the same as for b, but the nose of the blade was slightly different in shape, which gave a different type of bubble formation. This shows the effect of a slight change in design.

Sample Problem 5.9
A liquid ($s = 0.86$) with a vapor pressure of 3.8 psia flows through the horizontal constriction in Fig. S5.9. Atmospheric pressure is 26.8 inHg. Find the maximum theoretical flow rate, i.e., at what minimum Q does cavitation occur in the *throat* (narrowest section)? Neglect head loss.

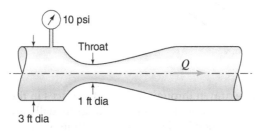

FIGURE S5.9

Solution
Since the standard atmosphere is equivalent to 29.92 inHg and 14.70 psia (Sec. 3.5),

$$p_{atm} = \frac{26.8}{29.92}(14.70) = 13.16 \text{ psia}$$

From Eq. (5.43):
$$\left(\frac{p_{crit}}{\gamma}\right)_{gage} = -\left[\frac{13.16 - 3.8}{0.86(62.4)}\right]144 = -25.1 \text{ ft}$$

Eq. (4.7):
$$V_1 = \frac{Q}{A_1} = \frac{4Q}{\pi D_1^2} = \frac{4Q}{\pi 3^2} = \frac{Q}{7.07}; \quad V_2 = \frac{4Q}{\pi 1^2} = \frac{Q}{0.785}$$

Eq. (5.29):
$$\frac{10(144)}{0.86(62.4)} + 0 + \left(\frac{Q}{7.07}\right)^2 \frac{1}{2(32.2)} = -25.1 + 0 + \left(\frac{Q}{0.785}\right)^2 \frac{1}{2(32.2)}$$

$$Q = 45.7 \text{ cfs} \quad ANS$$

5.11   Definition of Hydraulic Grade Line and Energy Line

When dealing with flow problems involving liquids, the concepts of energy line and hydraulic grade line are usually advantageous. Even with gas flow, these concepts can be useful.

We refer to the quantity $p/\gamma + z$ as the *piezometric head,* because it represents the level to which liquid will rise in a *piezometer tube,* which has its connecting end in a plane parallel to the flow (Fig. 5.7 and Sec. 3.5). The *piezometric line,* or *hydraulic grade*

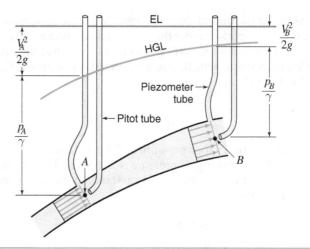

FIGURE 5.7   Ideal fluid.

line (HGL), is a line drawn through the liquid surfaces in the piezometers (Fig. 5.7); on the HGL the pressure is that of the surroundings, usually atmospheric. Let us consider point A on the stream tube in Fig. 5.7. The z portion of the piezometric head represents the elevation of point A. Therefore, the vertical distance from point A to the corresponding point on the HGL represents the remainder, which is the *static pressure head* p/γ in the flow at point A (see also Sec. 5.4). This is the difference in pressure *head* between the streamline and the surroundings, and of course this also indicates the difference in *pressure*.

A *pitot tube,* a small tube with its open end in the flow pointing upstream (Fig. 5.7 and Sec. 9.3), will intercept the kinetic energy of the flow in addition to the piezometric head, and so its liquid level indicates the *total energy head,* $p/\gamma + z + u^2/2g$. Therefore, the vertical distance between the liquid surface in the piezometer tube and that in the pitot tube is $u^2/2g$, from which we can easily compute the local flow velocity, u. The line drawn through the pitot-tube liquid surfaces (Fig. 5.7) is known as the *energy line* (EL). For the flow of an ideal fluid, as depicted in Fig. 5.7, the energy line is horizontal, because there is no head loss; for a real fluid, the energy line must slope downward in the direction of flow because of head loss due to fluid friction.

Because the local velocity u usually varies across a flow cross section, as shown in Fig. 5.8, the reading given by a pitot tube will depend on the precise location of its submerged open end. So a pitot tube will indicate the true level of the energy line only when we place it in the flow at a point where $u^2/2g = \alpha(V^2/2g)$ or, in other words, where $u = \sqrt{\alpha}V$. If we assume α (Sec. 5.1) has a value of 1.0, then, to indicate the true energy line, we must place the tube in the flow at a point where $u = V$. We rarely know ahead of time where in the flow $u = V$ (or $= \sqrt{\alpha}V$); so the correct positioning of a pitot tube, in order that it indicate the true position of the energy line, is generally unknown.

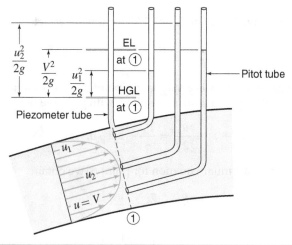

FIGURE 5.8   Real fluid.

Sample Problem 5.10

Water flows in a wide open channel as shown in Fig. S5.10. Two pitot tubes are connected to a differential manometer containing a liquid ($s = 0.82$). Find u_A and u_B.

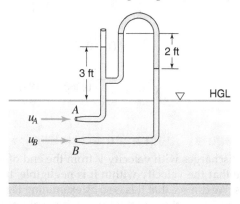

FIGURE S5.10

Solution

The water surface in the channel is the HGL, and the water surface in the pitot tube is at the EL, so the difference is:

$$\frac{u_A^2}{2g} = 3 \text{ ft}$$

from which $u_A = \sqrt{2(32.2)3} = 13.90 \text{ fps}$ *ANS*

From Sec. 3.5, Fig. 3.14b, and Eq. (3.13a) for the manometer,

$$\frac{p_A}{\gamma} - \frac{p_B}{\gamma} = z_B - z_A + \left(1 - \frac{s_M}{s_F}\right)R_m \qquad (1)$$

The tip of piezometer A is a stagnation point (Sec. 4.10) where $V = 0$, so, considering Eq. (5.29) for the approaching streamline, with y_A as the depth of point A, we obtain

$$y_A + z_A + \frac{u_A^2}{2g} = \frac{P_A}{\gamma} + z_A + 0, \quad \text{i.e.,} \quad \frac{p_A}{\gamma} = \frac{u_A^2}{2g} + y_A$$

and, subtracting a similar equation for point B, we obtain

$$\frac{p_A}{\gamma} - \frac{p_B}{\gamma} = \frac{u_A^2}{2g} - \frac{u_B^2}{2g} + y_A - y_B$$

Substituting for $\Delta p/\gamma$ from manometer Eq. (1),

$$z_B - z_A + \left(1 - \frac{s_M}{s_F}\right)R_m = \frac{u_A^2}{2g} - \frac{u_B^2}{2g} + y_A - y_B$$

and, noting that $y_A + z_A = y_B + z_B =$ the elevation of the HGL, this simplifies to

$$\frac{u_A^2}{2g} - \frac{u_B^2}{2g} = \left(1 - \frac{s_M}{s_F}\right)R_m$$

i.e.,

$$3 - \frac{u_B^2}{2g} = (1 - 0.82)2 = 0.360 \text{ ft}$$

from which

$$u_B = \sqrt{2(32.2)(3 - 0.360)} = 13.04 \text{ fps} \qquad ANS$$

5.12   Loss of Head at Submerged Discharge

When a fluid discharges with velocity V from the end of a pipe into a tank or reservoir that is so large that the velocity within it is negligible, the entire kinetic energy of the flow dissipates. We can see that this is so by examining Fig. 5.9. In the pipe up to point (a) the kinetic energy of the flowing fluid per unit weight of fluid is $V^2/2g$, but at point (b)

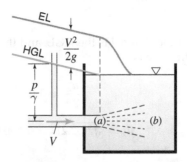

FIGURE 5.9   Fluid discharging into a tank.

in the tank the velocity is zero and hence the kinetic energy per unit weight of fluid is also zero. Because p and z are the same at points (a) and (b), the loss of head in this case, with **submerged discharge,** must be $V^2/2g$. The loss occurs after the fluid leaves the end of the pipe. This is a situation where fast-moving fluid impinges on stationary fluid. It is an impact situation not unlike that in which a fast-moving mudball collides with an immovable wall. The loss of head at submerged discharge into still water is $V^2/2g$, regardless of whether the fluid is ideal or real, compressible or incompressible. We shall consider this topic in more detail in Sec. 7.21.

5.13 Application of Hydraulic Grade Line and Energy Line

Familiarity with the concepts of the energy line and hydraulic grade line is useful in the solution of flow problems involving incompressible fluids. In the piezometer tube at B in Fig. 5.10, the liquid in it will rise to a height BB' equal to the pressure head existing at that point. If the end of the pipe at E were closed so that no flow would take place, the water would rise in this column to M. The drop from M to B' when flow occurs is due to two factors, one being that part of the pressure head has transformed into the velocity head which the liquid has at B, and the other being that there is a loss of head due to fluid friction between A and B.

As noted in Sec. 5.11, if we connected a series of piezometers all along the pipe, the liquid would rise in them to various levels along what is called the **hydraulic grade line** (Figs. 5.7 and 5.10). We can see that the hydraulic grade line represents what would be the free surface if one could exist and maintain the same conditions of flow.

The hydraulic grade line indicates the *pressure* along the pipe, since at any point the vertical distance from the pipe to the hydraulic grade line is the pressure head at that point, assuming the profile is drawn to scale. At C this distance is zero, indicating that the absolute pressure in the pipe there is atmospheric. At D the pipe is above the hydraulic grade line, indicating that the pressure head there is $-DN$, or a vacuum of DN ft (or m) of liquid.

If we draw the profile of a pipeline to scale, not only does the hydraulic grade line enable us to determine the pressure head (and so the pressure) at any point by measuring

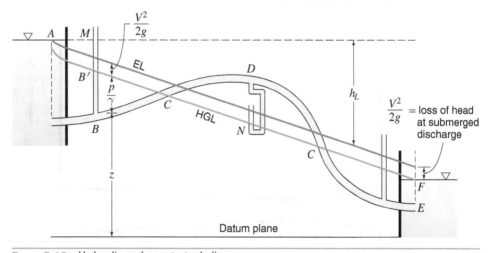

FIGURE 5.10   Hydraulic and energy grade lines.

---

136 **Chapter Five**

the diagram, but it also shows by mere inspection how the pressure varies over the entire length of the pipe. The hydraulic grade line is a straight line only if the pipe is straight and of uniform diameter and roughness (friction). But for the gradual curvatures that are often found in long pipelines, the deviation from a straight line will be small. Of course, if there are local losses of head, in addition to those due to normal pipe friction, there may be abrupt drops in the hydraulic grade line. Changes in diameter with resulting changes in velocity will also cause abrupt changes in the hydraulic grade line.

If the velocity head is constant, as in Fig. 5.10, the drop in the hydraulic grade line between any two points is equal to the loss of head between those two points, and so the slope of the hydraulic grade line is a measure of the rate of loss. In Fig. 5.11, for example, the rate of loss in the larger pipe (lower velocity) is much less than in the smaller pipe (higher velocity). If the velocity changes, the hydraulic grade line might actually rise in the direction of flow, as in Figs. 5.11 and 5.12.

The vertical distance from the level of the surface at A in Fig. 5.10 down to the hydraulic grade line for any point represents the h_L from A to the point in question plus $V^2/2g$ at that point. Thus the position of the grade line does not depend on the position

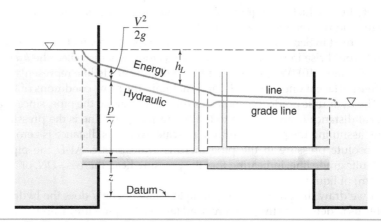

FIGURE 5.11   Profile of a pipeline plotted to scale.

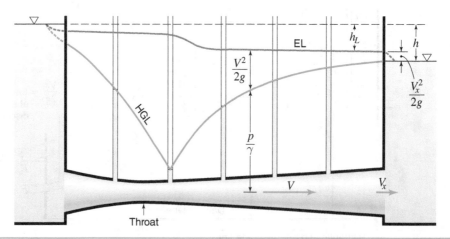

FIGURE 5.12   Profile of a converging-diverging pipe, plotted to scale.

of the pipe. Therefore we need not compute pressure heads at various points in the pipe to plot the hydraulic grade line. Instead, we can set off values of $V^2/2g + h_L$ from A to various points, below the horizontal line through A, and this procedure is often more convenient. If the pipe diameter is uniform, we need only locate a few points, and often two are sufficient.

If Fig. 5.10 represents to scale the profile of a pipe of uniform diameter, we can draw the hydraulic grade line as follows. At the intake to the pipe there will be a drop below the surface at A, which we should set off equal to $V^2/2g$ plus a local entrance loss. (This latter we explain in Sec. 7.20.) At E the pressure is EF, and hence the grade line must end at the surface at F. If the pipe discharged freely into the air at E, the line would pass through E. We can compute the location of other points, such as B' and N, if desired. In the case of a *long* pipe of uniform diameter the error is very small if we draw the hydraulic grade line as a straight line from A to F for a submerged discharge, or from A to E for a free discharge into the atmosphere.

If we set off values of h_L below the horizontal line through A, the resulting line represents values of the total energy head H measured above any arbitrary datum plane inasmuch as the line is above the hydraulic grade line by an amount equal to $V^2/2g$. This line is the *energy grade line,* usually known as simply the *energy line* (see also Sec. 5.11). It shows the rate at which the energy decreases, and it must always drop downward in the direction of flow unless energy is added by a pump. The energy line *also* does not depend on the position of the pipeline.

Energy lines are shown in Figs. 5.10–5.12. The last one, plotted to scale, shows that the chief loss of head is in the diverging portion and just beyond the *throat* (section of minimum diameter). In all three of these cases the discharge is submerged and so the velocity head is lost at discharge (Sec. 5.12). But note in Fig. 5.12 how the conical *diffuser* (diverging pipe) greatly reduces this loss, because the enlarged discharge area reduces the velocity at discharge. The large pressure changes that occur in converging–diverging pipes similar to Fig. 5.12 provide a very convenient means of measuring flow rates, which we will discuss in Sec. 9.7.

5.14 Method of Solution of Liquid Flow Problems

For the solutions of problems of (incompressible) liquid flow, there are two fundamental equations: the equation of continuity given by Eq. (4.17) and the energy equation in one of the forms from Eqs. (5.22)–(5.29). We may employ the following procedure:

1. Choose a datum plane through any convenient (lower) point.

2. Note at what sections we know or must assume the velocity. If at any point the section area is great compared with its value elsewhere, the velocity head is so small that we may disregard it.

3. Note at what points we know or must assume the pressure. In a body of liquid at rest with a free surface we know the pressure at every point within the body. The pressure in a jet is the same as that of the medium surrounding the jet.

4. Note whether or not there is any point where we know all three terms, pressure, elevation, and velocity.

5. Note whether or not there is any point where there is only one unknown quantity.

Generally we can write an energy equation that will fulfill conditions 4 and 5. If there are two unknowns in the equation then we must also use the continuity equation. For the application of these principles, see Sample Problems 5.11 and 5.12.

Sample Problem 5.11

In a fire fighting system, a pipeline with a pump leads to a nozzle as shown in Fig. S5.11. Find the flow rate when the pump develops a head of 80 ft, given that we may express the friction head loss in the 6-in-diameter pipe by $h_f = 5V_6^2/2g$, and the friction head loss in the 4-in-diameter pipe by $h_f = 12V_4^2/2g$. (a) Sketch the energy line and hydraulic grade line. (b) Find the pressure head at the suction side of the pump. Find (c) the power delivered to the water by the pump, and (d) the power of the jet.

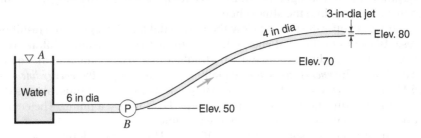

Figure S5.11

Solution

(a) Select the datum as the elevation of the water surface in the reservoir. If V_3 is the jet velocity, note from continuity equation given by Eq. (4.17) that

$$V_6 = \left(\frac{3}{6}\right)^2 V_3 = 0.25V_3, \qquad V_4 = \left(\frac{3}{4}\right)^2 V_3 = 0.563\,V_3$$

Writing energy equation given by Eq. (5.27) from the surface of the reservoir (point 1) to the jet (point 3),

$$\left(\frac{p_1}{\gamma} + z_1 + \frac{V_1^2}{2g}\right) - h_{f_6} + h_p - h_{f_4} = \frac{p_3}{\gamma} + z_3 + \frac{V_3^2}{2g}$$

$$0 + 70 + 0 - 5\frac{V_6^2}{2g} + 80 - 12\frac{V_4^2}{2g} = 0 + 80 + \frac{V_3^2}{2g}$$

Use the continuity relations to express all velocities in terms of V_3:

$$70 - \frac{5(0.25V_3)^2}{2g} + 80 - 12\frac{(0.563V_3)^2}{2g} = 80 + \frac{V_3^2}{2g}$$

from which $V_3 = 29.7$ fps

Eq. (4.17):
$$Q = A_3 V_3 = \frac{\pi}{4}\left(\frac{3}{12}\right)^2 29.7 = 1.458 \text{ cfs} \quad \textbf{\textit{ANS}}$$

Friction head loss in suction pipe:
$$h_{f_6} = 5\frac{V_6^2}{2g} = \frac{5(0.25V_3)^2}{2g} = \frac{0.312V_3^2}{2g} = 4.28 \text{ ft}$$

Friction head loss in discharge pipe:
$$h_{f_4} = 12\frac{V_4^2}{2g} = \frac{12(0.563V_3)^2}{2g} = 52.0 \text{ ft}$$

$$\frac{V_3^2}{2g} = 13.70 \text{ ft}, \quad \frac{V_4^2}{2g} = 4.33 \text{ ft}, \quad \frac{V_6^2}{2g} = 0.856 \text{ ft}$$

Drawing the energy line and hydraulic grade line on the figure to scale:

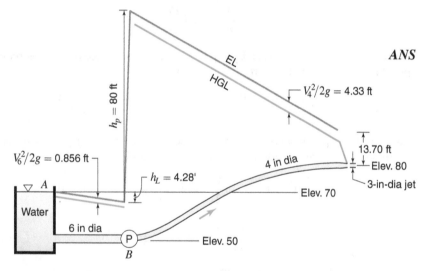

(b) From the figure we see that the pressure head on the suction side of the pump is
$$p_B/\gamma = 70 - 50 - 4.28 - 0.856 = 14.86 \quad \textbf{\textit{ANS}}$$

Likewise, we can find the pressure head at any point in the pipe if the figure is to scale.

(c) Eq. (5.40): $\quad P_{\text{deliv. by pump}} = \dfrac{\gamma Q h_p}{550} = \dfrac{62.4(1.458)80}{550} = 13.23 \text{ hp} \quad \textbf{\textit{ANS}}$

(d) Eq. (5.40) and Sec. 5.9: $\quad P_{\text{jet}} = \dfrac{\gamma Q(H_j - z_j)}{550} = \dfrac{\gamma Q}{550}\left[\dfrac{V_3^2}{2g}\right]$

$$= \frac{62.4(1.458)13.70}{550} = 2.27 \text{ hp} \quad \textbf{\textit{ANS}}$$

Sample Problem 5.12

For a system nearly identical to that of Sample Problem 5.11, but discharging to a reservoir at C, determine the flow rate through the system.

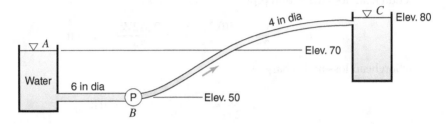

Solution

Following a similar approach to Sample Problem 5.11, the continuity equation [Eq. (4.17)] allows the two unknown velocities to be expressed as a single unknown:

$$V_6 = \frac{A_4}{A_6} V_4 = \left(\frac{4}{6}\right)^2 V_4 = 0.444 V_4$$

Writing energy equation given by Eq. (5.27) from the surface of the upstream reservoir (point A) to the downstream reservoir (point C):

$$\left(\frac{p_A}{\gamma} + z_A + \frac{V_A^2}{2g}\right) - h_{f_6} + h_p - h_{f_4} = \frac{p_C}{\gamma} + z_C + \frac{V_C^2}{2g}$$

Pressure at an open reservoir is atmospheric by definition, so $p_A = p_C = 0$ (gage). Velocities in reservoirs are considered negligibly small, so $V_A = V_C = 0$.

$$0 + 70 + 0 - 5\frac{V_6^2}{2g} + 80 - 12\frac{V_4^2}{2g} = 0 + 80 + 0$$

$$5\frac{(0.444V_4)^2}{2(32.2)} + 12\frac{V_4^2}{2(32.2)} = 70$$

$$V_4^2 = 347.1 \quad \text{or} \quad V_4 = 18.63 \text{ ft/s}$$

Eq. (4.17): $Q = VA = V_4 A_4 = (18.63 \text{ ft/s}) \cdot \frac{\pi}{4}\left(\frac{4}{12}\right)^2 = 1.626 \text{ ft}^3/\text{s}$ *ANS*

Sample Problem 5.13

Find the flow rate per meter width for the two-dimensional channel flow shown in Fig. S5.13. Assume no head loss.

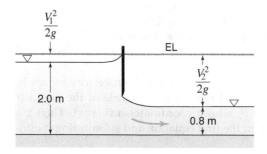

FIGURE S5.13

Solution

Select the datum as the (effectively horizontal) channel bed. The water surface represents the hydraulic grade line in the region where the streamlines are parallel. The energy line is a distance $V^2/2g$ above the water surface, assuming $\alpha = 1.0$. If there is a no head loss, the energy line is horizontal. Writing the energy equation given by Eq. (5.29) from section 1 to 2, we have

$$0 + 2.0 + \frac{V_1^2}{2g} = 0 + 0.8 + \frac{V_2^2}{2g} \tag{1}$$

Note that this applies either (*a*) between points on the water surface, with $p_1 = p_2 = 0$, $z_1 = 2.0$, and $z_2 = 0.8$, or (*b*) between points on the bed, with $z_1 = z_2 = 0$, $p_1/\gamma = 2.0$, and $p_2/\gamma = 0.8$.

But from the continuity equation given by Eq. (4.17), for 1 m of channel width perpendicular to the figure,

$$(2 \times 1)V_1 = (0.8 \times 1)V_2 \tag{2}$$

Substituting Eq. (1) into Eq. (2), and using $g = 9.81$ m/s², we obtain

$$V_1 = 2.12 \text{ m/s}, \quad V_2 = 5.29 \text{ m/s}, \quad \frac{V_1^2}{2g} = 0.229 \text{ m}, \quad \frac{V_2^2}{2g} = 1.429 \text{ m}$$

and $Q = A_1V_1 = (2 \times 1)2.12$ m/s $= 4.24$ m³/s (for 1 m of channel width)   *ANS*

5.15   Jet Trajectory

A free liquid jet in air will describe a ***trajectory,*** or path under the action of gravity, with a vertical velocity component that is continually changing. The trajectory is a stream-line, and consequently, if we neglect air friction, we can apply Bernoulli's theorem to it, with all the pressure terms zero. Thus the sum of the elevation and velocity head is the same for all points of the curve. The energy grade line is a horizontal line at distance $V_0^2/2g$ above the nozzle, where V_0 is the initial velocity of the jet as it leaves the nozzle (Fig. 5.13).

We can obtain the equation for the trajectory by applying Newton's equations of uniformly accelerated motion to a particle of the liquid traveling in time t from the nozzle to point P, whose coordinates are (x, z). Then $x = V_{x0}t$ and $z = V_{z0}t - \frac{1}{2}gt^2$. Solving for t from the first equation and substituting it into the second gives

$$z = \frac{V_{z0}}{V_{x0}}x - \frac{g}{2V_{x0}^2}x^2 \tag{5.44}$$

By setting $dz/dx = 0$, we find that z_{max} occurs when $x = V_{x0}V_{z0}/g$. Substituting this value for x in Eq. (5.44) gives $z_{max} = V_{z0}^2/2g$ Thus, Eq. (5.44) is that of an inverted parabola having its vertex at $x = V_{x0}V_{z0}/g$ and $z = V_{z0}^2/2g$. Since the velocity at the top of the trajectory is horizontal and equal to V_{x0}, the distance from this point to the energy line is evidently $V_{x0}^2/2g$. We can obtain this in another way by considering that $V_0^2 = V_{x0}^2 + V_{z0}^2$. Dividing each term by $2g$ gives the relations shown in Fig. 5.13.

If the jet is initially horizontal, as in the flow from a vertical orifice, $V_{x0} = V_0$ and $V_{z0} = 0$. Equation (5.44) then readily reduces to an expression for the initial jet velocity in terms of the coordinates from the vena contracta (Fig. 9.14) to any point of the trajectory, z now being positive downward:

$$V_0 = x\sqrt{\frac{g}{2z}} \tag{5.45}$$

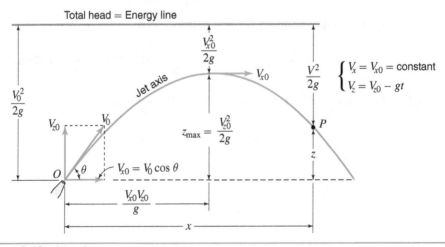

FIGURE 5.13   Jet trajectory.

Sample Problem 5.14

If a water jet is inclined upward 30° from the horizontal, what minimum initial velocity will enable it to reach over a 10-ft wall at a horizontal distance of 60 ft, neglecting friction?

Solution

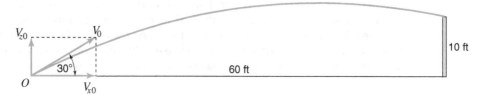

$$V_{x0} = V_0 \cos 30° = 0.866 V_0$$

$$V_{z0} = V_0 \sin 30° = 0.5 V_0$$

From Newton's laws,

$$x = 0.866 V_0 t = 60 \tag{1}$$

$$z = 0.5 V_0 t - 0.5 g t^2 = 10 \tag{2}$$

From (1), $t = 69.3 / V_0$. Substituting this into (2),

$$0.5 V_0 \frac{69.3}{V_0} - \frac{32.2}{2} \left(\frac{69.3}{V_0} \right)^2 = 10$$

from which $\qquad V_0^2 = 3140, \quad or \quad V_0 = 56.0 \text{ fps} \qquad ANS$

Plumes

When one fluid (specific gravity s_1) discharges into a second fluid (s_2) with a similar density, a plume of the first fluid forms. We are familiar with smoke and steam plumes; similar plumes form when treated sewage effluent discharges under the ocean from outfall sewers. Such plumes rise because $s_1 < s_2$. To a first approximation, neglecting fluid friction and mixing, if the second fluid is not moving then we may compute the path of the plume as a jet trajectory. However, we must then replace the gravitational acceleration in the trajectory equations by the force per unit mass on the plume fluid, which is

$$g' = \frac{(\rho_1 - \rho_2) g \mathcal{V}}{\rho_1 \mathcal{V}} = \frac{(s_1 - s_2) g}{s_1} = g \left(1 - \frac{s_2}{s_1} \right) \tag{5.46}$$

For a rising plume, both g' and z will be negative.

5.16   Flow in a Curved Path

The energy equations we developed previously apply fundamentally to flow along a streamline or along a stream of large cross section if we use certain average values. Now we will investigate conditions in a direction normal to a streamline. Figure 5.14 represents an element of fluid moving in a *horizontal* plane[9] with a velocity V along a curved path of radius r. The element has a linear dimension dr in the plane of the paper and an area dA normal to the plane of the paper. The mass of this fluid element is $\rho \, dA \, dr$, and the normal component of acceleration is V^2/r. Thus the centripetal force acting on the element toward the center of curvature is $\rho \, dA \, dr \, V^2/r$. As the radius increases from r to $r + dr$, the horizontal pressure will change from p to $p + dp$. Therefore the resultant force in the direction of the center of curvature is $dp \, dA$. Equating these two forces and dividing by dA,

$$dp = \rho \frac{V^2}{r} dr \qquad (5.47)$$

When horizontal flow is in a straight line for which r is infinite, the value of dp is zero. This demonstrates that no difference in pressure can exist in the horizontal direction perpendicular to horizontal flow in a straight line.

Because dp is positive if dr is positive, Eq. (5.47) shows that pressure increases from the concave to the convex side of the stream, but the exact way in which it increases depends on the way in which V varies with the radius. If we can express V as a function of r, or if V is constant, we can integrate Eq. (5.47) to find $p_{outer} - p_{inner}$. Usually V is *not* constant. In Secs. 5.17 and 5.18 we will consider two important practical cases in which V varies with the radius in two different ways.

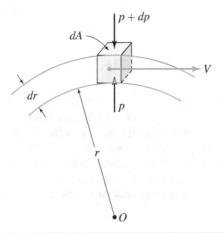

FIGURE 5.14   Circular motion in a horizontal plane.

---

[9] A more generalized analysis of flow along a curved path in a vertical or inclined plane leads to a result that includes z terms.

5.17 Forced or Rotational Vortex

In theory, we can make a fluid rotate as a solid body without relative motion between particles, either by rotating the containing vessel or by stirring the contained fluid. Thus, in one way or another, we apply an external torque. Common examples are the rotation of liquid within a centrifugal pump and that of gas in a centrifugal compressor.

Cylindrical Forced Vortex

If the entire body of fluid rotates as a solid then V varies directly with r; that is, $V = r\omega$, where ω (omega) is the imposed angular velocity. Substituting this for V in Eq. (5.47), for the case of rotation in any horizontal plane about a vertical axis, we have

$$dp = \rho\omega^2 r\,dr = \frac{\gamma}{g}\omega^2 r\,dr$$

Between any two radii r_1 and r_2, we can integrate this to give

$$\frac{p_2}{\gamma} - \frac{p_1}{\gamma} = \frac{\omega^2}{2g}(r_2^2 - r_1^2) \tag{5.48}$$

which is the pressure head difference between two points on the same horizontal plane. If p_0 is the pressure when $r_1 = 0$, Eq. (5.48) becomes

$$\frac{p}{\gamma} = \frac{\omega^2}{2g}r^2 + \frac{p_0}{\gamma} \tag{5.49}$$

which we recognize as the equation of a parabola. In Fig. 5.15a we see that if the fluid is a liquid then the pressure head p/γ at any point is equal to z, the depth of the point below the free surface. Therefore we may also write the preceding equations as

$$z_2 - z_1 = \frac{\omega^2}{2g}(r_2^2 - r_1^2) \tag{5.50}$$

and

$$z = \frac{\omega^2}{2g}r^2 + z_0 \tag{5.51}$$

where z_0 is the depth when $r_1 = 0$. Equations (5.50) and (5.51) define the free surface, if one exists, and in general they define *any* surface of equal pressure in the liquid; these surfaces are a series of paraboloids of the same shape as the free surface, such as the dashed curves in Fig. 5.15a.

For the *open* vessel of Fig. 5.15a, the pressure head at any point is equal to its depth below the free surface. If the liquid is *confined* within a vessel, as in Fig. 5.15b, the pressure will vary along any radius in just the same way as if there were a free surface. Therefore the two are equivalent.

In this discussion we assumed the axis of the vessel was vertical; however, the axis might be inclined. Since pressure varies with elevation z as well as radius, a more general equation applicable to fluid in a *closed* tank with an inclined axis is

$$\frac{p_2}{\gamma} - \frac{p_1}{\gamma} + z_2 - z_1 = \frac{\omega^2}{2g}(r_2^2 - r_1^2) \tag{5.52}$$

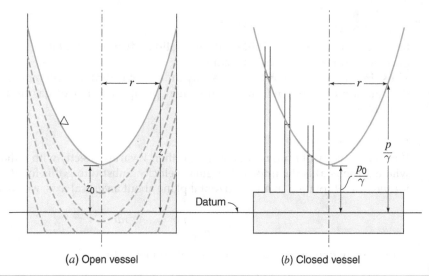

(a) Open vessel   (b) Closed vessel

FIGURE 5.15   Forced vortex.

Equation (5.48) is the special case where $z_1 = z_2$ (closed tank with vertical axis), and Eq. (5.50) is the special case where $p_1 = p_2$ (open tank with vertical axis). Note that Eqs. (5.48)–(5.52) are *not* energy equations, since they represent conditions *across* streamlines rather than along a streamline.

Spiral Forced Vortex

So far we have confined this discussion to the rotation of all particles in concentric circles. Suppose that now we superimpose a flow with a velocity having radial components either outward or inward. If the height of the walls of the open vessel in Fig. 5.15a were less than that of the liquid surface, and if we supplied liquid to the center at the proper rate by some means, then it is obvious that liquid would flow outward. If, on the other hand, liquid flowed into the tank over the rim from some source at a higher elevation and flowed out at the center, the flow would be inward. The combination of this approximately radial flow with the circular flow would result in path lines that were some form of spirals.

If the closed vessel in Fig. 5.15b is arranged with suitable openings near the center and also around the periphery, and if it is provided with vanes, as shown in Fig. 5.16, it becomes either a centrifugal pump impeller or a turbine runner, as the case may be. The vanes constrain the flow of the liquid and determine both its relative magnitude and its direction. If the area of the passages normal to the direction of flow is A, the equation of continuity fixes the relative velocities, since

$$Q = A_1 v_1 = A_2 v_2 = \text{constant}$$

This relative flow is the flow as it would appear to an observer or a camera revolving with the rotor. Neglecting friction losses and assuming a vertical axis of rotation, we can use the energy equation to find that the pressure head difference due to this superimposed flow alone is $p_2/\gamma - p_1/\gamma = (v_1^2 - v_2^2)/2g$.

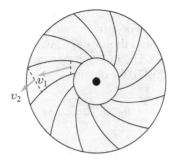

FIGURE 5.16 Flow through a rotor.

For the case of rotation with flow (i.e., spiral forced vortex), we can find the total pressure head difference between two points by adding together the pressure head differences due to the two flows considered separately. That is, for the case of a vertical axis,

$$\frac{p_2}{\gamma} - \frac{p_1}{\gamma} = \frac{\omega^2}{2g}(r_2^2 - r_1^2) + \frac{v_1^2 - v_2^2}{2g} \qquad (5.53)$$

Of course, friction losses will modify this result to some extent. If the axis is inclined, we must add z terms to the equation. We see that Eq. (5.48) is a special case of Eq. (5.53) when $v_1 = v_2$, either when both are finite or when both are zero.

For a forced vortex with spiral flow, a pump adds energy to the fluid and a turbine extracts energy from it. In the limiting case of zero flow, when all path lines become concentric circles (i.e., a cylindrical forced vortex), a real fluid still needs energy input from some external source to maintain the rotation.

5.18   Free or Irrotational Vortex

In a free vortex there is no expenditure of energy whatever from an outside source, and the fluid rotates because of some rotation previously imparted to it or due to some internal action. Some examples are a whirlpool in a river, the rotary flow that often arises in a shallow vessel when liquid flows out through a hole in the bottom (as we often see when water empties from a bathtub), and the flow in a centrifugal-pump casing just outside the impeller or that in a turbine casing as the water approaches the guide vanes.

As the fluid receives no additional energy, it follows that, neglecting friction, H is constant throughout; that is, $p/\gamma + z + V^2/2g =$ constant.

Cylindrical Free Vortex

The angular momentum with respect to the center of rotation of a particle of mass m moving along a circular path of radius r at a velocity V_u is $mV_u r$, where V_u is the velocity along the circular path (i.e., tangential velocity).[10] Newton's second law states that, for

---

[10] In this chapter we use V_u to represent the tangential component of velocity. Other symbols commonly used to represent tangential velocity are V_t and V_θ.

the case of rotation, the torque is equal to the time rate of change of angular momentum. Hence torque $= d(mV_u r)/dt$. For a free vortex (frictionless fluid) no torque is applied; therefore $mV_u r =$ constant, and thus $V_u r = C$, where we can determine the value of C by knowing the value of V at some radius r. Assuming a vertical axis of rotation and substituting $V_u = C/r$ in Eq. (5.47), we obtain

$$dp = \rho \frac{C^2}{r^2} \frac{dr}{r} = \frac{\gamma}{g} \frac{C^2}{r^3} dr$$

We can integrate this between any two radii r_1 and r_2, to get

$$\frac{p_2}{\gamma} - \frac{p_1}{\gamma} = \frac{C^2}{2g}\left(\frac{1}{r_1^2} - \frac{1}{r_2^2}\right) = \frac{V_{u1}^2}{2g}\left[1 - \left(\frac{r_1}{r_2}\right)^2\right] \qquad (5.54)$$

If there is a free surface, the pressure head p/γ at any point is equal to the depth below the surface. Also, at any radius the pressure varies in the vertical direction according to the hydrostatic law. So this equation is merely a special case in which $z_1 = z_2$.
As $H = p/\gamma + z + V_u^2/2g =$ constant, it follows that at any radius r

$$\frac{p}{\gamma} + z = H - \frac{V_u^2}{2g} = H - \frac{C^2}{2gr^2} = H - \frac{V_{u1}^2}{2g}\left(\frac{r_1}{r}\right)^2 \qquad (5.55)$$

Assuming a vertical axis, we can find the pressure along the radius from this equation by taking z constant; and for any constant pressure p, we can find values of z determining a surface of equal pressure. If p is zero, the values of z determine the free surface (Fig. 5.17a), if one exists.

Equation (5.55) indicates that H is the asymptote that $p/\gamma + z$ approaches as r approaches infinity and V_u approaches zero. On the other hand, as r approaches zero,

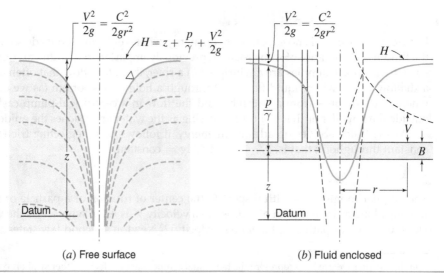

(a) Free surface   (b) Fluid enclosed

FIGURE 5.17   Free vortex.

V_u approaches infinity, and $p/\gamma + z$ approaches minus infinity. Since this is physically impossible, the free vortex cannot extend to the axis of rotation. In reality, since high velocities are attained as the axis is approached, the friction losses, which vary as the square of the velocity, become of increasing importance and are no longer negligible. In this region, then, the assumption that H is constant no longer holds; the core of the vortex tends to rotate as a solid body, as in the central part of Fig. 5.17b.

Spiral Free Vortex

If we superimpose a radial flow upon the concentric flow just discussed, the path lines will then be spirals. If the flow passes out through a circular hole in the bottom of a shallow vessel, the surface of the liquid takes the form shown in Fig. 5.17a, with an air *core* sucked down the hole. If there is an outlet symmetric with the axis as in Fig. 5.17b, we might have a flow component either radially inward or radially outward. If the two confining plates shown are a constant distance B apart, the radial flow component with velocity V_r is then across a series of concentric cylindrical surfaces whose area is $2\pi rB$. Thus

$$Q = 2\pi rBV_r = \text{constant}$$

from which we see that $rV_r = \text{constant}$. Therefore the radial velocity varies in the same way with r that the circumferential velocity did in the preceding discussion of the cylindrical free vortex. The pressure variation in a spiral free vortex (Fig. 5.17b) is given by

$$\frac{p_2}{\gamma} - \frac{p_1}{\gamma} = \frac{V_1^2}{2g} - \frac{V_2^2}{2g} \tag{5.56}$$

where $V = \sqrt{V_r^2 + V_u^2}$, the velocity of flow.

Sample Problem 5.15

A centrifugal pump with a 12-in-diameter impeller is inside a casing that has a constant height of 1.5 in between sections 1 and 2 and that then enlarges into a volute at 3 (Fig. S5.15). Water leaves the impeller with a velocity of 60 fps at an angle of 15° with the tangent. (a) At what rate is water flowing through the pump? (b) Neglecting friction, what will be the magnitude and direction of the velocity at section 2 and what will be the gain in pressure head from section 1 to 2?

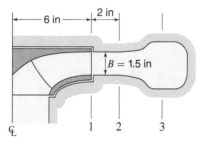

FIGURE S5.15

Solution

(a) Flow through the pump $Q = A_1(V_r)_1 = 2\pi r_1 B(V_r)_1$, where $(V_r)_1 = 60\sin 15° = 15.53$ fps

$$Q = 2\pi(6/12)(1.5/12)15.53 = 6.10 \text{ cfs} \qquad \textit{ANS}$$

(b) From continuity: $Q = 2\pi r_1 B(V_r)_1 = 2\pi r_2 B(V_r)_2$,

so
$$\frac{(V_r)_2}{(V_r)_1} = \frac{r_1}{r_2} = \frac{6}{8}$$

Because torque $= 0$ in the space between sections 1 and 2, angular momentum must be conserved. Thus

$$m(V_u)_1 r_1 = m(V_u)_2 r_2$$

so
$$\frac{(V_u)_2}{(V_u)_1} = \frac{r_1}{r_2} = \frac{6}{8}$$

The region between sections 1 and 2 is a spiral free vortex.

Because we have found that V_u and V_r both decrease in the same proportion as flow moves from section 1 to section 2, the angle α does not change, and

$$V_2/V_1 = \frac{6}{8}; \quad V_2 = \left(\frac{6}{8}\right)60 = 45 \text{ fps} \quad \text{at } 15° \text{ with the tangent} \qquad \textit{ANS}$$

Finally, writing the energy equation along the flow lines gives

$$\frac{p_2}{\gamma} - \frac{p_1}{\gamma} = \frac{V_1^2}{2g} - \frac{V_2^2}{2g} = \frac{60^2 - 45^2}{2(32.2)} = 24.5 \text{ ft} \qquad \textit{ANS}$$

CHAPTER 6

Momentum and Forces in Fluid Flow

Previously, we met two important fundamental concepts of fluid mechanics: the continuity equations and the energy equation. In this chapter we will develop a third basic concept, the ***momentum principle***. This concept is particularly important in flow problems where we need to determine forces. Such forces occur whenever the velocity of a stream of fluid changes in either direction or magnitude. By the law of action and reaction, the fluid exerts an equal and opposite force on the body producing the change. After developing the momentum principle, we will discuss its application to a number of important engineering problems.

Applications of the momentum principle to rotating machines like fans, propellers, windmills, sprinklers, pumps, and turbines are not included in this chapter. While such rotating machines are common and important, applications involving them can be found in earlier editions of this text; Chap. 11 includes pumps and turbines.

6.1   Development of the Momentum Principle

We will derive the momentum principle from Newton's second law. The flow may be compressible or incompressible, real (with friction) or ideal (frictionless), steady or unsteady, and the equation is not limited to flow along a streamline. In Chap. 5 when applying the energy equation to real fluids we found that the energy loss must be computed. We do not encounter this difficulty in momentum analysis.

We can express Newton's second law as

$$\sum \mathbf{F} = \frac{d(m\mathbf{V})_S}{dt} \tag{6.1}$$

This states that the sum of the external forces $\mathbf{F}$ on a body of fluid or system S is equal to the rate of change of linear momentum $m\mathbf{V}$ of that body or system. The boldface symbols $\mathbf{F}$ and $\mathbf{V}$ represent vectors, and so the change in momentum must be in the same direction as the force. Because we can also express Eq. (6.1) as $\sum(\mathbf{F})dt = d(m\mathbf{V})_S$, i.e., impulse equals change of momentum, we sometimes use the terminology ***impulse-momentum principle.***

Using the principles of Sec. 4.6, let us consider the linear momentum of the *fluid system* and *control volume* defined within the stream tube of Fig. 6.1a, just as we did for energy in Sec. 5.5. The fixed control volume (*CV*) lies between sections 1 and 2, and the moving fluid system (S) consists of the fluid mass contained at time t in the control volume.

151

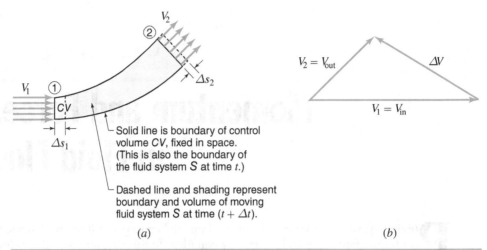

FIGURE 6.1   (a) Control volume for steady flow with control surface cutting a constant-velocity stream at right angles. (b) Velocity relations.

During a short time interval Δt, we shall assume that the fluid moves a short distance Δs_1 at section 1 and Δs_2 at section 2. Recalling the analysis of Sec. 4.6, and letting the general property X now be the momentum $m\mathbf{V}$, Eq. (4.9) becomes

$$\frac{d(m\mathbf{V})_S}{dt} = \frac{d(m\mathbf{V})_{CV}}{dt} + \frac{d(m\mathbf{V})_{CV}^{\text{out}}}{dt} - \frac{d(m\mathbf{V})_{CV}^{\text{in}}}{dt} \tag{6.2}$$

where, as before, subscript S denotes the moving fluid system and subscript CV denotes the fixed control volume. So, setting this equal to Eq. (6.1),

Unsteady flow: $$\sum \mathbf{F} = \frac{d(m\mathbf{V})_{CV}}{dt} + \frac{d(m\mathbf{V})_{CV}^{\text{out}}}{dt} - \frac{d(m\mathbf{V})_{CV}^{\text{in}}}{dt} \tag{6.3}$$

On the right side of this equation, the first term represents the rate of change or accumulation of momentum within the fixed control volume, whereas the second and third terms respectively represent the rates at which momentum enters and leaves the control volume. The entire Eq. (6.3) states that the resultant force acting on a fluid mass is equal to the rate of change of momentum of the fluid mass. It is perfectly general. It applies to compressible or incompressible, real or ideal, and steady or unsteady flow.

In the case of steady flow, conditions within the control volume do not change, so $d(m\mathbf{V})_{CV}/dt = 0$, and the equation becomes

Steady flow: $$\sum \mathbf{F} = \frac{d(m\mathbf{V})_{CV}^{\text{out}}}{dt} - \frac{d(m\mathbf{V})_{CV}^{\text{in}}}{dt} \tag{6.4}$$

Thus, for steady flow the net force on the fluid mass is equal to the net rate of *outflow* of momentum across the control surface.

Since Eqs. (6.1)–(6.4) are vector equations, we can also express them as scalar equations in terms of forces and velocities in the x, y, and z directions, respectively.

It helps if we select a control volume so that the control surface is normal to the velocity where it cuts the flow. Consider such a situation in Fig. 6.1a. Also, let us specify that the velocity is constant where it cuts across the control surface, and let us restrict ourselves to steady flow so that Eq. (6.4) is applicable. Since

$$\frac{d(m\mathbf{V})_1}{dt} = \frac{dm_1}{dt}\mathbf{V}_1 = \dot{m}_1\mathbf{V}_1 = \rho_1 Q_1 \mathbf{V}_1$$

and the same relations hold for section 2, we can write Eq. (6.4) as

Steady flow: $$\sum \mathbf{F} = \dot{m}_2\mathbf{V}_2 - \dot{m}_1\mathbf{V}_1 = \rho_2 Q_2 \mathbf{V}_2 - \rho_1 Q_1 \mathbf{V}_1 \tag{6.5}$$

But since the flow we are considering is steady, from continuity, $\dot{m}_1 = \dot{m}_2 = \dot{m} = \rho_1 Q_1 = \rho_2 Q_2 = \rho Q$. Also, using the vector relations of Fig. 6.1b, let us for convenience write $\Delta \mathbf{V} = \mathbf{V}_2 - \mathbf{V}_1 = \mathbf{V}_{out} - \mathbf{V}_{in}$. Using these, Eq. (6.5) becomes

Steady flow: $$\sum \mathbf{F} = \dot{m}(\Delta \mathbf{V}) = \rho Q(\Delta \mathbf{V}) = \rho Q(\mathbf{V}_2 - \mathbf{V}_1) \tag{6.6}$$

The direction of $\sum \mathbf{F}$ must be the same as that of the velocity change, $\Delta \mathbf{V}$. Note that the $\sum \mathbf{F}$ represents the vector sum of *all* forces acting *on the fluid mass* in the control volume, including gravity forces, shear forces, and pressure forces including those exerted by fluid surrounding the fluid mass under consideration as well as the pressure forces exerted by the solid boundaries in contact with the fluid mass. Often the force sought is just *one* of these many forces. Frequently it is not even one of them, but instead it is *opposite* to one of them, being the force of the liquid acting on a boundary. The right side of Eq. (6.6) represents the change in momentum per unit time.

Since Eq. (6.6) is a vector equation, we can express it by the following scalar (component) equations:

Steady flow:
$$\sum F_x = \dot{m}(\Delta V_x) = \rho Q(\Delta V_x) = \rho Q(V_{2x} - V_{1x}) \tag{6.7a}$$
$$\sum F_y = \dot{m}(\Delta V_y) = \rho Q(\Delta V_y) = \rho Q(V_{2y} - V_{1y}) \tag{6.7b}$$
$$\sum F_z = \dot{m}(\Delta V_z) = \rho Q(\Delta V_z) = \rho Q(V_{2z} - V_{1z}) \tag{6.7c}$$

In Sec. 6.4 and succeeding sections we will apply these equations to several situations that are commonly encountered in engineering practice. If the flow in a single stream tube splits up into several stream tubes, we just compute the ρQV values of each stream tube separately and then substitute them into Eqs. (6.5)–(6.7) (see Sample Problem 6.2). The great advantage of the momentum principle is that we need not know the details of what is occurring within the flow; only the conditions at the end sections of the control volume govern the analysis.

6.2   Navier-Stokes Equations

We can derive a set of differential equations that describe the motion of a real fluid for the general case by considering the forces acting on a small element or control volume of fluid like Fig. 3.2. The forces include gravitational, viscous (frictional), and pressure forces.

Before we can incorporate Newton's equation of viscosity [Eq. (2.9)] for one-dimensional flow we must generalize it to three-dimensional flow.

The full derivation of these equations is lengthy and involved, and beyond the scope of this text. However, for an incompressible fluid with constant viscosity, in rectangular coordinates with z increasing vertically upwards, the result is

$$-\frac{\partial p}{\partial x} + \mu\left(\frac{\partial^2 u}{\partial x^2} + \frac{\partial^2 u}{\partial y^2} + \frac{\partial^2 u}{\partial z^2}\right) = \rho\left[\frac{\partial u}{\partial t} + u\frac{\partial u}{\partial x} + v\frac{\partial u}{\partial y} + w\frac{\partial u}{\partial z}\right] \qquad (6.8a)$$

$$-\frac{\partial p}{\partial y} + \mu\left(\frac{\partial^2 v}{\partial x^2} + \frac{\partial^2 v}{\partial y^2} + \frac{\partial^2 v}{\partial z^2}\right) = \rho\left[\frac{\partial v}{\partial t} + u\frac{\partial v}{\partial x} + v\frac{\partial v}{\partial y} + w\frac{\partial v}{\partial z}\right] \qquad (6.8b)$$

$$-\rho g - \frac{\partial p}{\partial z} + \mu\left(\frac{\partial^2 w}{\partial x^2} + \frac{\partial^2 w}{\partial y^2} + \frac{\partial^2 w}{\partial z^2}\right) = \rho\left[\frac{\partial w}{\partial t} + u\frac{\partial w}{\partial x} + v\frac{\partial w}{\partial y} + w\frac{\partial w}{\partial z}\right] \qquad (6.8c)$$

These fundamental general equations of motion are known as the *Navier-Stokes equations.* We name them after the French scientist, C.L.M.H. Navier (1785–1836), who today we would describe as a civil engineer, and the English physicist, Sir George Stokes (1819–1903), both of whom first derived them. They are second-order nonlinear partial differential equations (Appendix B) that no one has analytically solved in general, although they have obtained analytical and numerical solutions for certain specific situations. Their complete derivations, for rectangular, cylindrical, and spherical coordinates, are given in advanced fluid mechanics texts.

The Navier-Stokes equations are in fact just a differential form of the linear momentum principle. Thus, on the left sides of Eqs. (6.8), we have the body force per unit volume (the term in g) and the surface force per unit volume (pressure force represented by terms in p, and viscous force represented by terms with μ and parentheses). These are equal to the time rate of change of momentum on the right side, consisting in the square brackets of the local acceleration (the derivatives with respect to time) and the convective acceleration (the other terms).

When we write Eqs. (6.8a), (6.8b), and (6.8c) in terms of the normal stresses σ and the shear stresses τ, we call them the *Cauchy equations.* For an ideal fluid ($\mu = 0$), they reduce to a set of three-dimensional equations known as the *Euler equations* of motion, which are the same as Eqs. (6.8a), (6.8b), and (6.8c) but with the terms containing μ and second-order derivatives eliminated.

A major area of endeavor has used the Navier-Stokes equations, together with various numerical computation methods, in efforts to solve flow fields with challenging features. The entire flow region is divided into many small elements, and the equations applied to each element. We call such procedures *computational fluid dynamics* (CFD). In Fig. 6.2 we can compare the excellent results of such a numerical computation with a photograph of an actual flow field.

6.3   Momentum Correction Factor

If the velocity is not uniform over a section (e.g., Secs. 2.11, 4.1, and 5.11), we shall find that the momentum per unit time crossing that section is greater than that computed by using the mean velocity. The rate of momentum transfer (*momentum flux*) across an elementary area dA, where the local velocity is u, is $\dot{m}u = (\rho u\,dA)u = \rho u^2\,dA$, and the

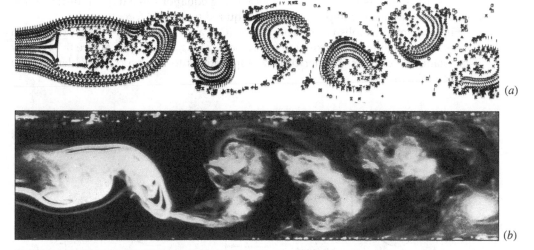

FIGURE 6.2   Comparison of flow around a square block. (a) Computed at **R** = 500. (b) Visualization at
R = 550. (Courtesy Ronald W. Davis, Chemical Science & Technology Laboratory, NIST)

rate of momentum transfer across the entire section is $\rho \int_A u^2 dA$, while that computed by using the mean velocity is $\rho QV = \rho AV^2$. Thus the ***momentum correction factor*** β (beta), which we should multiply ρQV by to obtain the true momentum per unit time, is

$$\beta = \frac{1}{AV^2} \int_A u^2 dA \qquad (6.9)$$

For laminar flow in a circular pipe, $\beta = \frac{4}{3}$, but for turbulent flow in circular pipes, it usually ranges from 1.005 to 1.05, as we can see from Eq. (7.45b). For open-channel flow, it may be greater. Mathematically, it cannot be less than 1.0, and, unless otherwise specified, we will take the value of β in the following discussion as 1.0.

6.4   Applications of the Momentum Principle

In a common application of the momentum principle, we use it to find forces that flowing fluid exert on structures open to the atmosphere, like gates (e.g., Fig. 9.34a) and overflow spillways (e.g., Fig. 9.32).

It is very important to remember that the momentum principle deals only with forces that act *on* the fluid mass in a *designated control volume* (CV); fluid forces acting on a structure are equal and opposite to boundary pressure forces acting on the fluid. To avoid confusion over signs, we strongly advise students to first solve for the magnitude and direction of the (reaction) force of the structure on the fluid, and only in the last step to find the equal and opposite force of the fluid on the structure. (We will use additional subscripts to designate forces *not* acting on the control volume.)

The force of a fluid on a structure is usually distributed as varying pressure forces over the surface. We are normally interested in the resultant of this distribution, and we usually only consider pressures that differ from atmospheric. If we need the total resultant force, we usually obtain this by writing equations like (6.7), in order to first find the perpendicular components of the required (opposite) force.

First, we must establish a control volume (CV), and, as noted in Sec. 6.1, we should do this by cutting the flow normal to the velocity, along a boundary where the velocity is constant. We can best further discuss applications to this type of problem with a sample problem.

Sample Problem 6.1
The water passage in Fig. S6.1 is 10 ft (3 m) wide normal to the plane of the figure. Determine the horizontal force acting on the shaded structure. Assume ideal flow.

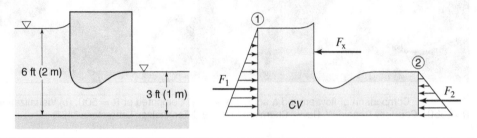

FIGURE S6.1

Solution (BG units)
We first establish a control volume (CV) as in Fig. S6.1.
In free-surface flow such as this where the streamlines are parallel, the water surface is coincident with the hydraulic grade line. Writing an energy equation from the upstream section to the downstream section, where $V = V_x$,

$$6 + \frac{V_1^2}{2g} = 3 + \frac{V_2^2}{2g} \tag{1}$$

From continuity, $6(10)\,V_1 = 3(10)V_2$ (2)

Substituting Eq. (2) into Eq. (1) yields

$$V_1 = 8.02 \text{ fps}, \quad V_2 = 16.05 \text{ fps}$$
$$Q = A_1V_1 = A_2V_2 = 481 \text{ cfs}$$

Next take a free-body diagram of the control volume (CV) of water shown in the figure and apply the momentum equation given by Eq. (6.7a),

$$F_1 - F_2 - F_x = \rho Q(V_2 - V_1)$$

where F_x represents the force of the structure on the water (CV) in the horizontal direction, and the F's and V's are understood to have no y components.
From Eq. (3.16), we have $F_1 = \gamma h_{c1} A_1$ and $F_2 = \gamma h_{c2} A_2$. Hence

$$62.4(3)(10 \times 6) - 62.4(1.5)(10 \times 3) - F_x = 1.94(481)(16.05 - 8.02)$$

and $F_x = +936 \text{ lb} = 936 \text{ lb} \leftarrow$

The positive sign means that the assumed direction is correct. The force of the water on the structure is equal and opposite, namely,

$$(F_{W/S})_x = 936\,\text{lb} \rightarrow \quad ANS$$

Note that the momentum principle will not permit us to obtain the vertical component of the force of the water on the shaded structure, because the pressure distribution along the bottom of the channel is unknown. We can estimate the pressure distribution along the boundary of the structure and along the bottom of the channel by sketching a flow net and applying Bernoulli's principle. Then we can find the horizontal and vertical components of the force by computing the integrated effect of the pressure-distribution diagram.

Solution   (SI units)

Energy:
$$2 + \frac{V_1^2}{2(9.81)} = 1 + \frac{V_2^2}{2(9.81)} \tag{3}$$

Continuity:
$$2(3)V_1 = 1(3)V_2 \tag{4}$$

Substituting Eq. (4) into Eq. (3) yields

$$V_1 = 2.56\,\text{m/s}, \quad V_2 = 5.11\,\text{m/s}$$

$$Q = A_1 V_1 = A_2 V_2 = 15.34\,\text{m}^3/\text{s}$$

Applying the momentum equation, given by Eq. (6.7a), to the free-body diagram,

$$F_1 - F_2 - F_x = \rho Q(V_2 - V_1)$$

$$9.81(1)(2)(3) - 9.81(0.5)(1)(3) - F_x = 1.0(15.34)(5.11 - 2.56)$$

$$F_x = +4.91\,\text{kN} = 4.91\,\text{kN} \leftarrow$$

So
$$(F_{W/S})_x = 4.91\,\text{kN} \rightarrow \quad ANS$$

There are numerous other fluid-flow situations where the momentum principle is useful. In the Sec. 6.5 we shall apply it to find forces exerted on pressure conduits such as at bends and nozzles. The same approach can be used to find the force a jet exerts on a stationary object, such as a vane or blade, with the main difference being that with a jet in the open atmosphere the gage pressures p in the jet are zero and the pA forces disappear. We will also use the momentum principle to develop an expression for the head loss in a pipe expansion (Sec. 7.24), and for the conjugate depths of a hydraulic jump (Sec. 8.18).

6.5   Force on Pressure Conduits

Consider the case of horizontal flow to the right through the reducer of Fig. 6.3a. A free-body diagram of the forces acting on the fluid mass contained in the reducer (the control volume, *CV*) is shown in Fig. 6.3b. We shall apply Eq. (6.7a) to this fluid mass to examine the forces that are acting in the x direction. The forces $p_1 A_1$ and $p_2 A_2$ represent pressure forces that fluid located just upstream and just downstream exerts on the

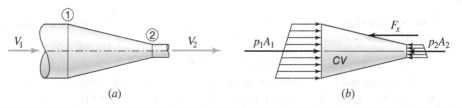

FIGURE 6.3   (a) Horizontal flow through a reducer. (b) Forces acting on the control volume.

control volume. The force F_x represents the force exerted *by the reducer on the fluid* (CV) in the x direction. Neglecting shear forces at the boundary of the reducer, the force F_x is the resultant (integrated) effect of the normal pressure forces that the wall of the reducer exerts on the fluid. The intensity of pressure at the wall will decrease as the diameter decreases because of the increase in velocity head, in accordance with Bernoulli's theorem [Eq. (5.30)]. Figure 6.4 is a typical pressure diagram.

The effect that atmospheric pressure has on such analyses can be confusing. The pressures shown in Fig. 6.4 are gage pressures (Sec. 3.4). If we used absolute pressures, we would have to increase all the pressures shown by a constant amount, p_{atm} (about 14.70 psi or 101.3 kPa at sea level). This would increase $p_1 A_1 - p_2 A_2$, F, and the equal and opposite force exerted *by the fluid on the reducer*, $F_{F/R}$. However, this increased force on the inside of the reducer would exactly balance the force of the atmosphere on the outside. Therefore the atmospheric pressure does not affect the *net* force on the reducer, which results from the fluid flow and which tends to move the reducer. It is this *net* force which interests us, and we can most easily obtain it by excluding atmospheric pressure, i.e., by using gage pressures. Therefore, customarily we use *gage* pressures for p_1 and p_2.

Applying Eq. (6.7a) and assuming the fluid is ideal with F_x directed as shown, since the entry and exit velocities are parallel to the x direction, we get

$$\Sigma F_x = p_1 A_1 - p_2 A_2 - F_x = \rho Q(V_2 - V_1) \tag{6.10}$$

In Eq. (6.10) each term can be evaluated independently from the given flow data, except F_x, which is the quantity we wish to find. Rewriting Eq. (6.10), the result is

$$F_x = p_1 A_1 - p_2 A_2 - \rho Q(V_2 - V_1) \tag{6.11}$$

This gives the net force of the *reducer on the fluid* (the CV) in the x direction. This force acts to the left as assumed in Fig. 6.3b and as applied in Eq. (6.10). The force of the *fluid on the reducer* ($F_{F/R}$) is, of course, equal and opposite to that of the reducer on the fluid. If the flow were to the left in Fig. 6.3, a similar analysis would apply, but we need to be

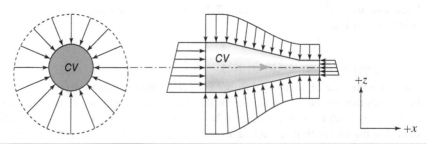

FIGURE 6.4   Gage pressure distribution on the fluid flowing through a reducer.

consistent in regard to plus and minus signs. Conventionally we usually take the flow direction as the positive direction.

By considering the weight of fluid between sections 1 and 2 in Fig. 6.3a we must conclude that pressures are larger on the bottom half of the pipe than on the upper half. Recall (Sec. 6.1) that it is the conditions at the end sections of the control volume that govern the analysis. How flow moves between sections 1 and 2 is unimportant to the determination of forces. Figure 6.4 gives a schematic representation of the gage pressure distribution on the fluid within the reducer. The integrated effect of the pressures exerted by the reducer itself is equivalent in the x direction to F_x and in the z direction to the weight of fluid between sections 1 and 2.

If the fluid undergoes a change in both direction and velocity, as in the reducing pipe bend in Fig. 6.5, the procedure is similar to that of the preceding case, except that we find it convenient to deal with components. Assuming the flow is in a horizontal plane so that we can neglect the weight, applying Eq. (6.7a) by summing up x-forces acting on the fluid in the CV, and equating them to the change in fluid momentum in the x direction, gives

$$\Sigma F_x = p_1 A_1 - p_2 A_2 \cos\theta - F_x = \rho Q(V_{2x} - V_{1x}) \tag{6.12}$$

which, after noting that $V_{2x} = V_2 \cos\theta$ and $V_{1x} = V_1$, when rewritten for the force we wish to find, becomes

$$F_x = p_1 A_1 - p_2 A_2 \cos\theta - \rho Q(V_2 \cos\theta - V_1) \tag{6.13}$$

Similarly, in the y direction,

$$\Sigma F_y = 0 - p_2 A_2 \sin\theta + F_y = \rho Q(V_{2y} - V_{1y}) \tag{6.14}$$

which, after noting that $V_{2y} = V_2 \sin\theta$ and $V_{1y} = 0$, when rewritten, becomes

$$F_y = p_2 A_2 \sin\theta + \rho Q V_2 \sin\theta \tag{6.15}$$

In a specific case, if the numerical values of F_x and F_y determined from these equations are positive then the assumed directions are correct. A negative value for either one merely indicates that that component is in the direction opposite to that assumed.

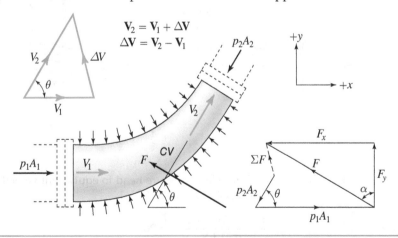

FIGURE 6.5   Forces on the fluid flowing through a reducing pipe bend. (ΣF is parallel to ΔV.)

Note that $\Sigma\mathbf{F} = \rho Q\Delta\mathbf{V}$ is the resultant of *all* the forces acting on the fluid in the control volume, which *includes* the pressure forces on the two ends and the force $\mathbf{F}$ exerted by the bend on the fluid. The directions of $\Sigma\mathbf{F}$ and $\Delta\mathbf{V}$ must be the same (see Fig. 6.5). The value of F is $\sqrt{F_x^2 + F_y^2}$, and we can obtain its direction α from the force diagram shown in Fig. 6.5.

The total force $\mathbf{F}_{F/B}$ exerted by the fluid on the bend is equal in magnitude but opposite in direction to the force $\mathbf{F}$ of the bend on the fluid. The force of the fluid on the bend tends to move the portion of the pipe under consideration. Hence, to prevent damage where such changes in velocity or alignment occur, a large pipe will usually be "anchored" by attaching it to a concrete block of sufficient size and/or weight to provide the necessary resistance.

If the flow in Fig. 6.5 had been in a vertical plane, i.e., y was vertical, we would have to calculate the weight of the fluid between sections 1 and 2 and include it in Eqs. (6.14) and (6.15). Also, we could include the effects of shear stresses due to fluid friction in the problem; however, these effects are usually small. If there are multiple inlets or exits, the principle remains the same: $\Sigma\mathbf{F} = \Sigma(\rho Q\mathbf{V})_{\text{out}} - \Sigma(\rho Q\mathbf{V})_{\text{in}}$; this is illustrated in Sample Problem 6.4.

Sample Problem 6.2

A hydrant test is being conducted to ensure a water main has the capacity to deliver the flow required for firefighting. The discharge diameter is 6.4 cm (2.5 inch). A pitot gage has measured the discharge velocity as equivalent to 240 kPa (35 psi). Neglect the velocity within the hydrant. Determine the moment caused by the discharging water about the flange at A and the shear force, F_A, that the flange bolts exert.

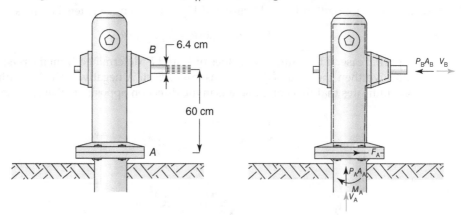

Solution

Use the relations of Eq. (5.35) to relate pressure head to equivalent velocity head, using $\rho = 1000$ kg/m³ and $\gamma = 9.81$ kN/m³ from Table A.1:

Sec. 5.1:
$$\frac{p}{\gamma} = \frac{240 \text{ kPa}}{9.81 \text{ kN/m}^3} = 20.46 \text{ m} = \frac{V_B^2}{2g}$$

$$V_B = \sqrt{2(9.81 \text{ m/s}^2)(20.46 \text{ m})} = 21.91 \text{ m/s}$$

Eq. (4.3): $Q = AV = \dfrac{\pi}{4}(0.064 \text{ m})^2 \cdot 21.91\text{m/s} = 0.0705 \text{ m}^3/\text{s}$

Using the dashed line in Fig. S6.2 to indicate the control volume, and recognizing for discharge to the atmosphere that $p_B = 0$, the left side of Eq. (6.7a) gives:

$$\Sigma F_x = F_A$$

The right side of Eq. (6.7a) is:

$$(\rho Q_B V_B - 0)$$

Equate the left and right sides and solve for F_A.

$$F_A = (1000 \text{ kg/m}^3)(0.0705 \text{ m}^3/\text{s})(21.91 \text{ m/s}) = 1545 \text{ N} \qquad \textbf{\textit{ANS}}$$

Summing moments about A (assuming clockwise is positive), the only moment about A is caused by the change in momentum at B:

$$\Sigma M_A = M_A - 0.6 \text{ m}(\rho Q_B V_B) = 0$$

$$M_A = 0.6 \text{ m } (1545 \text{ N}) = 927 \text{ N·m} \qquad \textbf{\textit{ANS}}$$

Sample Problem 6.3

A 30-cm diameter water pipe is laid in a horizontal trench as shown, with a bend of $\theta = 45°$. It carries a flow, Q, of 140 L/s (0.14 m³/s), with a design pressure of $p = 1000$ kPa. Ignore friction through the bend. Determine the required reaction force, F required to counteract the forces due to the bend, and the required area of the concrete thrust block ($a \times b$ in the cross-section) to provide that reaction force if the bearing capacity of the soil is 95 kN/m².

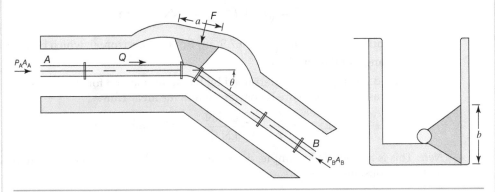

Figure S6.3

Solution

$$\text{Eq. (6.12):} \quad V_A = V_B = \frac{Q}{A} = \frac{0.14 \text{ m}^3/\text{s}}{\frac{\pi}{4}(0.30 \text{ m})^2} = 1.98 \text{ m/s}$$

The reaction F will be considered by its components F_x ($\leftarrow$) and F_y ($\downarrow$).

The left side of Eq. (6.12): $\Sigma F_x = -F_x + p_A A_A - p_B A_B \cos 45°$

The right side of Eq. (6.12): $\rho Q(V_{Bx} - V_{Ax}) = \rho Q(V \cos 45° - V)$

Equate the left and right sides and solve for F_x, recognizing 1 kPa = 1000 Pa:

$$F_x = p_A A_A - p_B A_B \cos 45° - \rho Q(V \cos 45° - V) =$$

$$10^6 \text{ Pa}\left(\frac{\pi}{4} 0.3^2 \text{ m}^2\right) - 10^6 \text{ Pa}\left(\frac{\pi}{4} 0.3^2 \text{ m}^2\right) \cos 45° -$$

$$(1000 \text{ kg/m}^3)(0.14 \text{ m}^3/\text{s})(1.98 \text{ m/s})(\cos 45° - 1)$$

$$F_x = 20\,787 \text{ N} = 20.8 \text{ kN} \leftarrow$$

Similarly for the y-direction, adapting Eq. (6.14) for the current configuration:

The left side: $\Sigma F_y = -F_y + p_B A_B \sin 45°$

The right side: $\rho Q(V_{By} - 0) = \rho Q(-V \sin 45°)$

Equate the left and right sides and solve for F_y:

$$F_y = p_B A_B \sin 45° - \rho Q(-V \sin 45°) =$$

$$10^6 \text{ Pa}\left(\frac{\pi}{4} 0.3^2 \text{ m}^2\right) \sin 45° - (1000 \text{ kg/m}^3)(0.14 \text{ m}^3/\text{s})(-1.98 \text{ m/s})(\sin 45°)$$

$$F_y = 50\,188 \text{ N} = 50.2 \text{ kN} \downarrow$$

$$F = \sqrt{F_x^2 + F_y^2} = 54.3 \text{ kN} \qquad \textit{ANS}$$

The line of action for F is $\tan^{-1}\left(\frac{F_y}{F_x}\right) = 67.5°$ up from the horizontal, which by inspection bisects the angle of the pipe bend.

The required bearing area is determined by the definition of a force as pressure × area [Eq. (3.14)], assuming pressure is constant across the area. Thus,

$$A_{\text{bearing}} = \frac{F}{p} = \frac{54.3 \text{ kN}}{95 \text{ kN/m}^2} = 0.572 \text{ m}^2 \qquad \textit{ANS}$$

Note: This is a simplified example; design of a thrust block may include a safety factor and other considerations.

Sample Problem 6.4

Water flows through the double nozzle as shown in Fig. S6.2. Determine the magnitude and direction of the resultant force the water exerts on the nozzle. The velocity of both nozzle jets is 12 m/s. The axes of the pipe and both nozzles lie in a horizontal plane. $\gamma = 9.81$ kN/m³. Neglect friction.

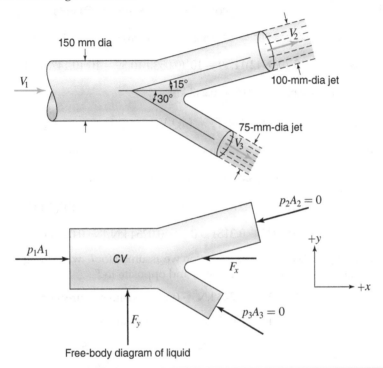

FIGURE S6.4

Solution

Continuity:
$$A_1 V_1 = A_2 V_2 + A_3 V_3$$

$$15^2 V_1 = 10^2 (12) + 7.5^2 (12), \qquad V_1 = 8.33 \text{ m/s}$$

$$Q_1 = \frac{\pi}{4}(0.15)^2 8.33 = 0.1473 \text{ m}^3/\text{s}, \quad Q_2 = 0.0942 \text{ m}^3/\text{s}, \quad Q_3 = 0.0530 \text{ m}^3/\text{s}$$

Jets 2 and 3 are "free," i.e., in the atmosphere, so $p_2 = p_3 = 0$. Writing the energy equation, given by Eq. (5.29), along a streamline:

$$\frac{p_1}{\gamma} + z + \frac{8.33^2}{2(9.81)} = 0 + z + \frac{12^2}{2(9.81)}$$

So
$$\frac{p_1}{\gamma} = 3.80 \text{ m}, \quad p_1 = 37.3 \text{ kN/m}^2, \quad p_1 A_1 = 0.659 \text{ kN}$$

Eq. (6.7a): $\qquad \sum F_x = p_1 A_1 - 0 - F_x = (\rho Q_2 V_{2x} + \rho Q_3 V_{3x}) - \rho Q_1 V_{1x}$

Where $\qquad\qquad \rho = \dfrac{\gamma}{g} = \dfrac{9.81 \text{ kN/m}^3}{9.81 \text{ m/s}^2} = 1.0 \dfrac{\text{kN·s}^2}{\text{m}^4} = 10^3 \dfrac{\text{kg}}{\text{m}^3}$

And $\qquad\qquad V_{2x} = V_2 \cos 15° = 12(0.966) = 11.59 \text{ m/s}$

And $\qquad V_{3x} = V_3 \cos 30° = 12(0.866) = 10.39 \text{ m/s}, \quad V_{1x} = V_1 = 8.33 \text{ m/s}$

So $\quad 0.659 - F_x = 10^3(0.0942)11.59 + 10^3(0.0530)10.39 - 10^3(0.1473)8.33 = 0.416 \text{ kN}$

$\qquad\qquad F_x = 0.659 - 0.416 = 0.243 \text{ kN} \leftarrow$

Eq. (6.7b): $\qquad \sum F_y = 0 - 0 + F_y = (\rho Q_2 V_{2y} + \rho Q_3 V_{3y}) - \rho Q_1 V_{1y}$

Where $\qquad\qquad V_{2y} = V_2 \sin 15° = 12 (0.259) = 3.11 \text{ m/s}$

And $\qquad\quad V_{3y} = -V_3 \sin 30° = -12 (0.50) = -6.00 \text{ m/s}, \quad V_{1y} = 0$

So $\qquad F_y = 10^3 (0.0942) 3.11 + 10^3 (0.0530)(-6.00) - 10^3(0.1473)(0)$

$\qquad\qquad = 0.2927 - 0.3181 - 0 = -0.0254 \text{ kN} \uparrow = 0.0254 \text{ kN} \downarrow$

The minus sign indicates that the direction we assumed for F_y was wrong. Therefore F_y acts in the negative y direction. $F_{L/N}$ is equal and opposite to F.

$\qquad\qquad (F_{L/N})_x = 0.243 \text{ kN} \rightarrow \text{(in the positive } x \text{ direction)}$

$\qquad\qquad (F_{L/N})_y = 0.0254 \text{ kN} \uparrow \text{(in the positive } y \text{ direction)}$

$\qquad\qquad F_{L/N} = 0.244 \text{ kN at } 5.97° \angle \qquad ANS$

Note: As the calculation for F_y shows, when finding the difference between two similar numbers, significant figures can be lost; it can help to retain more significant figures at intermediate steps, so the final answer retains three significant figures.

CHAPTER 7

Steady Incompressible Flow in Pressure Conduits

I n this chapter we will discuss some aspects of steady flow in pressure conduits. We shall limit our discussion to *incompressible fluids,* that is, to those that have $\rho \approx$ constant. This includes all liquids. When gases flow with very small pressure changes we can consider them incompressible, for then $\rho \approx$ constant. In this chapter we will assume isothermal conditions so as to eliminate thermodynamic effects.

7.1 Laminar and Turbulent Flow

If we measure the head loss in a given length of uniform pipe at different velocities, we will find that, as long as the velocity is low enough to secure laminar flow (Sec. 4.2), the head loss, due to friction, is directly proportional to the velocity, as shown in Fig. 7.1. But with increasing velocity, at some point B, where visual observation of dye injected in a transparent tube would show that the flow changes from laminar to turbulent (Sec. 4.2), there will be an abrupt increase in the rate at which the head loss varies. If we plot the logarithms of these two variables on linear scales or, in other words, if we plot the values directly on log–log paper, we will find that, after passing a certain transition region (BCA in Fig. 7.1), the lines will have slopes ranging from about 1.75 to 2.00.

Thus we see that for laminar flow the drop in energy due to friction varies as V, while for turbulent flow the friction varies as V^n, where n ranges from about 1.75 to 2. The lower value of 1.75 for turbulent flow occurs for pipes with very smooth walls; as the wall roughness increases, the value of n increases up to its maximum value of 2.

The points in Fig. 7.1 are plotted directly from measurements made by Osborne Reynolds (1842–1912), an English physicist and professor who conducted pioneering work in fluid mechanics. Figure 7.1 shows decided curves in the transition zone where values of n are even greater than 2. If we gradually reduce the velocity from a high value, the points will not return along the line BC. Instead, the points will lie along curve CA. We call point B the **higher critical point,** and A the **lower critical point.**

However, velocity is not the only factor that determines whether the flow is laminar or turbulent. Additional factors include a characteristic length, usually the diameter, D, for circular pipe, fluid properties of density, ρ, and viscosity, μ, and if the pipe is not smooth, a characteristic roughness. The interrelationship among these variables for a

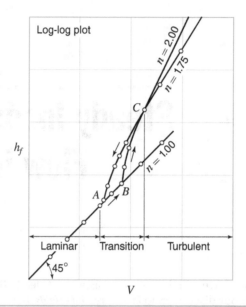

FIGURE 7.1 Log–log plot for flow in a uniform pipe ($n = 2.00$, rough-wall pipe; $n = 1.75$, smooth-wall pipe).

specific phenomenon, such as characterization of turbulence, can be expressed through combining them into dimensionless groups. This is discussed in Sec. 7.2.

7.2 Dimensional Analysis and the Reynolds Number

In dimensional analysis, from a general understanding of fluid phenomena, we first predict the physical parameters, or variables, that will influence the flow, and then we group these parameters into dimensionless combinations which enable a better understanding of the flow phenomena. The number of independent dimensionless groups is less than the number of variables, simplifying experimental design and analysis. A more generalized method of dimensional analysis developed by E. Buckingham[1] and others is widely used. Because Buckingham used an upper case Π (pi) to represent the product of variables in each group, we now call this method the *pi theorem* or the Buckingham pi theorem, described below.

Let $X_1, X_2, X_3, \ldots, X_n$ represent n *dimensional* variables, such as velocity, density, and viscosity, which are involved in some physical phenomenon. We can write the dimensionally homogeneous equation relating these variables as

$$f(X_1, X_2, X_3, \ldots, X_n) = 0$$

in which the dimensions of each *term* are the same. It follows that we can rearrange this equation into

$$\phi(\Pi_1, \Pi_2, \ldots, \Pi_{n-k}) = 0$$

---

[1] E. Buckingham, Model Experiments and the Form of Empirical Equations, *Trans. ASME*, Vol. 37, pp. 263–296, 1915.

where ϕ (phi) is another function and each Π is an independent *dimensionless* product of some of the Xs. The reduction k in the number of terms (from n to $n - k$), is usually equal to and sometimes less than the number of fundamental dimensions m involved in all the variables.

We need to follow a series of seven steps when applying the pi theorem. As we review these, we will apply them to the concept of turbulence and head loss (Fig. 7.1). Since the head loss in Fig. 7.1 was derived by measuring pressure difference across a length of pipe, using a differential manometer (Sec. 3.5), we will use pressure loss per unit length, $p_L = \Delta p / \Delta L$ as the variable describing head loss.

Step 1: Visualize the physical problem, consider the factors that are of influence, and list and count the n variables.

In our example we include the physical factors identified above that influence pressure loss due to turbulence in a circular pipe: the diameter of the pipe, D, must enter the problem; also, the flow velocity, V, must be important. The fluid properties involved are the density ρ and the viscosity μ. Thus we can write

$$f(p_L, D, V, \rho, \mu) = 0$$

Here f stands for "some function." We see that $n = 5$. Note that the procedure cannot work if any relevant variables are omitted. Experimentation with the procedure and experience will help determine which variables are relevant.

Step 2: Choose a dimensional system, MLT (mass-length-time) or FLT (force-length-time), and list the dimensions of each variable. Find m, the number of fundamental dimensions involved in all the variables.

In our example, choosing the MLT system, the dimensions are, respectively

$$\frac{M}{L^2 T^2}, \quad L, \quad \frac{L}{T}, \quad \frac{M}{L^3}, \quad \frac{M}{LT}$$

We can get help in identifying these dimensions from Sec. 1.5 and from the units of quantities in the appendices, etc. Since M, L, and T are all involved, $m = 3$.

Step 3: Find the reduction number k. Usually this equals m, which it cannot exceed, but rarely it is less than m. To check this, try to find m dimensional variables that *cannot* be formed into a dimensionless group. If m are found, $k = m$; if not, reduce k by one and retry.

In our example we find that three of the dimensional variables, namely, ρ, D, and V, with dimensions M/L^3, L, and L/T, will *not* form into a dimensionless Π group because M and T cannot cancel among them, so $k = 3$. The rare circumstances that reduce k arise when some of the dimensions (usually M and T) occur in the parameters only in fixed combinations.

Step 4: Determine $n - k$, the number of dimensionless Π groups needed.
In our example this is $5 - 3 = 2$.

Step 5: From the list of dimensional variables, select k of them to be so-called primary (repeating) variables. These must contain all of the m fundamental dimensions, and must not form a Π among themselves (see step 3). Generally it helps to choose primary variables that relate to mass, geometry, and kinematics (flow without forces or energy). Form the Π groups by multiplying the product of the primary variables, with unknown exponents, by each of the remaining variables, one at a time.

We choose ρ, D, and V (as in step 3) as the primary variables for our example. Then the Π terms are

$$\Pi_1 = \rho^{a_1} D^{b_1} V^{c_1} \mu$$

$$\Pi_2 = \rho^{a_2} D^{b_2} V^{c_2} p_L$$

Step 6: To satisfy dimensional homogeneity, equate the exponents of each dimension on both sides of each pi equation, and so solve for the exponents and the forms of the dimensionless groups. Since the Πs are dimensionless, we can replace them with $M^0 L^0 T^0$. Experience in fluid mechanics tells us that these dimensionless groups commonly take the form of a Reynolds number (see below), Froude number (Sec. 8.10), or Mach number (Sec. 9.3). So, we should always watch out for them when using dimensional analysis.

For our example, working with Π_1,

$$M^0 L^0 T^0 = (ML^{-3})^{a_1} L^{b_1} (LT^{-1})^{c_1} (ML^{-1}T^{-1})$$

M: $0 = a_1 + 1$

L: $0 = -3a_1 + b_1 + c_1 - 1$

T: $0 = -c_1 - 1$

Solving for a_1, b_1, and c_1,

$$a_1 = -1, \quad b_1 = -1, \quad c_1 = -1.$$

Thus,

$$\Pi_1 = \rho^{-1} D^{-1} V^{-1} \mu = \frac{\mu}{\rho DV} = \left(\frac{\rho DV}{\mu}\right)^{-1}$$

Note that $(\rho DV/\mu)$ is the Reynolds number,[2] **R**, discussed further below.

So $\Pi_1 = \mathbf{R}^{-1}$

Working in a similar fashion with Π_2, we get

$$\Pi_2 = \frac{p_L D}{\rho V^2}$$

Check that all Πs are in fact dimensionless.

Step 7: Rearrange the pi groups as desired. The pi theorem states that the Πs are related, and may be expressed as $f_1(\Pi_1, \Pi_2, \ldots, \Pi_{n-k}) = 0$ or as $\Pi_1 = f_2(\Pi_2, \Pi_3, \ldots, \Pi_{n-k})$, etc. It does not predict the functional form of f_1 or f_2; we must determine these relations experimentally. Further, we can raise each Π parameter to any power, as this will not affect their dimensionless status. Note that if there is only one dimensionless group, $f(\Pi) = 0$ and so we must have $\Pi = $ constant.

In our example, since $f_1(p_L D/\rho V^2, \mathbf{R}) = 0$ this indicates that $p_L D/\rho V^2$ depends only on **R**. Alternatively, since we are interested in p_L, if we write $\Pi_2 = \phi(\Pi_1^{-1})$ we get

$$\frac{p_L D}{\rho V^2} = \phi(\mathbf{R}) \quad \text{so that} \quad p_L = \frac{\rho V^2}{D} \phi(\mathbf{R})$$

---

[2] It is standard practice to represent Reynolds, Froude, and Mach number by bold face **R**, **F**, and **M**, respectively as this simplifies its use with subscripts. *Re* and N_R are also sometimes used for Reynolds number.

We must emphasize that dimensional analysis does not provide a complete solution to fluid problems. It provides a partial solution only. The success of dimensional analysis depends entirely on the ability of the individual using it to define the parameters that are applicable. If we omit an important variable, the results are incomplete, and this may lead to incorrect conclusions.

The first Π group identified in the example above is a form of the Reynolds number, **R**, the most widely used dimensionless number in fluid mechanics. It is named in honor of Osborne Reynolds who presented this in a publication of his experimental work in 1882. **R** is the ratio of *inertia forces to viscous forces*. The ratio of these two forces is generally written for a characteristic length, L, for example a diameter of a pipe or sphere (Sec. 7.4), a measure of soil pore space in an aquifer, or the hydraulic radius of a river or channel (Sec. 8.3). Using relationships for viscous and inertial forces from Secs. 2.11 and 6.1, the ratio is

$$\mathbf{R} = \frac{F_I}{F_V} = \frac{L^2 V^2 \rho}{L V \mu} = \frac{L V \rho}{\mu} = \frac{L V}{\nu}$$

For any consistent system of units, **R** is a dimensionless number. For a circular pipe flowing full the characteristic length is the diameter of the pipe, D, resulting in

Pipe flow:
$$\mathbf{R} = \frac{D V \rho}{\mu} = \frac{D V}{\nu} \tag{7.1}$$

7.3   Critical Reynolds Number

The upper critical Reynolds number, corresponding to point B of Fig. 7.1, is really indeterminate and depends on the care taken to prevent any initial disturbance from affecting the flow. Its value is normally about 4000, but experimenters have maintained laminar flow in circular pipes up to values of **R** as high as 50,000. However, in such cases this type of flow is inherently unstable, and the least disturbance will transform it instantly into turbulent flow. On the other hand, it is practically impossible for turbulent flow in a straight pipe to persist at values of **R** much below 2000, because any turbulence that occurs is damped out by viscous friction. This lower value is thus much more definite than the higher one, and is the real dividing point between the two types of flow. So, we define this lower value as the ***true critical Reynolds number.*** However, it may vary slightly. It will be higher in a converging pipe and lower in a diverging pipe than in a straight pipe. Also, it will be less for flow in a curved pipe than in a straight one, and even for a straight uniform pipe it may be as low as 1000, where there is excessive roughness. However, for normal cases of flow in straight pipes of uniform diameter and usual roughness, we can take the critical value as

$$\mathbf{R}_{\text{crit}} = 2000 \tag{7.2}$$

For water at 59 °F (15 °C), using Eqs. (7.1) and (7.2) we find:

when $D = 1$ in (25 mm), $V_{\text{crit}} = 0.30$ fps (0.091 m/s)

and when $V = 3$ fps (0.91 m/s), $D_{\text{crit}} = 0.10$ in (2.5 mm)

Sample Problem 7.1

In a refinery, oil ($s = 0.85$, $\nu = 1.8 \times 10^{-5}$ m^2/s) flows through a 100-mm diameter pipe at 0.50 L/s. Is the flow laminar or turbulent?

Solution

Eq. (4.7):

$$V = \frac{4Q}{\pi D^2} = \frac{4(0.0005 \text{ m}^3/\text{s})}{\pi (0.1 \text{ m})^2} = 0.0637 \text{ m/s}$$

Eq. (7.1):

$$\mathbf{R} = \frac{DV}{\nu} = \frac{0.10 \text{ m}(0.0637 \text{ m/s})}{1.8 \times 10^{-5} \text{ m}^2/\text{s}} = 354$$

Using Eq. (7.2), since $\mathbf{R} < \mathbf{R}_{\text{crit}} = 2000$, the flow is laminar.   *ANS*

We seldom encounter velocities or pipe diameters as small as these with water flowing in practical engineering, though they may occur in certain laboratory instruments. Therefore, for such fluids as water and air, practically all cases of engineering importance are in the turbulent-flow region. But if the fluid is a viscous oil, laminar flow often occurs.

7.4   Hydraulic Radius, Hydraulic Diameter

For conduits having noncircular cross sections, we need to use some value other than the diameter for the linear dimension in the Reynolds number. The characteristic dimension we use is the *hydraulic radius*, defined as

$$R_h = \frac{A}{P} \tag{7.3}$$

where A is the cross-sectional area of the flowing fluid, and P is the *wetted perimeter*, that portion of the perimeter of the cross section where the fluid contacts the solid boundary, and therefore where friction resistance is exerted on the flowing fluid. For a circular pipe flowing full,

Full-pipe flow:

$$R_h = \frac{\pi r^2}{2\pi r} = \frac{r}{2} = \frac{D}{4} \tag{7.4}$$

Thus R_h is not the radius of the pipe, and so the term "radius" is misleading. If a circular pipe is exactly half full, both the area and the wetted perimeter are half the preceding values; so R_h is $r/2$, the same as if it were full. But if the depth of flow in a circular pipe is 0.8 times the diameter, for example, $A = 0.674D^2$ and $P = 2.21D$, then $R_h = 0.304D$, or $0.608r$. We discuss part-full pipe flow further in Sec. 8.7.

The hydraulic radius is a convenient means for expressing the shape as well as the size of a conduit, since, for the same cross-sectional area, the value of R_h will vary with the shape.

In evaluating the Reynolds number for flow in a noncircular conduit (Sec. 7.7), customarily we substitute $4R_h$ for D in Eq. (7.1).

Workers in some engineering fields define $4R_h$ to be the **hydraulic diameter** D_h. Then, for a pipe, $D_h = D$, which is fine, but $D_h = 4R_h$, which seems strange.

7.5 Friction Head Loss in Conduits of Constant Cross Section

This discussion applies to either laminar or turbulent flow and to any shape of cross section.

Consider steady flow in a conduit of uniform cross section A, not necessarily circular (Fig. 7.2). The pressures at sections 1 and 2 are p_1 and p_2, respectively. The distance between the sections is L. For equilibrium in steady flow, the summation of forces acting on any fluid element must be equal to zero (i.e., $\Sigma F = ma = 0$). Thus, in the direction of flow,

$$p_1 A - p_2 A - \gamma LA \sin \alpha - \bar{\tau}_0(PL) = 0 \tag{7.5}$$

where we define $\bar{\tau}_0$, the **average shear stress** (average shear force per unit area) at the conduit wall, by

$$\bar{\tau}_0 = \frac{\int_0^P \tau_0 dP}{P} \tag{7.6}$$

in which τ_0 is the local shear stress[3] acting over a small incremental portion dP of the wetted perimeter, P.

Noting that $\sin \alpha = (z_2 - z_1)/L$ and dividing each term in Eq. (7.5) by γA gives

$$\frac{p_1}{\gamma} - \frac{p_2}{\gamma} - z_2 + z_1 = \bar{\tau}_0 \frac{PL}{\gamma A} \tag{7.7}$$

---

[3] The local shear stress varies from point to point around the perimeter of all conduits (regardless of whether the wall is smooth or rough), except for the case of a circular pipe flowing full where the shear stress at the wall is the same at all points on the perimeter.

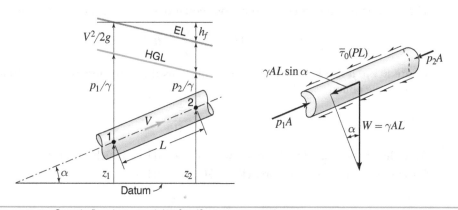

FIGURE 7.2   Steady flow in a conduit of uniform cross-section.

From the left-hand sketch of Fig. 7.2, we can see that the head loss due to friction at the wetted perimeter is

$$h_f = \left(z_1 + \frac{p_1}{\gamma}\right) - \left(z_2 + \frac{p_2}{\gamma}\right)$$

This equation indicates that h_f depends only on the values of z and p on the centerline, and so it is the same regardless of the size of the cross-sectional area A. Substituting h_f for the right-hand side of this expression and replacing A/P in Eq. (7.7) by R_h from Eq. (7.3), we get

$$h_f = \bar{\tau}_0 \frac{L}{R_h \gamma} \tag{7.8}$$

This equation is applicable to any shape of uniform cross section, regardless of whether the flow is laminar or turbulent. Its derivation is very similar to that in Sec. 5.3, in which the conduit cross section was not constant, however.

For a smooth-walled conduit, where we can neglect wall roughness (discussed in Sec. 7.10), we might assume that the average fluid shear stress $\bar{\tau}_0$ at the wall is some function of ρ, μ, and V and some characteristic linear dimension, which we will here take as the hydraulic radius R_h. Thus

$$\bar{\tau}_0 = f(\rho, \mu, V, R_h) \tag{7.9}$$

Using the pi theorem of dimensional analysis (Sec. 7.2) to better determine the form of this relationship, we choose ρ, R_h, and V as primary variables, so that

$$\Pi_1 = \mu \rho^{a_1} R_h^{b_1} V^{c_1}$$

$$\Pi_2 = \bar{\tau}_0 \rho^{a_2} R_h^{b_2} V^{c_2}$$

With the dimensions of the variables being $ML^{-1}T^{-1}$ for μ, $ML^{-1}T^{-2}$ for $\bar{\tau}_0$, ML^{-3} for ρ, L for R_h, and LT^{-1} for V, the dimensions for Π_1 are

For M: $0 = 1 + a_1$
For L: $0 = 1 - 3a_1 + b_1 + c_1$
For T: $0 = -1 - c_1$

The solution of these three simultaneous equations is $a_1 = b_1 = c_1 = -1$, from which

$$\Pi_1 = \frac{\mu}{\rho R_h V} = \mathbf{R}^{-1}$$

where $R_h V \rho / \mu$ is a Reynolds number with R_h as the characteristic length. In a similar manner, we obtain

$$\Pi_2 = \frac{\bar{\tau}_0}{\rho V^2}$$

According to Sec. 7.2, step 7, we can write $\Pi_2 = \phi(\Pi_1^{-1})$, which results in $\bar{\tau}_0 = \rho V^2 \phi(\mathbf{R})$. Setting the dimensionless term $\phi(\mathbf{R}) = \frac{1}{2}C_f$ this yields

$$\bar{\tau}_0 = C_f \rho \frac{V^2}{2} \tag{7.10}$$

Inserting this value of $\bar{\tau}_0$ into Eq. (7.8), and noting from Eq. (2.1) that $\gamma = \rho g$,

$$h_f = C_f \frac{L}{R_h} \frac{V^2}{2g} \tag{7.11}$$

which we can apply to any shape of smooth-walled cross section. From this equation, we may easily obtain an expression for the slope of the energy line,

$$S = \frac{h_f}{L} = \frac{C_f}{R_h} \frac{V^2}{2g} \tag{7.12}$$

which we also know as the *energy gradient.*

Later, in Sec. 7.13, we shall see that Eqs. (7.11) and (7.12) also apply to rough-walled conduits.

7.6 Friction in Circular Conduits

In Sec. 7.4 we saw that for a circular pipe flowing full $R_h = D/4$. Substituting this value into Eqs. (7.11) and (7.12), we obtain (for both smooth-walled and rough-walled conduits) the well-known equation for *pipe-friction head loss,*

Circular pipe,
flowing full
(laminar or
turbulent flow):

$$h_f = f \frac{L}{D} \frac{V^2}{2g} \tag{7.13}$$

and

$$\frac{h_f}{L} = S = \frac{f}{D} \frac{V^2}{2g} \tag{7.14}$$

where

$$f = 4C_f = 8\phi(\mathbf{R}) \tag{7.15}$$

Equation (7.13) is known as the *pipe-friction equation,* and as the *Darcy-Weisbach equation.*[4] Like the coefficient C_f, the *friction factor* f (also sometimes called the *Darcy friction factor*) is dimensionless and some function of Reynolds number. Much research has gone into determining how f varies with $\mathbf{R}$ and also with pipe roughness (see Sec. 7.13). The pipe-friction equation states that the head lost in friction in a given pipe is proportional to the velocity head. The equation is dimensionally homogeneous, and we may use it with any consistent system of units.

Dimensional analysis gives us the proper form for an equation, but does not yield a numerical result, since it does not deal with abstract numerical factors. It also shows

---

[4] In a slightly different form where the hydraulic radius R_h replaces D, so that f changes, Eq. (7.13) is known as the *Fanning equation,* which chemical engineers use widely.

that Eq. (7.13) is a rational expression for pipe friction. But we must determine the exact form of $\phi(\mathbf{R})$ and numerical values for C_f and f by experiment or other means.

For a circular pipe flowing full, by substituting $R_h = r_0/2$, where r_0 is the radius of the pipe, we can write Eq. (7.8) as

$$h_f = \bar{\tau}_0 \frac{L}{R_h \gamma} = \frac{2\tau_0 L}{r_0 \gamma} \tag{7.16}$$

where the local shear stress at the wall, τ_0, is equal to the average shear stress $\bar{\tau}_0$ because of symmetry.

For a cylindrical fluid body of radius r concentric with the pipe, if we follow a development similar to that of Eqs. (7.5)–(7.8) and note that $A = \pi r^2$ and $P = 2\pi r$, we can show that

$$h_f = \frac{2\tau L}{r\gamma} \tag{7.17}$$

where τ is the shear stress in the fluid at radius r; this also follows from Eq. (5.12). Since h_f is the same for all r (Sec. 7.4), equating the right sides of Eqs. (7.16) and (7.17) yields

$$\tau = \tau_0 \frac{r}{r_0} \tag{7.18}$$

or the shear stress is zero at the center of the pipe and increases linearly with the radius to a maximum value τ_0 at the wall as in Fig. 7.3. This is true regardless of whether the flow is laminar or turbulent.

From Eqs. (7.8) and (7.13) and substituting $R_h = D/4$ for a circular pipe, we obtain

$$\tau_0 = \frac{f}{4}\rho\frac{V^2}{2} = \frac{f}{4}\gamma\frac{V^2}{2g} \tag{7.19}$$

With this equation, we can compute τ_0 for flow in a circular pipe for any experimentally determined value of f.

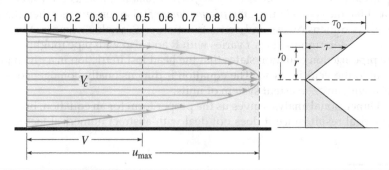

FIGURE 7.3   Velocity profile in laminar flow, and distribution of shear stress.

7.7 Friction in Noncircular Conduits

Most closed conduits we use in engineering practice are of circular cross section; however, we do occasionally use rectangular ducts and cross sections of other geometry. We can modify many of the equations in this chapter for application to noncircular sections by using the concept of hydraulic radius.

In Sec. 7.4 we defined the hydraulic radius as $R_h = A/P$, where A is the cross-sectional area and P is the wetted perimeter. Some equations include R_h, as in Sec. 7.19, in which case we can simply determine R_h for that conduit. But, because circular pipes are so common, many more equations instead use the pipe diameter, D. If the conduit of interest is not circular, we have no diameter to use in such equations, but we can find its hydraulic radius, R_h. Then we notice for a circular pipe flowing full, that

$$R_h = \frac{A}{P} = \frac{\pi D^2/4}{\pi D} = \frac{D}{4} \tag{7.20}$$

or

$$D = 4R_h \tag{7.21}$$

This provides us with an *equivalent* diameter, which we can substitute into Eq. (7.13) for example to yield

$$h_f = f\frac{L}{4R_h}\frac{V^2}{2g} \tag{7.22}$$

and into Eq. (7.1) for Reynolds number to give

Pipe flow: $$\mathbf{R} = \frac{(4R_h)V\rho}{\mu} = \frac{(4R_h)V}{\nu} \tag{7.23}$$

and into any other pipe flow equations and charts that use D instead of R_h. We will meet many such equations and an important such chart later in this chapter.

This approach gives reasonably accurate results for turbulent flow, but the results are poor for laminar flow, because in such flow viscous action causes frictional phenomena throughout the body of the fluid, while in turbulent flow the frictional effect occurs largely in the region close to the wall; i.e., it depends on the wetted perimeter.

7.8 Laminar Flow in Circular Pipes

In Sec. 2.11 we noted that for laminar flow $\tau = \mu\, du/dy$, where u is the value of the velocity at a distance y from the boundary. As $y = r_0 - r$, we also see that $\tau = -\mu\, du/dr$; in other words, the minus sign indicates that u decreases as r increases. The coefficient of viscosity μ is a constant for any particular fluid at a constant temperature, and therefore if the shear varies from zero at the center of the pipe to a maximum at the wall, it follows that the velocity profile must have a zero slope at the center and have a continuously steeper velocity gradient approaching the wall.

To determine the velocity profile for laminar flow in a circular pipe, we substitute the expression $\tau = \mu\,du/dy$ into Eq. (7.17). Thus

$$h_f = \frac{2\tau L}{r\gamma} = \mu\frac{du}{dy}\frac{2L}{r\gamma} = -\mu\frac{du}{dr}\frac{2L}{r\gamma}$$

From this,
$$du = -\frac{h_f\gamma}{2\mu L}r\,dr$$

Integrating and determining the constant of integration from the fact that $u = u_{max}$ when $r = 0$, we obtain

$$u = u_{max} - \frac{h_f\gamma}{4\mu L}r^2 = u_{max} - kr^2 \tag{7.24}$$

From this equation we see that the velocity profile is a parabola, as shown in Fig. 7.3. Note that $k = h_f\gamma/4\mu L$.

At the wall we have the no-slip boundary condition (Sec. 2.11) that $u = 0$ when $r = r_0$. Substituting this into the second expression of Eq. (7.24) and noting that $u_{max} = V_c$, the centerline velocity, we find $k = V_c/r_0^2$. Thus we can express Eq. (7.24) as

Laminar
pipe flow:
$$u = V_c - \frac{V_c}{r_0^2}r^2 = V_c\left(1 - \frac{r^2}{r_0^2}\right) \tag{7.25}$$

Combining Eqs. (7.24) and (7.25), we get an expression for the centerline velocity as follows

$$V_c = u_{max} = \frac{h_f\gamma}{4\mu L}r_0^2 = \frac{h_f\gamma}{16\mu L}D^2 \tag{7.26}$$

We can multiply Eq. (7.24) by a differential area $dA = 2\pi r\,dr$ and integrate the product from $r = 0$ to $r = r_0$ to find the rate of discharge. From Eq. (4.3), the rate of discharge is equivalent to the volume of a solid bounded by the velocity profile. In this case the solid is a paraboloid with a maximum height of u_{max}. The mean height of a paraboloid is one-half the maximum height, and hence the mean velocity V is $0.5u_{max}$. Thus

$$V = \frac{h_f\gamma}{32\mu L}D^2 \tag{7.27}$$

From this last equation, noting that $\gamma = g\rho$ and $\mu/\rho = \nu$, the loss of head due to friction is

Laminar
flow:
$$h_f = 32\frac{\mu}{\gamma}\frac{L}{D^2}V = 32\nu\frac{L}{gD^2}V \tag{7.28}$$

which is the **Hagen-Poiseuille law** for laminar flow in tubes. G. Hagen, a German engineer, experimented with water flowing through small brass tubes, and published his results in 1839. J. L. Poiseuille, a French scientist, experimented with water flowing through capillary tubes in order to determine the laws of flow of blood through the veins of the body, and published his studies in 1840.

From Eq. (7.28) we see that in laminar flow the loss of head is proportional to the first power of the velocity. This is verified by experiment, as shown in Fig. 7.1. The striking feature of this equation is that it involves no empirical coefficients or experimental factors of any kind, except for the physical properties of the fluid such as viscosity and density (or specific weight). From this, it would appear that in laminar flow the friction is independent of the roughness of the pipe wall. Experiments also bear out that this is true.

Dimensional analysis shows that we may express the friction loss by Eq. (7.13). Equating (7.13) and (7.28) and solving for the friction factor f, we obtain for laminar flow under pressure in a circular pipe,

Laminar
flow:
$$f = \frac{64\nu}{DV} = \frac{64}{\mathbf{R}}$$
(7.29)

Thus, if $\mathbf{R}$ is less than 2000, we may use Eq. (7.28) to find pipe friction head loss, or we may use the pipe-friction equation given by Eq. (7.13) with the value of f as given by Eq. (7.29).

7.9 Entrance Conditions in Laminar Flow

In the case of a pipe leading from a reservoir, if the entrance is rounded so as to avoid any initial disturbance of the entering stream, all particles will start to flow with the same velocity, except for a very thin film in contact with the wall. Particles in contact with the wall have zero velocity (the no-slip boundary condition, Sec. 2.11), but the velocity gradient there is extremely steep, and, with this slight exception, the velocity is uniform across the diameter, as shown in Fig. 7.4. As the fluid progresses along the pipe, friction originating from the wall slows down the streamlines in the vicinity of the wall, but since Q is constant for successive sections, the velocity in the center must accelerate, until the final velocity profile is a parabola, as shown in Fig. 7.3. Theoretically, this requires an infinite distance, but both theory and observation have

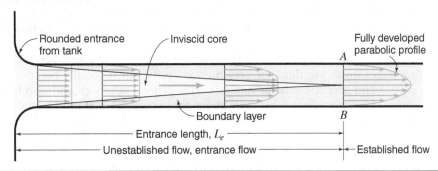

FIGURE 7.4 Velocity profiles and development of the boundary layer along a pipe in laminar flow.

established that the maximum velocity in the center of the pipe will reach 99% of its ultimate value in a distance[5]

Laminar flow: $$L_e = 0.0575\mathbf{R}D \qquad\qquad (7.30)$$

We call this distance the ***entrance length*** (Fig. 7.4). For the critical (maximum) value of $\mathbf{R} = 2000$ (Sec. 7.3), the entrance length L_e equals 115 pipe diameters. In other cases of laminar flow with Reynolds numbers less than 2000, the distance L_e will be correspondingly less, in accordance with Eq. (7.30).

Within the entrance length the flow is ***unestablished;*** that is, the velocity profile is changing. In this region (Fig. 7.4), we can visualize the flow as consisting of a central ***inviscid core*** in which there are no frictional effects, i.e., the flow is uniform, and an outer, annular zone extending from the core to the pipe wall. This outer zone increases in thickness as it moves along the wall, and is known as the ***boundary layer.*** Viscosity in the boundary layer acts to transmit the effect of boundary shear inwardly into the flow. At section AB the boundary layer has grown until it occupies the entire cross section of the pipe. At this point, for laminar flow, the velocity profile is a perfect parabola. Beyond section AB, for the same straight pipe the velocity profile does not change, and the flow is known as (laminar) ***established flow,*** or (laminar) ***fully developed flow.*** The flow will continue as fully developed so long as no change occurs to the straight pipe surface. When a change occurs, such as at a bend or other pipe fitting, the velocity profile will deform and will require some more flow length to return to established flow. Usually such fittings are so far apart that fully developed flow is common; but when they are close enough it is possible that established flow never occurs.

The concept of a boundary layer within which viscosity is important, and outside of which friction is unimportant and we can consider the fluid to be ideal, originated in 1904 with Ludwig Prandtl (1875–1953), a German engineering professor. Perhaps the single most significant contribution to fluid mechanics, this concept is particularly important with turbulent flow; we will discuss it further in Secs. 7.11–7.13.

As we saw in Sample Problem 5.1 for a circular pipe, the kinetic energy of a stream with a parabolic velocity profile is $2V^2/2g$ (because $\alpha = 2$), where V is the mean velocity. At the entrance to the pipe the velocity is uniformly V across the diameter, except for an extremely thin layer next to the wall. Thus at the entrance to the pipe the kinetic energy per unit weight is practically $V^2/2g$. So in the distance L_e there is a continuous increase in kinetic energy accompanied by a corresponding decrease in pressure head. Therefore, at a distance L_e from the entrance with laminar flow, the piezometric head is less than the static value in the reservoir by $2V^2/2g$ plus the friction loss in this distance.

We have dealt with laminar flow rather fully, not merely because it is important in problems involving fluids of very high viscosity, but especially because it permits a simple and accurate rational analysis. This general approach will also help with the study of *turbulent* flow, where conditions are so complex that rigid mathematical treatment is impossible.

---

[5] H. L. Langhaar, Steady Flow in the Transition Length of a Straight Tube, *J. Appl. Mech.,* Vol. 10, p. 55, 1942.

Sample Problem 7.2

For the case of Sample Problem 7.1, find the centerline velocity, the velocity at $r = 20$ mm, the friction factor, the shear stress at the pipe wall, and the head loss per meter of pipe length.

Solution

From the solution to Sample Problem 7.1: The flow is laminar, with

$$V = 0.0637 \text{ m/s and } \mathbf{R} = 354 \text{ and } \rho = 0.85(1000 \text{ kg/m}^3) = 850 \text{ kg/m}^3.$$

Therefore, per Sec. 7.8: $\qquad V_c = 2V = 0.1273 \text{ m/s} \qquad ANS$

Eq. (7.24): $\qquad u = u_{max} - kr^2 \qquad u_{max} = V_c = 0.1273 \text{ m/s}$

When $r = r_0 = 50$ mm, $u = 0$; hence $0 = 0.1273 - k(0.05)^2$, from which

$$k = 50.9/(\text{m·s}) \qquad u_{20 \text{ mm}} = 0.1273 - 50.9(0.02)^2 = 0.1070 \text{ m/s} \qquad ANS$$

Eq. (7.29): $\qquad f = \dfrac{64}{\mathbf{R}} = \dfrac{64}{354} = 0.1810 \qquad ANS$

Eq. (7.19): $\qquad \tau_0 = \dfrac{f}{4}\rho\dfrac{V^2}{2} = \dfrac{0.1810}{4}(850 \text{ kg/m}^3)\dfrac{(0.0637 \text{ m/s})^2}{2}$

$$= 0.0779 \text{ kg/(m·s}^2)$$

$$\tau_0 = 0.0779\dfrac{\text{kg}}{(\text{m·s}^2)}\dfrac{\text{N·s}^2}{\text{kg· m}} = 0.0779 \text{ N/m}^2 \qquad ANS$$

Eq. (7.14): $\qquad S = \dfrac{h_f}{L} = f\dfrac{1}{D}\dfrac{V^2}{2g} = 0.1810\dfrac{1}{(0.10 \text{ m})}\dfrac{(0.0637 \text{ m/s})^2}{2(9.81 \text{ m/s}^2)}$

$$= 0.000374 \text{ m/m} \qquad ANS$$

7.10 Turbulent Flow

In Sec. 4.2 we saw that in laminar flow the fluid particles move in straight lines while in turbulent flow they follow random paths. Consider the case of laminar flow shown in Figs. 7.5(a, b), where the velocity u increases with y. Even though the fluid particles are moving horizontally to the right, because of molecular motion, molecules will cross line ab and transport momentum with them. On the average, the velocities of the molecules in the slower-moving fluid below the line will be less than those of the faster-moving fluid above; the result is that the molecules that cross from below tend to slow down the faster-moving fluid. Likewise, the molecules that cross the line ab from above tend to speed up the slower-moving fluid below. The result is a shear stress τ (tau) along the surface whose trace is ab. As given in Sec. 2.11, $\tau = \mu\, du/dy$. This equation is applicable to laminar flow only.

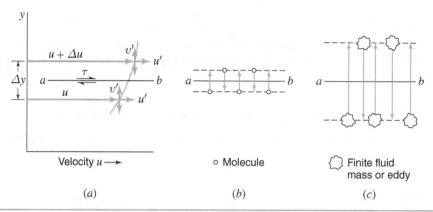

FIGURE 7.5   (a) Velocity profile. (b) Laminar flow (transfer of molecules across *ab*). (c) Turbulent flow (transfer of finite fluid masses across *ab*).

Let us examine some of the characteristics of turbulent flow to see how it differs from laminar flow. In turbulent flow the velocity at a point in the flow field fluctuates in both magnitude and direction.[6] We may observe these fluctuations in accurate velocity measurements (Sec. 9.4), and we commonly see their effects on pressure gages and manometers. The fluctuations result from a multitude of small eddies (Sec. 4.2), created by the viscous shear between adjacent particles. These eddies grow in size and then disappear as their particles merge into adjacent eddies. Thus, there is a continuous mixing of particles, with a consequent transfer of momentum. Viscosity dissipates mechanical energy, generating small amounts of heat.

First Expression

In the modern conception of turbulent flow, we assume a mechanism similar to that just described for laminar flow. However, we replace the molecules by minute but finite masses or eddies (Fig. 7.5c). So, by analogy, for turbulent flow we may define the shear stress along the plane through *ab* in Fig. 7.5 as

$$\text{Turbulent shear stress} = \eta \frac{du}{dy} \tag{7.31}$$

But unlike μ, the **eddy viscosity** η (eta) is not a constant for a given fluid at a given temperature, but it depends on the turbulence of the flow. We may view it as a coefficient of momentum transfer, expressing the transfer of momentum from points where the velocity is low to points where it is higher, and vice versa. Its magnitude may range from zero to many thousand times the value of μ. However, its numerical value is of less interest than its physical concept. In dealing with turbulent flow, it is sometimes convenient to use **kinematic eddy viscosity** ε (epsilon) $= \eta/\rho$, which is a property of the flow alone, analogous to kinematic viscosity.

---

[6] We can best visualize the velocity at a point in a so-called steady turbulent flow as a vector that fluctuates in both direction and magnitude. The mean temporal velocity at that point corresponds to the "average" of those vectors.

In general, the total shear stress in turbulent flow is the sum of the laminar shear stress plus the turbulent shear stress, i.e.,

$$\tau = \mu \frac{du}{dy} + \eta \frac{du}{dy} = \rho(\nu + \varepsilon)\frac{du}{dy} \tag{7.32}$$

With turbulent flow the second term of this equation is usually many times larger than the first term.

For turbulent flow we saw in Sec. 4.5 and Fig. 4.6 that the local axial velocity has fluctuations of plus and minus u' and there are also fluctuations of plus and minus v' and w' normal to u as shown in Fig. 7.6b. As it is obvious that there can be no values of v' next to and perpendicular to a smooth wall, turbulent flow cannot exist there. Hence, near a smooth wall, the shear is due to laminar flow alone, and $\tau = \mu \, du/dy$. Note that the shear stress always acts to cause the velocity distribution to become more uniform.

At some distance from the wall, such as $0.2r$, the value of du/dy becomes small in turbulent flow, and so the viscous shear becomes negligible in comparison with the turbulent shear. The latter can be large, even though du/dy is small, because of the possibility of η being very large. This is because of the great turbulence that may exist at an appreciable distance from the wall. But at the center of the pipe, where du/dy is zero, there can be no shear at all. So, in turbulent flow as well as in laminar flow, the shear stress is a maximum at the wall and decreases linearly to zero at the axis, as shown in Fig. 7.3 and proved in Sec. 7.6.

Second Expression

We can obtain another expression for turbulent shear stress that is different from Eq. (7.31). In Fig. 7.5a, if a mass m of fluid below ab, where the temporal mean axial velocity is u, moves upward into a zone where the temporal mean axial velocity is $u + \Delta u$, its initial momentum in the axial direction must increase by $m\Delta u$. Conversely, when a mass m moves from the upper zone to the lower its axial momentum will decrease by $m\Delta u$. So this transfer of momentum back and forth across ab will produce a shear in the plane through ab proportional to Δu. This shear is possible only because of the velocity profile shown. If the latter were vertical, Δu would be zero and there could be no shear.

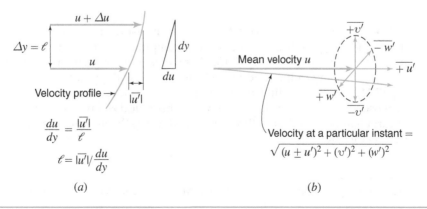

$$\frac{du}{dy} = \frac{|\overline{u'}|}{\ell}$$

$$\ell = |\overline{u'}| / \frac{du}{dy}$$

(a)

(b)

FIGURE 7.6   (a) Mixing length ℓ. (b) Instantaneous local velocity in turbulent flow.

If the distance Δy in Fig. 7.5a is chosen so that the average value of $+u'$ in the upper zone over a time period long enough to include many velocity fluctuations is equal to Δu, i.e., $\Delta u = \overline{|u'|}$ the distance between the two streams will be what is known as the *mixing length* ℓ, which we will discuss shortly. Consider, over a short time interval, a mass moving upward from below ab with a velocity v'; it will transport into the upper zone, where the velocity is $u + u'$, a momentum per unit time which is on the average equal to $\rho(v'dA)(u)$. The slower-moving mass from below ab will tend to retard the flow above ab; this creates a shear force along the plane of ab. We can find this force by applying the momentum principle, Eq. (6.6),

$$F = \tau dA = \rho Q(\Delta V) = \rho(v'dA)(u + u' - u) = \rho u'v'dA$$

So, over a period of time sufficiently long to include a large number of velocity fluctuations, the shear stress is given by

$$\tau = F/dA = -\rho\overline{u'v'} \tag{7.33}$$

where $\overline{u'v'}$ is the temporal average of the product of u' and v' This is an alternate form for Eq. (7.31), and in modern turbulence theory $-\rho\overline{u'v'}$ is known as the *Reynolds stress.*

The minus sign appears in Eq. (7.33) because the product $\overline{u'v'}$ on the average is negative. By inspecting Fig. 7.5a, we can see that $+v'$ is associated with $-u'$ values more than with $+u'$ values. The opposite is true for $-v'$ Even though the temporal mean values of u' and v' are individually equal to zero, the temporal mean value of their product is not zero. This is because combinations of $+v'$ and $-u'$ and of $-v'$ and $+u'$ predominate over combinations of $+v'$ and $+u'$ and $-v'$ and $-u'$ respectively.

Prandtl (Sec. 7.9) reasoned that in any turbulent flow $\overline{|u'|}$ and $\overline{|v'|}$ must be proportional to each other and of the same order of magnitude. He also introduced the concept of mixing length ℓ, which is the distance transverse to the flow direction such that $\Delta u = \overline{|u'|}$. From Fig. 7.6a we can see that $\Delta u = \ell\, du/dy$ and so $\overline{|u'|} = \ell du/dy$. If $\overline{|u'|} \propto \overline{|v'|}$ and if we allow ℓ to account for the constant of proportionality, Prandtl[7] showed that $-\overline{u'v'}$ varies as $\ell^2(du/dy)^2$. Thus

$$\tau = -\rho\overline{u'v'} = \rho\ell^2\left(\frac{du}{dy}\right)^2 \tag{7.34}$$

This equation expresses terms that we can measure. Thus, in any experiment that determines the pipe friction loss, we can compute $\overline{\tau}_0$ from Eq. (7.8), and then we can find τ at any radius from Eq. (7.18). A traverse of the velocity across a pipe diameter will give u at any radius, and the velocity profile will give du/dy at any radius. Thus Eq. (7.34) allows us to find the mixing length ℓ as a function of the pipe radius. The purpose of all of this is to help us develop theoretical equations for the velocity profile in turbulent flow, and from this in turn to develop theoretical equations for f, the friction coefficient.

---

[7] H. Schlichting, *Boundary Layer Theory*, 7th ed., McGraw-Hill Book Co., 1987, p. 605.

7.11 Viscous Sublayer in Turbulent Flow

In Fig. 7.4 we saw that, for *laminar* flow, if the fluid enters a pipe with no initial disturbance, the velocity is uniform across the diameter except for an exceedingly thin film at the wall, because the velocity touching any wall is zero (the no-slip condition, Sec. 2.11). But as flow proceeds down the pipe, the velocity profile changes because of the growth of a laminar boundary layer, which continues until the boundary layers from opposite sides meet at the pipe axis and then there is fully developed (or established) laminar flow.

If the Reynolds number is above the critical value (Sec. 7.3), so that the developed flow is turbulent, the initial condition is much like that in Fig. 7.4. But as the laminar boundary layer increases in thickness, at a certain point a transition occurs and the boundary layer becomes turbulent (Fig. 7.7). This transition occurs where the length x_c of the laminar portion of the boundary layer is about equal to $500{,}000\nu/U$, where U is the uniform velocity (i.e., where $\mathbf{R}_x = Ux/\nu \approx 500{,}000$). After the transition, the *turbulent boundary layer* generally increases in thickness much more rapidly. So the length of the *inviscid core*, to where the two opposite layers meet, is relatively shorter.

The development of turbulent flow is considerably more complex than that of laminar flow (Sec. 7.9). Although the length of the inviscid core is relatively short with turbulent flow, it takes about four times this length for the velocity profile to become fully developed, and 8 to 12 times this length for the detailed structure of the turbulence to become fully developed. Only when all these aspects are complete do we have *fully developed turbulent flow.* These various developments depend on a variety of features, including the entrance conditions (Sec. 7.22) and the wall roughness (Secs. 7.11–7.13). While no single equation exists to predict the *entrance length*, L_e for turbulent flow, i.e., the length over which the flow is developing, two commonly used equations are $L_e/D = 1.6(\mathbf{R})^{1/4}$ and $L_e/D = 4.4(\mathbf{R})^{1/6}$, with some evidence the former may be preferred.[8] Regardless, in the common range of Reynolds numbers encountered in civil

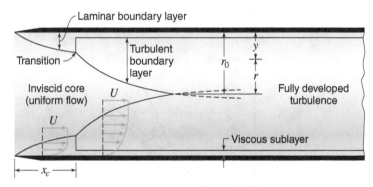

FIGURE 7.7 Development of boundary layer in a pipe where fully developed flow is turbulent (the horizontal scale is very compressed).

---

[8] Anselmet, F., et al., Axial Development of the Mean Flow in the Entrance Region of Turbulent Pipe and Duct Flows, *Comptes Rendus Mécanique*, 337(8):573–584, 2009.

engineering design, the velocity profile generally becomes fully developed in the range

Turbulent
flow:

$$20 < \frac{L_e}{D} < 40$$

In all that now follows we shall consider only *fully* developed turbulent flow.

As v' must be zero at a smooth wall, turbulence there is inhibited so that a laminar-like sublayer occurs immediately next to the wall. However, the adjacent turbulent flow does repeatedly induce random transient effects that momentarily disrupt this sub-layer, even though they fade rapidly. Because it is therefore not a true laminar layer, and because shear in this layer is predominantly due to viscosity alone, we call it a *viscous sublayer* (see Fig. 7.7). This viscous sublayer is extremely thin, usually only a few hundredths of a millimeter, but its effect is great because of the very steep velocity gradient within it and because $\tau = \mu \, du/dy$ in that region. At a greater distance from the wall the viscous effect becomes negligible, but the turbulent shear is then large. Between the two, there must be a transition zone where both types of shear are significant. It is evident that there cannot be sharp lines of demarcation separating these three zones, instead one must merge gradually into the other.

By plotting one velocity profile from the wall on the assumption that the flow is entirely laminar (Sec. 7.8) and plotting another velocity profile on the assumption that the flow is entirely turbulent (Sec. 7.12), the two will intersect, as shown in Fig. 7.8. It is obvious that there can be no abrupt change in profile at this point of intersection, but that one curve must merge gradually into the other with some kind of transition, which is in fact what happens as shown by the experimental points.

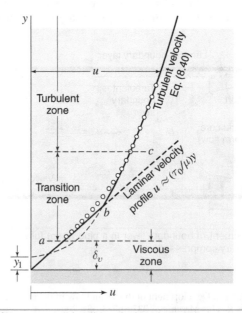

FIGURE 7.8   Velocity profile near a solid wall (vertical scale greatly exaggerated). Theoretical relations (solid lines) are compared with experimental data (small circles).

When studying such velocity profiles, we find that the quantity $\sqrt{\tau_0/\rho}$ frequently occurs. Because it has the dimensions of velocity, scientists have named it the *shear-stress velocity* (or *friction velocity*) u_*, although it is not actually a flow velocity.

When the flow in a circular pipe is entirely laminar, we have seen that the velocity profile is a parabola (Fig. 7.3). But when there is only an extremely thin film closest to the wall where viscous shear dominates, we can scarcely distinguish the velocity profile in it from a straight line. If we ignore the momentary fluctuations in this viscous sublayer, with such a linear velocity profile Eq. (2.9) at the wall becomes

$$\tau_0 = \mu\frac{u}{y} \quad \text{or} \quad \frac{\tau_0}{\rho} = \frac{\nu u}{y}$$

But from the definition of u_*, we have $u_*^2 = \tau_0/\rho$, so, by eliminating τ_0/ρ, we obtain

$$\frac{\nu u}{y} = u_*^2$$

$$\frac{u}{u_*} = \frac{y u_*}{\nu} \tag{7.35}$$

which is known as a *law of the wall*. This linear relation for $u(y)$ approximates experimental data well in the range $0 \le y u_*/\nu \le 5$. If we call this imprecise, but commonly accepted, upper limiting value of y the thickness of the viscous sublayer, δ_v, then

$$\delta_v = 5\nu/u_* \tag{7.36}$$

The transition zone appears to extend from a to c in Fig. 7.8. For the latter point, the value of y seems to be about $70\nu/u_*$ or $14\delta_v$. Beyond this, the flow is so turbulent that viscous shear is negligible.

Noting from Eq. (7.19) that

$$u_* = \sqrt{\frac{\tau_0}{\rho}} = V\sqrt{\frac{f}{8}} \tag{7.37}$$

and that the Reynolds number $\mathbf{R} = DV/\nu$, we see that when $y u_*/\nu = 5$, or $y = \delta_v$,

$$\delta_v = \frac{14.14\nu}{V\sqrt{f}} = \frac{14.14D}{\mathbf{R}\sqrt{f}} \tag{7.38}$$

From this, we see that the higher the velocity or the lower the kinematic viscosity, the thinner is the viscous sublayer. Thus, for a given constant pipe diameter, the thickness of the viscous sublayer decreases as the Reynolds number increases.

Now we can consider what is meant by a smooth wall and a rough wall. There is no such thing in reality as a mathematically smooth surface. But if the irregularities on any actual surface are small enough that the effects of the projections do not pierce through the viscous sublayer (Fig. 7.8), the surface is *hydraulically smooth* from the fluid-mechanics viewpoint. If the effects of the projections extend beyond the sublayer, the laminar

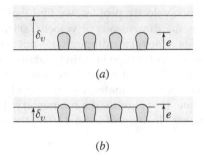

FIGURE 7.9 Turbulent flow near a boundary. (a) Low **R**, $\delta_v > e$; the pipe behaves as a smooth pipe. (b) Relatively high **R**, $\delta_v < e$; if $\delta_v < \frac{1}{14}e$, the pipe behaves as a fully rough pipe.

layer is broken up and the surface is no longer hydraulically smooth. If the surface roughness projections are large enough to protrude right through the transition layer, it is totally broken up. The resulting flow is completely turbulent, known as *fully rough (-pipe) flow,* and friction is independent of Reynolds number. We shall discuss the significance of this in Sec. 7.12. If the roughness projections protrude only partially into the transition layer, we say the flow is *transitionally rough,* and there is a moderate Reynolds number effect.

To be more specific, if e is the equivalent height of the roughness projections then for $eu_*/\nu < 5$ (or $e < \delta_v$) the viscous sublayer completely buries the surface roughness, the roughness has no effect on friction, and the pipe is hydraulically smooth. If $eu_*/\nu > 70$ (or $e > 14\delta_v$), the pipe will behave as fully rough. In the region between these values, i.e., when the roughness projections are such that $5 \leq eu_*/\nu \leq 70$ (or $\delta_v \leq e \leq 14\delta_v$), the pipe will behave in a transitional mode, neither hydraulically smooth nor fully rough. Most engineering pipe flows fall in this range.

Because the thickness of the viscous sublayer in a given pipe decreases with an increase in Reynolds number, we see that the same pipe may be hydraulically smooth at low Reynolds numbers and rough at high Reynolds numbers. Thus, even a relatively smooth pipe may behave as a rough pipe if the Reynolds number is high enough. It is also apparent that, with increasing Reynolds number, there is a gradual transition from smooth to rough-pipe flow. Figure 7.9 depicts these concepts schematically.

7.12 Velocity Profile in Turbulent Flow

Prandtl (Secs. 7.9 and 7.10) reasoned that turbulent flow in a pipe is strongly influenced by the flow phenomena near the wall. In the vicinity of the wall, $\tau \approx \tau_0$. He assumed that the mixing length ℓ (Sec. 7.10) near the wall was proportional to the distance from the wall, that is, $\ell = Ky$. Experiments have confirmed this, and have determined that $K = 0.40$. Using this relationship in Eq. (7.34), we get

$$\tau \approx \tau_0 = \rho\ell^2 \left(\frac{du}{dy}\right)^2 = \rho K^2 y^2 \left(\frac{du}{dy}\right)^2$$

or
$$du = \frac{1}{K}\sqrt{\frac{\tau_0}{\rho}}\frac{dy}{y} = \frac{u_*}{K}\frac{dy}{y}$$

Integrating, and inserting 0.4 for K, we obtain

$$u = 2.5u_* \ln y + C$$

We can evaluate the constant C by noting that $u = u_{max}$ = the centerline velocity when $y = r_0$ = the pipe radius. Substituting the expression for C, we get

$$\frac{u_{max} - u}{u_*} = 2.5 \ln \frac{r_0}{y} \qquad (7.39)$$

This is known as the **velocity defect law,** because we call $(u_{max} - u)$ the velocity defect. Replacing y by $r_0 - r$, and changing the base e logarithm (ln) to a base 10 logarithm (log), the equation becomes

$$u = u_{max} - 2.5u_* \ln \frac{r_0}{r_0 - r} = u_{max} - 5.76u_* \log \frac{r_0}{r_0 - r} \qquad (7.40)$$

Although this equation is derived by assuming certain relations very near to the wall, it holds almost as far out as the pipe axis.

Starting with the derivation of Eq. (7.34), this entire development is open to argument at nearly every step. But the fact remains that Eq. (7.40) agrees very closely with actual measurements of velocity profiles for both smooth and rough pipes. However, there are two zones in which the equation is defective. At the axis of the pipe, du/dy must be zero. But Eq. (7.40) is logarithmic and does not have a zero slope at $r = 0$, and hence the equation gives a velocity profile with a sharp point (or cusp) at the axis, whereas in reality it is rounded at the axis. This discrepancy affects only a very small area and causes only a very slight error when computing the rate of discharge using Eq. (7.40).

Equation (7.40) is also not applicable very close to the wall. In fact, it indicates that when $r = r_0$, the value of u is minus infinity. The equation indicates that $u = 0$, not at the wall, but at a small distance from it, shown as y_1 in Fig. 7.8. However, this discrepancy is well within the confines of the viscous sublayer, where the equation is not supposed to apply, and where we have Eq. (7.35). In the intervening transition or overlap zone (Fig. 7.8), where both viscous and turbulent shear are important, investigators have found that experimental velocity profile data follow a logarithmic relation

$$\frac{u}{u_*} = 2.5 \ln\left(\frac{yu_*}{\nu}\right) + 5.0 \qquad (7.41)$$

Although Eq. (7.40) is not perfect, it reliably fits the data except for the two small areas mentioned, where it is still close. So we can determine the rate of discharge Q with a high degree of accuracy by using the value of u given by Eq. (7.40) and integrating over the area of the pipe. Thus

$$Q = \int u\,dA = 2\pi \int_0^{r_0} urdr$$

Substituting from the first expression of Eq. (7.40) for u, integrating and dividing by the pipe area πr_0^2, the mean velocity is[9]

$$V = u_{max} - 2.5u_* \left[\ln r_0 - \frac{2}{r_0^2} \int_0^{r_0} r\ln(r_0 - r)\, dr \right]$$

Making use of Eq. (7.37), this equation reduces to

$$V = u_{max} - \tfrac{3}{2} \times 2.5\, u_* = u_{max} - 1.326V\sqrt{f} \tag{7.42}$$

From Eq. (7.42) we can obtain the **pipe factor**, which is the ratio of the mean to the maximum velocity. It is

$$\text{Pipe factor} = \frac{V}{u_{max}} = \frac{1}{1 + 1.326\sqrt{f}} \tag{7.43}$$

Using Eq. (7.43) to eliminate u_{max} from Eq. (7.40) and using Eq. (7.37) to eliminate u_*, the result is

$$u = (1 + 1.326\sqrt{f})V - 2.04\sqrt{f}V \log \frac{r_0}{r_0 - r} \tag{7.44}$$

which enables us to plot a velocity profile for any mean velocity and any value of f in turbulent flow. In Fig. 7.10 profiles for both a smooth and a rough pipe are plotted from this equation. The only noticeable difference between these and measured profiles is that the latter are more rounded at the axis of the pipe.[10] Of course, the measured turbulent profiles also exhibit turbulent fluctuations everywhere except near the walls.

Comparing the turbulent-flow velocity profiles with the laminar-flow velocity profile in Fig. 7.10 we see the turbulent-flow profiles are much flatter near the central portion of the pipe and steeper near the wall. We also notice that the turbulent profile for the smooth pipe is flatter near the center (i.e., blunter) than for the rough pipe. In contrast, the velocity profile in laminar flow is independent of pipe roughness.

As we have now derived a theoretical equation for the velocity profile for turbulent flow in circular pipes, we can also derive equations for the kinetic-energy- and momentum-correction factors (Secs. 5.1 and 6.3) using mean velocities. Respectively, these equations are[11]

$$\alpha = 1 + 2.7f \tag{7.45a}$$

$$\beta = 1 + 0.98f \tag{7.45b}$$

---

[9] The integral results in indeterminate values at $r = r_0$, as we should expect, since the equation for u does not really apply close to the wall. However, for all practical purposes, these reduce to negligible quantities.
[10] Although the preceding theory agrees very well with experimental data, it is not absolutely correct throughout the entire range from the axis to the pipe wall, and some slight shifts in the numerical constants could improve agreement with test data. Thus, in Eqs. (7.43) and (7.44), we can replace the value 1.326 by 1.44, and in Eq. (7.44) although many writers use 2 instead of 2.04, a better practical value seems to be 2.15.
[11] L. F. Moody, Some Pipe Characteristics of Engineering Interest, *Houille Blanche*, May–June, 1950.

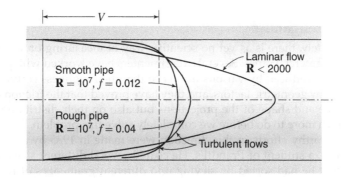

FIGURE 7.10   Velocity profiles across a pipe for equal flow rates. The turbulent profiles are plotted from Eq. (7.44).

Sample Problem 7.3

The pipe friction head loss in 200 ft of 6-in-diameter pipe is 25 ft·lb/lb when oil ($s = 0.90$) of viscosity 0.0008 lb·sec/ft^2 flows at 2.0 cfs. Determine the centerline velocity, the shear stress at the wall of the pipe, and the velocity at 2 in from the centerline.

Solution

First determine whether the flow is laminar or turbulent:

Eq. (4.7):
$$V = \frac{4Q}{\pi D^2} = \frac{4(20)}{\pi(0.5)^2} = 10.19 \text{ fps}$$

Eq. (7.1):
$$\mathbf{R} = \frac{DV\rho}{\mu} = \frac{0.5(10.19)(0.9 \times 1.940)}{0.0008} = 11,120$$

Since $\mathbf{R} > 2000$, the flow is turbulent.

From Eq. (7.13):
$$f = \frac{h_f D(2g)}{LV^2} = \frac{25(0.5)2(32.2)}{200(10.19)^2} = 0.0388$$

From Eq. (7.43), $u_{max} = 10.19(1 + 1.326\sqrt{0.0388}) = 12.85$ fps   *ANS*

Eq (7.19): $\tau_0 = \dfrac{f\rho V^2}{8} = \dfrac{0.0388(0.9 \times 1.940)(10.19)^2}{8} = 0.878$ lb/ft^2 *ANS*

Eq. (7.37):
$$u_* = V\sqrt{\frac{f}{8}} = 10.19\sqrt{\frac{0.0388}{8}} = 0.709 \text{ fps}$$

Finally, from Eq. (7.40),

$$u_{2 \text{ in}} = u_{max} - 5.76(0.709)\log\left(\frac{3}{1}\right) = 12.85 - 1.948 = 10.90 \text{ fps}   \textit{ANS}$$

Note: $u_{max}/V = 12.85/10.19 = 1.261$. If the flow had been laminar, the velocity profile would have been parabolic and u_{max}/V would have been 2 (Sec. 7.8).

7.13   Pipe Roughness

Unfortunately, there is as yet no scientific way of measuring or specifying the roughness of commercial pipes. Several experimenters have worked with pipes with artificial roughness produced by various means so that the roughness could be measured and described by geometric factors, and they have proved that the friction depends not only on the size and shape of the projections but also on their distribution or spacing. We have much more to do before we completely solve this problem.

Noteworthy efforts in this direction were made in 1933 by a German engineer, J. Nikuradse, a student of Prandtl's. He coated several different sizes of pipe with sand grains that he had sorted by sieving into different grain sizes of reasonably uniform diameters. Let us represent the diameters of the sand grains by e, which is known as the ***absolute roughness***. In Sec. 7.5 dimensional analysis of pipe flow showed that for a smooth-walled pipe the friction factor f is a function of Reynolds number. A more general approach, including e as a parameter, reveals that $f = \phi(\mathbf{R}, e/D)$. The term e/D is known as the ***relative roughness***. In his experimental work Nikuradse had values of e/D ranging from 0.000985 to 0.0333.

In the case of artificial roughness such as this, the roughness is uniform, whereas in commercial pipes it is irregular both in size and in distribution. However, we may describe the irregular roughness of commercial pipe by a single e value, if we understand this means that the pipe has the same value of f at a high Reynolds number that it would have for a smooth pipe coated with sand grains of uniform size e. This "equivalent" grain size e must be close to the mean size of the irregular roughness elements.

Investigators working with pipes have found that if the thickness of the viscous sublayer $\delta_v > e$ (i.e., $eu_*/\nu < 5$), the viscous sublayer completely submerges the effect of e (see Sec. 7.11). Prandtl (Secs. 7.9 and 7.12), using information from Eq. (7.40) and data from Nikuradse's experiments, developed an equation for the friction factor for such a case:

Smooth-pipe flow:
$$\frac{1}{\sqrt{f}} = 2 \log \left(\frac{\mathbf{R}\sqrt{f}}{2.51} \right) \qquad (7.46)$$

This equation applies to turbulent flow in any pipe as long as $\delta_v > e$; when this condition prevails, the flow is known as ***smooth-pipe flow***. Since e does not appear in the equation, it is as though $e = 0$. Users have found the equation to be reliable for smooth pipes for all values of $\mathbf{R}$ over 4000. For such pipes, i.e., drawn tubing, brass, glass, etc., we can extrapolate it with confidence for values of $\mathbf{R}$ far beyond any present experimental values because it is functionally correct, assuming the wall surface is so smooth that the effects of the projections do not pierce the viscous sublayer, which becomes increasingly thinner with increasing $\mathbf{R}$. That this is so is evident from the fact that the formula yields a value of $f = 0$ for $\mathbf{R} = \infty$. This agrees with the facts, because $\mathbf{R}$ is infinite for a fluid of zero viscosity, and for such a case f must be zero.

Because of the way that f appears in two places in Eq. (7.46), it is ***implicit*** in f (see Appendix B) and hard to solve; we must use iteration, a graph of f versus $\mathbf{R}$, or an equation solver. However, as suggested by Colebrook,[12] we can approximate it by the ***explicit*** equation

---

[12] C. F. Colebrook, Turbulent Flow in Pipes, with Particular Reference to the Transition Region between the Smooth and Rough Pipe Laws, *J. Inst. Civil Engrs. (London)*, Vol. 11, February 1939.

Smooth-pipe flow:
$$\frac{1}{\sqrt{f}} = 1.8 \log\left(\frac{\mathbf{R}}{6.9}\right)$$
(7.47)

which differs from Eq. (7.35) by less than ±1.5% for $4000 \leq \mathbf{R} \leq 10^8$.

Blasius[13] has shown that for Reynolds numbers between 3000 and 10^5 we can approximately express the friction factor for a smooth pipe as

Blasius, smooth pipe,
$3000 \leq \mathbf{R} \leq 10^5$:
$$f = \frac{0.316}{\mathbf{R}^{0.25}}$$
(7.48)

Sometimes we can use this very conveniently to simplify equations. Blasius also found that over the same range of Reynolds numbers, the velocity profile in a smooth pipe closely agrees with the expression

Blasius, smooth pipe,
$3000 \leq \mathbf{R} \leq 10^5$:
$$\frac{u}{u_{max}} = \left(\frac{y}{r_0}\right)^{1/7}$$
(7.49)

where $y = r_0 - r$, the distance from the pipe wall. We call this equation the *seventh-root law* for turbulent-velocity distribution. Though it is not absolutely accurate, it is useful because it is easy to work with mathematically. At Reynolds numbers above 10^5 we must use an exponent somewhat smaller than $\frac{1}{7}$ to give good results.

At high Reynolds numbers δ_v becomes much smaller, and the roughness elements protrude through the viscous sublayer as in Fig. 7.9b. If $\delta_v < \frac{1}{14}e$ (i.e., $eu_*/\nu > 70$), investigators have found that the flow behaves as *fully-rough-pipe flow*, i.e., the friction factor is independent of the Reynolds number. For such a case, Theodor von Kármán (1881–1963) found that we can express the friction factor as

Fully-rough-pipe
flow (f_{min}):
$$\frac{1}{\sqrt{f}} = 2 \log\left(\frac{3.7}{e/D}\right)$$
(7.50)

The values of f from this equation correspond to the right-hand side of the Moody chart (Fig. 7.11), where the curves become horizontal. We sometimes refer to these values as f_{min}.

In the interval where $e > \delta_v > \frac{1}{14}e$ (i.e., $5 < eu_*/\nu < 70$) neither smooth flow Eq. (7.46) nor fully rough flow Eq. (7.50) applies. In 1939 Colebrook[14] combined Eqs. (7.46) and (7.50) to yield

Turbulent flow, all
pipes (Colebrook):
$$\frac{1}{\sqrt{f}} = -2 \log\left(\frac{e/D}{3.7} + \frac{2.51}{\mathbf{R}\sqrt{f}}\right)$$
(7.51)

---

[13] H. Blasius, Das Ähnlichkeitsgesetz bei Reibungsvorgängen in Flüssigkeiten, *Forsch. Gebiete Ingenieurw.*, Vol. 131, 1913.
[14] See Footnote 12.

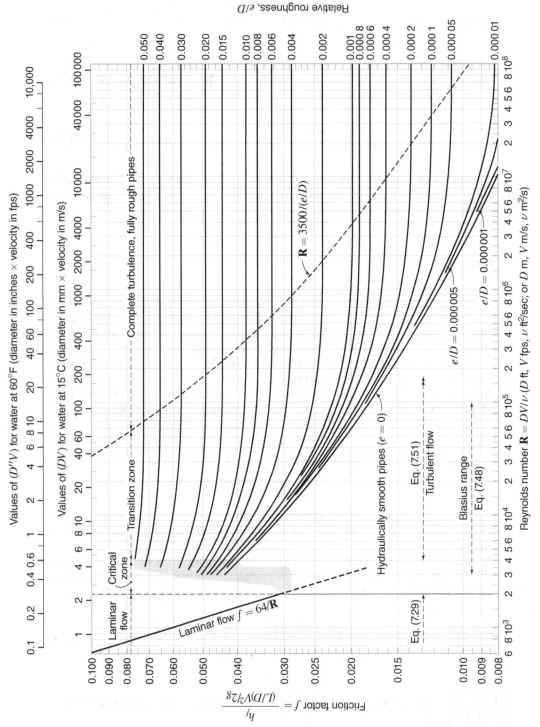

FIGURE 7.1.1   Moody chart for pipe friction factor (Stanton diagram). See Eq. (7.1) for **R**, Eq. (7.13) for f, Sec. 7.13 for e/D.

Besides providing a good approximation to conditions in the intermediate range, for $e = 0$ the Colebrook equation reduces to the smooth-pipe equation given by Eq. (7.46), and for large $\mathbf{R}$ it reduces to the fully-rough-pipe equation given by Eq. (7.50). Thus it applies to *all* turbulent flow conditions. Values of friction factor f that it predicts are generally accurate to within 10–15% of experimental data. This equation is so useful that engineers have long used it as the accepted design formula for turbulent flow; however, it has one major disadvantage. Like Eq. (7.46), it is implicit in f (Appendix B), which makes it inconvenient to use to manually evaluate f. In 1983 Haaland[15] combined Eqs. (7.47) and (7.50) to provide another approximation,

$$
\begin{matrix} \text{Turbulent flow,} \\ \text{all pipes (Haaland):} \end{matrix} \qquad \frac{1}{\sqrt{f}} = -1.8\log\left[\left(\frac{e/D}{3.7}\right)^{1.11} + \frac{6.9}{\mathbf{R}}\right] \qquad (7.52)
$$

which has the advantage of being explicit in f; it has the same asymptotic behavior as Eq. (7.51), from which it differs by less than $\pm 1.5\%$ for $4000 < \mathbf{R} < 10^8$. Many others have developed alternative explicit approximations for f that vary in complexity, most showing accuracy comparable to the Haaland equation.[16]

7.14   Chart for Friction Factor (Moody Diagram)

The preceding equations for f have been very inconvenient to use in a number of circumstances, which we will discuss further in coming sections, and this was especially true before Haaland's equation appeared. The inconvenience was largely overcome by reading numerical values from a chart (Fig. 7.11), prepared by Moody[17] in 1944. The chart, often called the *Moody diagram,*[18] is based on the best information available, and was plotted with the aid of Eqs. (7.29) and (7.51). All the quantities involved are dimensionless, so the chart may be used for both BG and SI unit systems. For convenience, BG values of DV (diameter times velocity) for water at 60°F and similar SI values for water at 15°C are given across the top of the chart to save the need to compute Reynolds number for those common cases.

The Moody chart, and the various flow conditions that it represents, divides into four zones: the *laminar-flow zone;* a *critical zone* where values are uncertain because the flow might be either laminar or turbulent; a *transition zone,* where f is a function of both Reynolds number and relative pipe roughness; and a *zone of complete turbulence* (fully-rough-pipe flow), where the value of f is independent of Reynolds number and depends solely upon the relative roughness, e/D.

There is no sharp line of demarcation between the transition zone and the zone of complete turbulence. The dashed line of Fig. 7.11 that separates the two zones was

---

[15] S. E. Haaland, Simple and Explicit Formulas for the Friction Factor in Turbulent Pipe Flow, *J. Fluids Eng.,* Vol. 105, March 1983.

[16] Brkić, D., Review of Explicit Approximations to the Colebrook Relation for Flow Friction, *J. Pet. Sci. Eng.,* 77(1), 34–48, 2011.

[17] L. F. Moody, Friction Factors for Pipe Flows, *ASME Trans.,* Vol. 66, 1944. Lewis F. Moody (1880–1953), an eminent American engineer and professor, also contributed greatly to our understanding of similitude and cavitation as applied to hydraulic machinery.

[18] Figure 7.11 is also called a *Stanton diagram,* because Stanton first proposed such a plot.

suggested by R. J. S. Pigott; the equation of this line is $\mathbf{R} = 3500/(e/D)$. On the right-hand side of the chart the given values of e/D correspond to the curves and *not* to the grid. Note how their spacing varies. The lowest of the curves in the transition zone is the smooth-pipe ($e = 0$) curve given by Eqs. (7.47) and (7.48); notice how many of the other curves blend asymptotically into the smooth-pipe curve. The use of the Moody diagram (Fig. 7.11) in solving pipe flow problems is illustrated in the sample problems in Secs. 7.16–7.18.

7.15   Single-Pipe Flow: Solution Basics

The methods of solution we will use for single-pipe flow are extensions of the method summarized in Sec. 5.14. We recommend that you review that section before proceeding, together with Secs. 5.11 and 5.13, which discuss the energy line and the hydraulic grade line.

Governing Equations

Four simultaneous equations govern flow at a point in a single pipe. Three of these are the equations of continuity [Eq. (4.7)], energy loss [Eq. (7.13)], and the Reynolds number [Eq. (7.1)]:

$$V = \frac{4Q}{\pi D^2}; \qquad h_f = f\frac{L}{D}\frac{V^2}{2g}; \qquad \mathbf{R} = \frac{DV}{\nu}$$

If the flow is turbulent, as is most common, the fourth is the Colebrook Eq. (7.51) or the Haaland Eq. (7.52) for the friction factor, as just presented in Sec. 7.13; if the flow is laminar, which is more rare, the friction factor is instead given by $f = 64/\mathbf{R}$, Eq. (7.29). The four unknowns are usually f, $\mathbf{R}$, D, or h_f, and Q or V. If the pipeline is uniform in size, shape, and roughness, then these equations will yield the same results at all points along it.

In some cases we will need h_f as a known quantity, and if it is not given we can often obtain it by rearranging energy Eq. (5.28) to first find the total head loss h_L as follows:

$$h_L = H_1 - H_2 = \left(\frac{p}{\gamma} + z + \frac{V^2}{2g}\right)_1 - \left(\frac{p}{\gamma} + z + \frac{V^2}{2g}\right)_2$$

$$= \frac{\Delta p}{\gamma} + \Delta z + \frac{V_1^2 - V_2^2}{2g}$$

If, for example, points 1 and 2 are in contact with the atmosphere, $p_1 = p_2 = \Delta p = 0$; if they are also on reservoir water surfaces, $V_1 = V_2 = 0$, so that $h_L = \Delta z$. If they are in a pipeline of constant diameter, $V_1 = V_2$. As we noted in Sec. 5.3, the pipe friction head loss h_f must be either equal to or less than h_L; the difference results from what are known as minor losses, which are often negligible, and which we will discuss more fully in Secs. 7.21–7.27.

Solution of Special Cases

Laminar Flow.   For laminar flow (see Fig. 7.11) we know from Eq. (7.29) that $f = 64/\mathbf{R}$. Substituting this into pipe-friction Eq. (7.13) to eliminate f, and using Eq. (7.1) for $\mathbf{R}$, yields Eq. (7.28). This equation is explicit (Appendix B) in each variable of interest, h_f, D, and V, so we can solve directly for any one of them. We must remember to check the validity of the solution we obtain by confirming that **R** is in the laminar range (normally < 2000, see Sec. 7.3).

Smooth-Pipe Flow. When e is so small that the roughness elements do not extend through the viscous sublayer or zone (i.e., $e < \delta_\nu$, see Sec. 7.11), so that the pipe behaves as a smooth pipe (Fig. 7.9a) even though most of the flow is turbulent, this is represented by the lowest curve on Fig. 7.11, for which $e = 0$. From Sec. 7.12 we know we may express f along this curve by the Blasius equation given by Eq. (7.48) provided 3000 $\leq$ **R** $\leq 10^8$. Combining Eq. (7.48) with pipe-friction Eq. (7.13) to eliminate f, and using Eq. (7.1) for **R**, yields

Smooth pipe,
$3000 \leq$ **R** $\leq 10^5$:
$$h_f = 0.1580\nu^{0.25}\frac{LV^{1.75}}{gD^{1.25}}$$
(7.53)

This equation has the advantage of being explicit in each variable of interest, h_f, D, or V. We must remember to check the validity of the solution it yields by confirming that **R** is in the required Blasius range.

Fully-Rough-Pipe Flow. When e is so large that the roughness elements extend through both the viscous zone and the transition zone (i.e., $e > 14\delta_\nu$; see Fig. 7.8), so that there is complete turbulence and the flow is fully rough (Secs. 7.11 and 7.14), this is represented by the horizontal lines at the right of Fig. 7.11, where **R** $> 3500/(e/D)$. Then $f = f_{min} =$ constant is given by Eq. (7.50), which when rearranged gives

Fully-rough-
pipe
$$f = f_{min} = \left[2\log\left(\frac{3.7}{e/D}\right)\right]^{-2}$$
(7.54)

Substituting this into pipe-friction Eq. (7.13) to eliminate f, yields

Fully-rough-
pipe (f_{min}):
$$h_f = \frac{1}{\left[2\log\left(\frac{3.7}{e/D}\right)\right]^2}\frac{L}{D}\frac{V^2}{2g}$$
(7.55)

We notice that for fully-rough-pipe flow h_f is directly proportional to V^2 and is independent of **R**. Equation (7.55) is explicit in h_f and V, but implicit in D (see Appendix B). The solution of implicit equations is discussed in Secs. 7.16–7.18 and 7.28. Values of e are given in Table 7.1. We must check the validity of a solution we obtain from this equation by confirming that **R** is sufficiently large.

In engineering applications, flows in any of these three special cases occur relatively rarely. In Fig. 7.11 we see that the only flow region not covered by these three cases is the transition zone. In addition to the fact that most engineering flows occur in the transition zone, flow solutions for this zone are more involved. We will discuss various ways to solve such flows, and flows in which the regimen (zone) is initially unknown, in Secs. 7.16–7.20 and 7.28.

Types of Single-Pipe Flow Problems. As we have already noted, often some of the equations governing pipe flow are implicit in form (Appendix B), and therefore they do not

Material	Feet	Millimeters
Glass, plastic (smooth)	0.0	0.0
Drawn tubing, brass, lead, copper, centrifugally spun cement, bituminous lining, transite	0.000005	0.0015
Commercial steel, wrought iron, welded-steel pipe	0.00015	0.046
Asphalt-dipped cast iron	0.0004	0.12
Galvanized iron	0.0005	0.15
Cast iron, ductile iron, average (unlined)	0.00085	0.25
PVC	0.000005	0.0015
Wood stave	0.0006–0.003	0.18–0.9
Concrete	0.001–0.01	0.3–3
Riveted steel	0.003–0.03	0.9–9

Note: $\dfrac{e}{D} = \dfrac{e \text{ in ft}}{D \text{ in ft}} = 12 \times \dfrac{e \text{ in ft}}{D \text{ in inches}} = \dfrac{e \text{ in mm}}{D \text{ in mm}}$

TABLE 7.1   Values of absolute roughness e for new commercial pipes

readily permit direct solution. For example, although Haaland's equation given by Eq. (7.52) is explicit in f, it is still implicit in V and D, since these are involved in both **R** and in f through Eq. (7.13). The classical way to overcome this difficulty has been to use trial-and-error procedures in combination with the Moody chart described in Sec. 7.14. With the availability of the Haaland equation and similar explicit approximations and the wide availability of equation solvers, we can now also perform these procedures using equations alone. We will explore both methods in this and subsequent sections, including applications to different types of problems.

To solve any problem (other than laminar flow or smooth-pipe flow) using the friction factor f, we will require a value of e, and we can obtain this from Table 7.1. As the ratio of e/D is dimensionless, we may use any units, provided they are the same for both e and D. Note also that we should use exact values of the internal pipe diameter. Exact diameters often differ from the nominal sizes, particularly for smaller pipes (see Table A.9). If the pipe is not of circular cross section, we can replace D in all instances by $4R_h$, as we noted in Sec. 7.7.

With regard to the values of e, note that those given in Table 7.1 are for new, clean pipes, and even in such cases there may be considerable variation in the values. Consequently, in practical cases, the value of f may be in error by ±5% for smooth pipes and by ±10% for rough ones. For old pipes, values of e may be much higher, but there is much variation in the degree with which pipe roughness increases with age, since so much depends on the nature of the fluid transported. In addition, deposits in small pipes materially reduce the internal diameter. The effect of the roughness of pipe joints may also increase the value of f substantially. So we must use judgment in estimating a value of e, and consequently of f. When solving pipe flow problems, it is important to have some idea what effects changes in e will have on h_f, Q, and D. From Eqs. (7.13) and (4.7), it follows that $h_f \propto f$, $Q \propto f^{-1/2}$, and $D \propto f^{1/5}$. This is discussed in the text following Sample Problem 7.4.

Most engineering problems fall within the transition zone (see Fig. 7.11). When solving single-pipe flow problems in this zone, where conditions depend on both e/D

and the Reynolds number, we need a definite value of $\mathbf{R}$. But when V or D is unknown, so is $\mathbf{R}$. The solution procedure varies with the type of problem, and we can categorize most single-pipe flow problems into one of the following three types:

Type	Find	Given	
1. Head-loss problem	h_f	D, Q or V,	and g, L, e, ν
2. Discharge problem	Q or V	D, h_f	and g, L, e, ν
3. Sizing problem	D	Q, h_f	and g, L, e, ν

Here g is the gravitational acceleration, L is the pipe length, Q is the flow rate, and ν is the kinematic viscosity. Of course, we may know μ and ρ instead of ν, in which case we may obtain ν from Eq. (2.11).

Experience has shown that new users of the Moody diagram (Fig. 7.11) frequently misread it. This most probably occurs because none of the scales are linear, and because the intervals between grid lines and chart curves keep changing. Take special care in reading the chart, and confirm any interpolated values by comparing them with nearby grid values or curves in both directions.[19]

Note that in all cases where iterations are used we should assume an f value whenever possible, rather than some other variable. This is because f varies the least, so trial calculations are less sensitive to its changes, and as a result they converge faster.

7.16   Single-Pipe Flow: Solution of Type 1 Problems (Head Loss Unknown)

Type 1 problems are the simplest. If Q is given, we can easily obtain V from continuity Eq. (4.7). Then we can obtain the Reynolds number and e/D from the given data.

Solving with the Moody diagram. With known values for e/D and $\mathbf{R}$, we enter Fig. 7.11 vertically with $\mathbf{R}$ and along a curve (possibly interpolated), following the bend of the curves, for e/D from the right, to identify an operating point for which we can read off the friction factor f horizontally to the left. With this value of f, we can directly compute the friction head loss h_f from Eq. (7.13).

Solving with equations.[20] To solve using equations most directly, we can compute f directly from Eq. (7.52) and find h_f using Eq. (7.13), as demonstrated in Sample Problem 7.4(c).

Alternatively, we can use the Colebrook Eq. (7.51) using trial and error or an equation solver. The latter approach benefits from rearranging Eq. (7.51) into the form

$$\frac{1}{\sqrt{f}} + 2\log\left(\frac{e/D}{3.7} + \frac{2.51}{\mathbf{R}\sqrt{f}}\right) = 0$$

---

[19] Charts involving the same functional relations have been plotted with different coordinates from those in Fig. 7.11 and may be more convenient for certain specific purposes, but we believe that the form shown is best both for purposes of instruction and for general use.

[20] Direct solutions with equations are particularly useful for more complex problems, such as when including minor losses (Sec. 7.28) and systems with multiple pipes (Secs. 7.30–7.33).

and graphing the left side of the equation versus f to see where the solution exists. Any equation solver can also rapidly produce a solution. Given a reasonable first guess of f, trial and error solutions generally converge rapidly. As with Moody diagram solutions, knowing f allows direct computation of the friction head loss h_f from Eq. (7.13).

Sample Problem 7.4

A 20-in-diameter galvanized iron pipe 2 miles long carries 4 cfs of water at 60°F. Find the friction head loss: (a) using Fig. 7.11 and the Reynolds number; (b) using Fig. 7.11 and its $D''V$ scale; (c) using only a basic scientific calculator, without Fig. 7.11.

Solution

This is a Type 1 problem, to find h_f. From Table 7.1 for galvanized iron:

$$e = 0.0005 \text{ ft:} \quad \text{so} \quad e/D = 0.0005(12)/20 = 0.0003$$

$$L = 2 \text{ mi}(5280 \text{ ft/mi}) = 10,560 \text{ ft}$$

Eq. (4.7):
$$V = \frac{4Q}{\pi D^2} = \frac{4(4)}{\pi(20/12)^2} = 1.833 \text{ fps}$$

Table A.1 for water at 60°F: $\nu = 1.217 \times 10^{-5} \text{ ft}^2/\text{sec}$

$$\mathbf{R} = \frac{DV}{\nu} = \frac{(20/12)1.833}{1.217 \times 10^{-5}} = 2.51 \times 10^5 \quad (> \mathbf{R}_{crit}, \text{ i.e., flow is turbulent})$$

(a)   Enter Fig. 7.11 at the right-hand side with $e/D = 0.0003$, by interpolating between 0.0002 and 0.0004; note that the e/D spacing varies. Follow this (unplotted) e/D curve to the left until it crosses a vertical line at $\mathbf{R} = 2.51 \times 10^5$ (*note:* this is between 10^5 and 10^6). For this operating point, reading horizontally to the left, $f = 0.0172$.

Eq. (7.13):
$$h_f = f \frac{L}{D} \frac{V^2}{2g} = 0.0172 \frac{(10560)}{(20/12)} \frac{1.833^2}{2(32.2)} = 5.69 \text{ ft} \quad ANS$$

(b) $D''V = 20(1.833) = 36.7$. Find this value on the scale across the top of Fig. 7.11; note that this scale is varying. Find where the (interpolated) curve for $e/D = 0.0003$ crosses the vertical line at $D''V = 36.7$. From this point, read horizontally to the left, to find $f = 0.0172$. Compute h_f as for part (a).

Note: From the operating point on Fig. 7.11, we see that flow conditions are in the transition zone of turbulent flow, which is typical.

(c)   Eq.(7.41):
$$\frac{1}{\sqrt{f}} = -1.8 \log \left[\left(\frac{0.0003}{3.7} \right)^{1.11} + \frac{6.9}{2.51 \times 10^5} \right] = 7.65$$

from which
$$f = 0.01709$$

Eq. (7.10):
$$h_f = 0.01709 \frac{10560}{(20/12)} \frac{(1.833)^2}{2(32.2)} = 5.65 \text{ ft} \quad ANS$$

Note: This differs from answer (a) by only 0.04 ft, or 0.70%.

We can demonstrate with Sample Problem 7.4, if e had been 20% larger then the h_f would have been 2.3% larger. This change in the head loss would be larger for larger f (or larger e/D), and vice versa. Percentage changes in Q and D will be smaller than those in h_f.

7.17 Single-Pipe Flow: Solution of Type 2 Problems (Q or V Unknown)

Solving with the Moody diagram. In *Type 2* problems, because V is unknown, the Reynolds number $\mathbf{R}$ is not known at the outset and so a direct solution using the Moody diagram is not possible. However, we notice in Fig. 7.11 that the value of f changes very slowly with large changes in $\mathbf{R}$. So we can usually solve the problem quite effectively by assuming an initial value of f, and then obtaining the final solution by successive trials (trial and error; see Sample Problem 3.8). Since D and hence e/D is known for a Type 2 problem, the fully-rough-pipe f value (f_{min}) given on the right-hand side of Fig. 7.11 or by Eq. (7.54) provides a good starting point. Each succeeding trial is started with the f value obtained from the previous trial. The value of f can be considered close enough when the first three significant figures of the required answer (Q or V) no longer change. Note that if we assume values of some other variable besides f, convergence is usually much slower.

It is convenient for *Type 2* problems to rearrange Eq. (7.13) into the form $V = K/\sqrt{f}$, where $K = \sqrt{2gDh_f/L}$ is known. Assuming an f (as just discussed) therefore yields a V, which enables us to calculate $\mathbf{R}$ and enter the diagram or to use Eq. (7.51) or (7.52) to obtain an improved value of f. If this is different from the assumed f, we must repeat the procedure assuming the just-obtained value, and successively repeat it until the two values converge. This usually only requires two or three trials, by which time all the values are correct, including the required value of V.

Solving with Equations. We can obtain a single equation for the direct calculation of velocity (or discharge) as follows. First, we rearrange Eq. (7.13) to obtain

$$\frac{1}{\sqrt{f}} = V\sqrt{\frac{L}{2gDh_f}}$$

Then, when we substitute this and $\mathbf{R} = DV/\nu$ into the Colebrook equation given by Eq. (7.51) and rearrange, we get

Turbulent flow, Type 2:
$$V = -2\sqrt{\frac{2gDh_f}{L}}\ \log\left(\frac{e/D}{3.7} + \frac{2.51\nu}{D}\sqrt{\frac{L}{2gDh_f}}\right) \qquad (7.56a)$$

This equation is explicit in V, which fortunately results because the Vs from $\mathbf{R}$ and $\sqrt{f}$ cancel in the last term of Eq. (7.51) [the same substitution into the Haaland equation given by Eq. (7.52) does not produce such a desirable result]. Having this explicit equation available will save much repetitive work in solving more involved problems, like those of branching pipes (Sec. 7.30). If Q is required rather than V, we can use continuity equation given by Eq. (4.7) to eliminate V, and Eq. (7.56a) becomes

Turbulent flow, Type 2:
$$Q = -2.221D^2\sqrt{\frac{gDh_f}{L}}\ \log\left(\frac{e/D}{3.7} + \frac{1.784\nu}{D}\sqrt{\frac{L}{gDh_f}}\right) \qquad (7.56b)$$

Using Eq. (7.56), we can calculate Q or V directly with a basic scientific calculator. Remember to use Eq. (7.1) to confirm that $\mathbf{R}$ is in the turbulent range. If $\mathbf{R}$ turns out to be in the laminar range, i.e., less than 2000, we must instead find V or Q from Eq. (7.28) rearranged.

Sample Problem 7.5

Water at 20°C flows in a 500-mm-diameter welded-steel pipe. If the friction loss gradient is 0.006, determine the flow rate: (*a*) using the Moody diagram (Fig. 7.11); (*b*) using only a basic scientific calculator, without Fig. 7.11.

Solution

This is a Type 2 problem, to find Q.

Table 7.1 for welded steel: $e = 0.046$ mm; $e/D = 0.046/500 = 0.000\,092$

Table A.1 at 20°C: $\nu = 1.003 \times 10^{-6}$ m²/s; $h_f/L = 0.006$ is given

Eq. (7.14):

$$S = \frac{h_f}{L} = \frac{f}{D}\frac{V^2}{2g}, \quad \text{i.e.,} \quad 0.006 = \frac{fV^2}{0.5(2)9.81}$$

from which $V = \dfrac{0.243}{f^{1/2}}$.

(*a*) Fig. 7.11 for $e/D = 0.000\,092$: $f_{min} \approx 0.0118$

Try $f = 0.0118$. Then $V = 0.243/(0.0118)^{1/2} = 2.23$ m/s.

Eq. (7.1):

$$\mathbf{R} = \frac{DV}{\nu} = \frac{0.5(2.23)}{1.003 \times 10^{-6}} = 1.114 \times 10^{6} \quad \text{(turbulent flow)}$$

Figure 7.11 with $e/D = 0.000\,092$ and $\mathbf{R} = 1.114 \times 10^{6}$: $f = 0.0131$. Assumed and obtained f values are different, so we must try again; use the obtained f for the next trial. Tabulating all the trials:

Try f	V, m/s	$\mathbf{R}$	Obtained f	
0.0118	2.23	1.114×10^6	0.0131	Try again
0.0131	2.12	1.059×10^6	0.0131	Converged!

The f values now agree, so we have the true operating point. Convergence is rapid!

$$Q = AV = (\pi/4)D^2V = (\pi/4)(0.5)^2 2.12 = 0.416 \text{ m}^3/\text{s} \quad \textit{ANS}$$

Caution: Take great care to read Fig. 7.11 correctly.

(*b*) Eq. (7.54) for $e/D = 0.000\,092$: $f_{min} = 0.011\,79$. Calculate V and $\mathbf{R}$ as in (*a*), then obtain an improved f from Eq. (7.52). Use the obtained f for the next trial. Tabulating all the trials:

Try f	V, m/s	$\mathbf{R}$	Obtained f	
0.011 79	2.23	1.114×10^6	0.013 09	Try again
0.013 09	2.12	1.057×10^6	0.013 15	Converged!

$$Q = AV = (\pi/4)(0.5)^2 2.12 = 0.416 \text{ m}^3/\text{s} \quad \textit{ANS}$$

Sample Problem 7.6
Solve Sample Problem 7.5 without trial and error, using equations rather than Fig. 7.11.

Solution
This is a Type 2 problem, to find Q. As in Sample Problem 7.5, $D = 0.5$ m, $h_f/L = 0.006$, $e/D = 0.000092$, and $\nu = 1.003 \times 10^{-6}$ m²/s. The quantity

$$\sqrt{\frac{2gDh_f}{L}} = \sqrt{2(9.81)0.5(0.006)} = 0.243 \text{ m/s},$$

so in turbulent-flow Eq. (7.56a):

$$V = -2(0.243)\log\left[\frac{0.000092}{3.7} + \frac{2.51(1.003 \times 10^{-6})}{0.5}\frac{1}{0.243}\right] = 2.11 \text{ m/s}$$

Check: $\mathbf{R} = \dfrac{DV}{\nu} = \dfrac{0.5(2.11)}{1.003 \times 10^{-6}} = 1.050 \times 10^6.$ This is $> \mathbf{R}_{crit} = 2000$

so flow *is* turbulent, validating the use of Eq. (7.56a). Finally,

$$Q = AV = (\pi/4)D^2V = 0.25\pi(0.5)^2 2.11 = 0.414 \text{ m}^3/\text{s} \qquad \textit{ANS}$$

7.18 Single-Pipe Flow: Solution of Type 3 Problems (D Unknown)

For *Type 3* problems, since D is unknown, neither e/D nor $\mathbf{R}$ are known initially, complicating the approaches to solving for D. As noted in Sec. 7.15, we must recognize that D is a theoretical (inside) diameter of pipe needed; only certain diameter pipes are commercially manufactured (Table A.9), so design recommendations should specify an appropriate diameter and check the system performance with that.

Solving with the Moody Diagram. Since D is not known, a value near the middle of the f range on Fig. 7.11, such as 0.03, makes a good initial guess. We substitute $V = 4Q/(\pi D^2)$ into Eq. (7.13) and rearrange it to obtain $D = (fK)^{1/5}$, where $K = 8LQ^2/(\pi^2gh_f)$ is known. Then assuming an f (as previously discussed) yields a D, which enables us to calculate e/D and $\mathbf{R}$, and determine a new estimate of f from the Moody diagram, proceeding similarly to the approach for Type 2. As with Type 2, the value of f can be considered close enough when the first three significant figures of the required answer (D) no longer change.

Solving with Equations. While no exact explicit formulation of the Colebrook and Darcy-Weisbach equations is possible for Type 3 problems, using an approximation we can eliminate iteration by reformulating our equations as follows. Substituting from Eqs. (7.13), (4.7), and (7.1) into the dimensionless quantity

$$\mathbf{N}_1 = f\mathbf{R}^5 = \frac{h_f D 2g}{L}\left(\frac{\pi D^2}{4Q}\right)^2 \left(\frac{D}{\nu}\frac{4Q}{\pi D^2}\right)^5 = \frac{128g}{\pi^3}\frac{h_f}{L}\frac{Q^3}{\nu^5} \qquad (7.57)$$

indicates that it is independent of the unknown D, so that N_1 is a known quantity. We can also transform the relative roughness into another known dimensionless quantity by dividing it by R, i.e.,

$$N_2 = \frac{e/D}{R} = \frac{e}{D}\left(\frac{\nu}{DV}\right) = \frac{e}{D}\left(\frac{\nu}{D}\right)\frac{\pi D^2}{4Q} = \frac{\pi e \nu}{4Q} \tag{7.58}$$

Now we can use N_1 to eliminate f from laminar flow Eq. (7.29) to obtain

Laminar flow, Type 3:
$$R = \left(\frac{N_1}{64}\right)^{0.25} \tag{7.59}$$

and we can use N_1 and N_2 to eliminate f and e/D from the Colebrook equation given by Eq. (7.51) for turbulent flow to obtain

$$R^{2.5} = -2N_1^{0.5} \log\left[\frac{N_2 R}{3.7} + \frac{2.51}{N_1^{0.5}}R^{1.5}\right] \tag{7.60}$$

Here R occurs in three places, so this equation is strongly implicit (Appendix B), and Eq. (7.60) appears to be not a very useful result. However, if we plot N_1 versus R we find it collapses the various flow curves on the Moody chart into a very narrow band that is closely approximated by the formula $R \approx 1.43 N_1^{0.208}$. We can substitute this into the right-hand side of Eq. (7.60) to create the more useful equation:

Turbulent flow Type 3:
$$R^{2.5} = -2N_1^{0.5} \log\left[\frac{N_2 N_1^{0.208}}{2.59} + \frac{4.29}{N_1^{0.188}}\right] \tag{7.61}$$

We see that this reformulation has converted the Colebrook equation into an explicit form.

Comparison of the two Reynolds numbers from Eqs. (7.59) and (7.61) with $R_{crit} = 2000$ indicates which equation (laminar or turbulent) is applicable for a given Type 3 problem.

Finally, from Eq. (7.1) combined with Eq. (4.7), we have

$$R = \frac{DV}{\nu} = \frac{D}{\nu}\left(\frac{4Q}{\pi D^2}\right) = \frac{4Q}{\pi \nu D}$$

and so, by rearrangement, we can obtain the required diameter D from

Type 3:
$$D = \frac{4Q}{\pi \nu R} \tag{7.62}$$

Other explicit approximations have also been developed of varying complexity and accuracy, a common one being that of Swamee and Jain,[21] limited to $5000 < R < 3 \times 10^8$

[21] Swamee P. K. and Jain A. K., Explicit Equations for Pipe Flow Problems, *J. Hydraul. Eng. ASCE*, 102(5):657–664, 1976.

and $10^{-6} < e/D < 10^{-2}$, which provides diameters within 5% of the direct solution of the Colebrook-White equation [Eq. (7.51)]:

$$D = 0.66 \left[e^{1.25} \left(\frac{LQ^2}{gh_f} \right)^{4.75} + \nu Q^{9.4} \left(\frac{L}{gh_f} \right)^{5.2} \right]^{0.04}$$

In addition, an equation solver can be used to obtain an exact solution to Eq. (7.56). Solving for D in this manner eliminates the need for approximations or iterations. As noted in Sec. 7.13 the Colebrook-White equation [Eq. (7.51)] and subsequently the Moody diagram (Fig. 7.11) are accurate to within 10–15% of experimental data, so we can justify the use of reasonably accurate approximations.

Sample Problem 7.7

A galvanized iron pipe 18,000 ft long must convey ethyl alcohol ($\nu = 2.3 10^{-5}$ ft^2/sec) at a rate of 135 gpm. If the friction head loss must be 215 ft, determine the pipe size theoretically required: (*a*) using the Moody diagram (Fig. 7.11); (*b*) using only a basic scientific calculator, without Fig. 7.11.

Solution

This is a Type 3 problem, to find D.

Appendices: $Q = 135$ gpm (2.23 cfs/1000 gpm) $= 0.301$ cfs

Table 7.1 for galvanized iron: e = 0.0005 ft.

$$\frac{e}{D} = \frac{0.0005}{D}$$

Eq. (4.7): $V = \dfrac{4Q}{\pi D^2} = \dfrac{4(0.301)}{\pi D^2} = \dfrac{0.383}{D^2}$

Eq. (7.1): $R = \dfrac{DV}{\nu} = \dfrac{D}{2.3 \times 10^{-5}} \left(\dfrac{0.383}{D^2} \right) = \dfrac{16,590}{D}$

Eq. (7.13): $215 = h_f = f \dfrac{18,000}{2D(32.2)} \left(\dfrac{0.383}{D^2} \right)^2$, from which $D = 0.718 f^{1/5}$

(*a*) Start by assuming a mid-range value of f

Try f	D, ft	e/D	R	Chart f	
0.0300	0.356	0.001 404	4.68×10^4	0.0253	Try again
0.0253	0.344	0.001 453	4.84×10^4	0.0253	Converged!

Values of f now agree, so we have the true operating point. Convergence is rapid!

$$D = 0.344 \text{ ft} = 4.13 \text{ in} \quad \textit{ANS}$$

(b) Start by assuming a mid-range value of f. Calculate V and $\mathbf{R}$ as before, then obtain an improved f from Eq. (7.52). Tabulating this and subsequent trials:

Try f	D, ft	e/D	$\mathbf{R}$	Eq. (7.52) f	
0.0300	0.356	0.001 404	46,800	0.025 01	Try again
0.0250	0.343	0.001 456	48,500	0.025 04	Converged!

The $\mathbf{R}$ value exceeds the minimum applicable value so Eq. (7.52) applies. If values below $\mathbf{R}_{crit}$ [Eq. (7.2)] occur, refer to Sec. 7.15 under *Laminar Flow*.

$$D = 0.343 \text{ ft} = 4.12 \text{ in} \quad \textbf{ANS}$$

Sample Problem 7.8
Solve Sample Problem 7.7 without trial and error, using equations.

Solution
This is a Type 3 problem, to find D. As in Sample Problem 7.7, $h_f = 215$ ft, $L = 18,000$ ft, $Q = 0.301$ cfs, $e = 0.0005$ ft, $g = 32.2$ ft/sec^2, and $\nu = 2.3 \times 10^{-5}$ ft^2/sec.

Eq. (7.57):
$$N_1 = \frac{128(32.2)}{\pi^3}\left(\frac{215}{18,000}\right)\frac{(0.301)^3}{(2.3 \times 10^{-5})^5} = 6.71 \times 10^{21}$$

Eq. (7.58):
$$N_2 = \frac{\pi e \nu}{4Q} = \frac{\pi(0.0005)2.3 \times 10^{-5}}{4(0.301)} = 3.00 \times 10^{-8}$$

Laminar flow Eq. (7.59):
$$\mathbf{R} = \left(\frac{N_1}{64}\right)^{0.25} = 101,300 > \mathbf{R}_{crit} = 2000$$

Turbulent-flow Eq. (7.61):
$$\mathbf{R}^{2.5} = -2(6.73 \times 10^{21})^{0.5}\log\left[\frac{3.0 \times 10^{-8}(6.71 \times 10^{21})^{0.208}}{2.59} + \frac{4.29}{(6.71 \times 10^{21})^{0.188}}\right]$$
$$= 5.13 \times 10^{11}$$

from which, $\mathbf{R} = 48,300$. Clearly, from the $\mathbf{R}$ values, the flow is turbulent, so the laminar flow equation is invalid. Thus, from Eq. (7.62),

$$D = \frac{4Q}{\pi \nu \mathbf{R}} = \frac{4(0.301)}{\pi(2.3 \times 10^{-5})48,300} = 0.345 \text{ ft} \quad \textbf{ANS}$$

Note: This differs by 0.000 55 ft or 0.16% from the accurate answer in Sample Problem 7.9.

: Programmed computing aids could help solve problems marked with this icon.

Sample Problem 7.9

Solve Sample Problem 7.8 using an equation solver.

Solution

As in Sample Problem 7.8, $h_f = 215$ ft, $L = 18,000$ ft, $Q = 0.301$ cfs, $e = 0.0005$ ft, $g = 32.2$ ft/sec^2, and $\nu = 2.3 \times 10^{-5}$ ft^2/sec.

Rearrange Eq. (7.56b) for use with an equation solver:

$$Q + 2.221D^2 \sqrt{\frac{gDh_f}{L}} \log \left(\frac{e/D}{3.7} + \frac{1.784\nu}{D} \sqrt{\frac{L}{gDh_f}} \right) = 0$$

Using known values for all variables except D, instruct the solver to change values of D until the equation is satisfied, and a root is found.

Result: $\qquad\qquad\qquad D = 0.344$ ft $\qquad\qquad\qquad$ *ANS*

While not necessary, a graph of the left side of the equation versus D can help envision the convergence toward a solution.

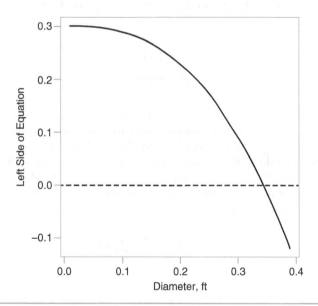

FIGURE S7.9

As always, check the **R** value to ensure the equation applies:

$$\mathbf{R} = \frac{4Q}{\pi\nu D} = \frac{4(0.301)}{\pi(2.3 \times 10^{-5})(0.344)} = 48,400$$

The **R** value exceeds the minimum applicable value so Eq. (7.56) applies. If values below $\mathbf{R}_{crit}$ [Eq. (7.2)] occur, refer to Sec. 7.15 under *Laminar Flow*.

7.19   Empirical Equations for Single-Pipe Flow

The presentation of friction loss in pipes given in Secs. 7.1–7.18 incorporates the best knowledge available on this subject, as far as application to Newtonian fluids (Sec. 2.11) is concerned. Admittedly, however, the trial-and-error type of solution, use of equation solvers, or direct use of complex direct or approximate solutions, especially when encumbered with computations for relative roughness and Reynolds number, have been challenging, especially prior to the wide availability of powerful computational tools. It is natural, therefore, that engineers developed simple and convenient-to-use design formulas, based on experiments and observations but limited to specific fluids and conditions. Such equations, in which the relations between the primary variables of interest (here V, R_h or D, and $S = h_f/L$) are based on observations rather than theory, are known as *empirical equations* (Appendix B). This is in contrast to the Darcy-Weisbach Eq. (7.13), for which we saw in Secs. 7.5 and 7.6 that these relations were developed from theory, primarily dimensional analysis; the fact that the roughness coefficients were determined experimentally does not affect the way in which the variable relations were developed.

Perhaps the best example of such an empirical equation is that of **Hazen and Williams**, applicable only to the flow of water in pipes larger than 2 in (50 mm) and at velocities less than 10 fps (3 m/s), but widely used in the waterworks industry. This formula takes the form

BG units: $$V = 1.318 C_{HW} R_h^{0.63} S^{0.54} \qquad\qquad (7.63a)$$

SI units: $$V = 0.849 C_{HW} R_h^{0.63} S^{0.54} \qquad\qquad (7.63b)$$

where R_h (ft or m) is the hydraulic radius (Sec. 7.3), and $S = h_f/L$, the energy gradient. The advantage of Eq. (7.63) over the standard pipe-friction formula is that the roughness coefficient C_{HW} is not a function of the Reynolds number, and so we need not make trial solutions. Values of C_{HW} range from 140 for very smooth, straight pipe down to 110 for new riveted-steel and vitrified pipe and to 90 or 80 for old and tuberculated pipe.

Another empirical formula, which we will discuss in detail in Sec. 8.2, is the **Manning** formula, which is

BG units: $$V = \frac{1.486}{n} R_h^{2/3} S^{1/2} \qquad\qquad (7.64a)$$

SI units: $$V = \frac{1}{n} R_h^{2/3} S^{1/2} \qquad\qquad (7.64b)$$

where n is a roughness coefficient, varying from 0.008 for the smoothest brass or plastic pipe, to 0.014 for average drainage tile or vitrified sewer pipe, to 0.021–0.030 for corrugated metal (Table 8.1). The Manning formula applies to about the same flow range as does the Hazen-Williams formula.

For some problems, it is more convenient to work with Eqs. (7.63) and (7.64) in the form of expressions for head loss (see Sec. 7.20). Because we can also express the equations in terms of V or Q, depending on which is given or sought, and in BG or SI units, the number of alternative forms is quite large. Because each variable occurs only once in the above empirical equations, they are always explicit regardless of how they are rearranged or which variable is unknown. This gives them their distinct advantage, that we can always solve them directly.

Engineers have developed nomographic charts and diagrams for the application of Eqs. (7.63) and (7.64). The lack of accuracy that results from using these formulas is not important in the design of water distribution systems, since we can seldom predict the capacity requirements with high precision, and because flows vary considerably throughout the day.

7.20 Nonrigorous Head-Loss Equations

If we rearrange the empirical Hazen-Williams and Manning's equations of Sec. 7.19 into the form of head-loss equations like Eq. (7.13), they will remain explicit in all unknowns and so easy to solve, but they will of course be less accurate than the rigorous equations that have a friction factor f that depends on the Reynolds number, $\mathbf{R}$. We can create a third, similar situation and equation if we use the Darcy-Weisbach Eq. (7.13) with a given, *constant* value of f.

To get good results with these nonrigorous equations the user must select proper values of the friction factor, either C_{HW}, Manning's n, or the constant f. This is more uncertain than estimating the e/D value for the Darcy-Weisbach equation with varying f. In Sec. 7.19 we noted that such less-than-rigorous methods and less accurate results are acceptable for the design of water distribution networks; they could also serve as first estimates for repetitive, rigorous solution procedures.

When the friction factor is constant, i.e., it does not vary with $\mathbf{R}$, we find in all three cases that we can conveniently represent the head-loss equations in the form

$$h_f = KQ^n \tag{7.65}$$

where we note that n is a constant exponent and not Manning's n, which we shall here represent by n_m. By rearranging Eq. (7.13) when f is constant and using Eq. (4.7) to replace V, we obtain

Darcy-Weisbach with constant f: $\quad K = \dfrac{8 fL}{\pi^2 gD^5}, \quad n = 2 \tag{7.66}$

By rearranging the empirical equation given by Eq. (7.63a), we obtain

Hazen-Williams BG units: $\quad K = \dfrac{4.727L}{C_{HW}^{1.852} D^{4.87}}, \quad n = 1.852 \tag{7.67}$

For SI units we must replace the value of 4.727 by 10.675 in the Hazen-Williams K; recall from Sec. 7.19 the restrictions on D and V. By rearranging the empirical equation given by Eq. (7.64a), we obtain

Manning, BG units: $\quad K = \dfrac{4.66n_m^2}{D^{5.33}}, \quad n = 2 \tag{7.68}$

For SI units, we must replace the value of 4.66 by 10.29 in the Manning K; recall the same restrictions on D and V.

We see that $h_f = KQ^2$ except when working with the Hazen-Williams coefficient, and we notice that K is a property of the pipe alone. When we need to solve for discharge, we can of course rearrange Eq. (7.65) into

$$Q = \left(\frac{h_f}{K}\right)^{1/n} \tag{7.69}$$

We will make use of these nonrigorous equations in Secs. 7.30–7.33.

7.21   Minor Losses in Turbulent Flow

Losses due to *local* disturbances of the flow in conduits such as changes in cross section, projecting gaskets, elbows, valves, and similar items we call **minor losses.** In the case of a very long pipe or channel, these losses are usually insignificant in comparison with the loss due to pipe (wall) friction (Secs. 5.3 and 7.6) in the length considered. But if the length of pipe or channel is very short, these so-called minor losses may actually be major losses. Thus, in the case of the suction pipe of a pump, the loss of head at entrance, especially if there is a strainer and a foot valve, may be very much greater than the friction loss in the short inlet pipe.

Whenever the average velocity of turbulent flow is altered either in direction or in magnitude, large *eddies* (Secs. 4.2 and 7.10) or eddy currents are set up causing a loss of energy in excess of the pipe friction in that same length.[22] Head loss in decelerating (i.e., diverging) flow is much larger than that in accelerating (i.e., converging) flow (Sec. 7.25). In addition, head loss generally increases with an increase in the geometric distortion of the flow. Though the causes of minor losses are usually confined to a very short length of the flow path, the effects may not disappear for a considerable distance downstream. Thus an elbow in a pipe may occupy only a small length, but the disturbance in the flow will extend for a long distance downstream.

In the following Secs. 7.22–7.27, we describe the most common sources of minor loss. There are two ways we can represent such losses. We can express them as $kV^2/2g$, where we must determine the **loss coefficient** k for each case. Or we can represent them as being equivalent to a certain length of straight pipe, usually expressed in terms of the number of pipe diameters, N. Since

$$k\frac{V^2}{2g} = \frac{fL}{D}\frac{V^2}{2g} = \frac{f(ND)}{D}\frac{V^2}{2g}$$

it follows that
$$k = Nf$$

To differentiate minor losses from other losses, we shall represent minor head losses by the symbol h' (with a prime). Because they are losses of energy, we will frequently relate them to the energy line (EL) and the hydraulic grade line (HGL), which we initially defined and discussed in Secs. 5.11 and 5.13. Note that these losses are really only degradations of mechanical energy to a less useful form (heat energy).

---

[22] In laminar flow these losses are insignificant, because irregularities in the flow boundary create a minimal disturbance to the flow and separation is essentially nonexistent.

7.22   Loss of Head at Entrance

Referring to Fig. 7.12, we can see that, as fluid from the reservoir enters the pipe, the streamlines continue to converge for a while, much as though this were a jet issuing from a sharp-edged orifice (Sec. 9.6). As a result, we find a cross section with maximum velocity and minimum pressure at *B*. This minimum flow area is known as the **vena contracta.** At *B*, surrounding the contracted flowing stream there is fluid in a state of turbulence but having very little forward motion. Between *B* and *C* the fluid is very disturbed because the stream expands (is less constrained) and the velocity decreases while the pressure rises. From *C* to *D* the flow is normal.

We see that the loss of energy at entrance occurs over the length *AC*, a distance of several diameters. The increased turbulence and vortex motion in this portion of the pipe cause the friction loss to be much greater than in a corresponding length where the flow is normal, as we can see from the drop of the total-energy line. Of this total loss, a small portion h_f would be due to the normal pipe friction (see Fig. 7.12). Hence the difference h_e' between the total entrance loss and h_f is the true value of the extra loss caused at entrance.

The loss of head at entrance we can express as

$$h_e' = k_e \frac{V^2}{2g} \qquad (7.70)$$

where V is the mean velocity in the pipe, and k_e is the loss coefficient, whose general values are given in Fig. 7.13.

The entrance loss results primarily from the turbulence created by the enlargement of the stream after it passes section *B*, and this enlargement in turn depends on how much the stream contracts as it enters the pipe. Thus it very much depends on the conditions at the entrance to the pipe. Experiments have determined values of the entrance loss coefficients. If the entrance to the pipe is well rounded or **bell-mouthed** (Fig. 7.13*a*),

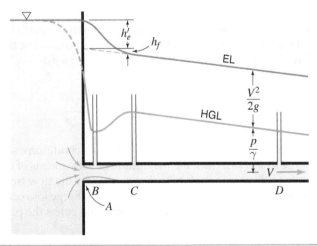

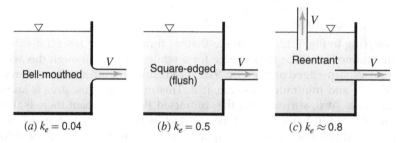

FIGURE 7.13 Entrance loss coefficients.

there is no contraction of the stream entering and the coefficient of loss is correspond-ingly small. For a *flush* or *square-edged entrance,* such as shown in Fig. 7.13b, k_e has a value of about 0.5. A *reentrant tube,* such as that shown in Fig. 7.13c, produces a maximum contraction of the entering stream, because the streamlines come from around the out-side wall of the pipe, as well as more directly from the fluid in front of the entrance. The degree of the contraction depends upon how far the pipe may project within the reser-voir and also upon how thick the pipe walls are, compared with its diameter. With very thick walls, the conditions approach that of a square-edged entrance. For these reasons, the loss coefficients for reentrant tubes vary; for very thin tubes, $k_e \approx 0.8$.

7.23   Loss of Head at Submerged Discharge[23]

Discharge into Still Water

When a fluid with a velocity V is discharged from the end of a pipe into a closed tank or reservoir which is so large that the velocity within it is negligible, the entire kinetic energy of the flow is dissipated. Thus the discharge loss is

$$h_d' = \frac{V^2}{2g} \qquad (7.71)$$

We can confirm that this is true by writing an energy equation between (a) and (c) in Fig. 7.14. Taking the datum plane through (a) and recognizing that the pressure head of the fluid at (a) is y, its depth below the surface, $H_a = y + 0 + V^2/2g$ and $H_c = 0 + y + 0$. Therefore we obtain

$$h_d' = H_a - H_c = \frac{V^2}{2g}$$

Thus the discharge loss coefficient $k_d = 1.0$ under all conditions; so the only way to reduce the discharge loss is to reduce the magnitude of V by means of a diverging tube. This is the reason for a diverging draft tube that discharges the flow from a reaction turbine.

As contrasted with entrance loss, note that discharge loss occurs *after* the fluid *leaves* the pipe,[24] while entrance loss occurs *after* the fluid *enters* the pipe.

---

[23] We first discussed this topic in Sec. 5.12.
[24] In a short pipe, where the discharge loss may be a major factor, greater accuracy is obtained by using the correction factor α, as explained in Sec. 5.1 [see also Eq. (7.45a)].

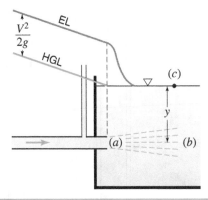

FIGURE 7.14 Loss due to submerged discharge into still water.

Discharge into Moving Water

Let us now find the head loss when a submerged discharge enters a body of water moving away in a channel, such as from a culvert with a submerged outlet (Sec. 8.23), or as in the tailrace of a reaction turbine.

At the outlet, the streamlines are straight and parallel, so, as demonstrated in Sec. 5.4, the pressure distribution across it must be hydrostatic. After a sufficient distance from the outlet, as at section *cf* in Fig. 7.15, the streamlines will again be essentially straight and parallel, and the flow velocity will be practically uniform ($= V_c$) over the full channel depth. So the pressure distribution at section *cf* must also be hydrostatic, and therefore per Eq. (3.6) (using gage pressures)

$$\frac{p_e}{\gamma} + z_e = \frac{p_f}{\gamma} + z_f = \frac{p_c}{\gamma} + z_c = 0 \tag{7.72}$$

For a general streamline such as *ef* in Fig. 7.15, from energy Eq. (5.28) we have

$$\left[\frac{p_e}{\gamma} + z_e + \frac{V^2}{2g}\right] - h'_d = \left[\frac{p_f}{\gamma} + z_f + \frac{V_c^2}{2g}\right] \tag{7.73}$$

Substituting from Eq. (7.72) into (7.73) and rearranging we obtain

$$h'_d = \frac{V^2}{2g} - \frac{V_c^2}{2g} \tag{7.74}$$

This equation states that the discharge loss is equal to the *difference* between the discharge and ultimate velocity heads. Equation (7.71) is a special case of this equation.

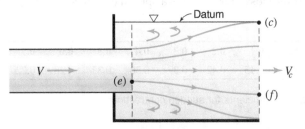

FIGURE 7.15 Loss due to submerged discharge into moving water.

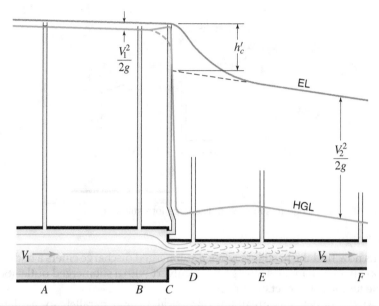

FIGURE 7.16 Loss due to sudden contraction. (*Plotted to scale.*)

7.24 Loss Due to Contraction

Sudden Contraction

The phenomena accompanying the sudden contraction of a flow are shown in Fig. 7.16. There is a marked drop in pressure due to the increase in velocity and to the loss of energy in turbulence. Note that in the corner upstream at section C there is a rise in pressure because the streamlines here are curving, so that the centrifugal action causes a greater pressure at the pipe wall than in the center of the stream. The dashed line indicates the pressure variation along the central streamline from sections B to C.

From C to E, the conditions are similar to those described for entrance (Sec. 7.22). We can represent the loss of head for a sudden contraction by

$$h_c' = k_c \frac{V_2^2}{2g} \tag{7.75}$$

where k_c has the values given in Table 7.2.

The entrance loss of Sec. 7.22 is a special case where $D_2/D_1 = 0$.

D_2/D_1	0.0	0.1	0.2	0.3	0.4	0.5	0.6	0.7	0.8	0.9	1.0
k_c	0.50	0.45	0.42	0.39	0.36	0.33	0.28	0.22	0.15	0.06	0.00

TABLE 7.2 Loss coefficients for sudden contraction

Gradual Contraction

In order to reduce the foregoing losses, we should avoid abrupt changes of cross section. We could achieve this by changing from one diameter to the other with a smoothly curved transition or with the frustrum of a cone. With a smoothly curved transition, a loss coefficient k_c as small as 0.05 is possible. For conical reducers, a minimum k_c of about 0.10 is possible, with a total cone angle of 20–40°. Smaller or larger total cone angles result in higher values of k_c.

The nozzle at the end of a pipeline (see Fig. S7.11b) is a special case of gradual contraction. An equation like Eq. (7.75) also governs the head loss through a nozzle at the end of a pipeline, where k_c becomes the ***nozzle loss coefficient*** k_n whose value commonly ranges from 0.04 to 0.20 and $V_2 = V_j$ is the jet velocity.[25] However, we *cannot* regard the head loss through a nozzle h_n as a minor loss, because the jet velocity head is usually quite large (see, e.g., Sample Problem 7.12). More details on the flow through nozzles is presented in Sec. 9.6.

7.25 Loss Due to Expansion

Sudden Expansion

The conditions at a sudden expansion are shown in Fig. 7.17. There is a rise in pressure because of the decrease in velocity, but this rise would be even greater if there were not the loss in energy. There is excessive turbulence in the flow from C to F, beyond which the flow is normal. The drop in pressure just beyond section C, which was measured by a piezometer not shown in the figure, is due to the fact that the pressures at the wall of the pipe are in this case less than those in the center of the pipe because of centrifugal effects.

Figures 7.16 and 7.17 are both drawn to scale from test measurements for the same diameter ratios and the same velocities, and they show that the loss due to sudden expansion is greater than the loss due to a corresponding contraction. This is so because

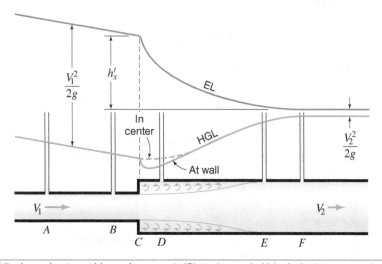

FIGURE 7.17 Loss due to sudden enlargement. (*Plotted to scale. Velocity is the same as in Fig. 7.16.*)

---

[25] See also Eq. (9.14).

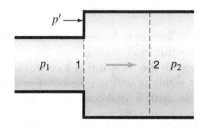

FIGURE 7.18   Detail of a sudden enlargement in a pipe.

of the inherent instability of flow in an expansion, where diverging flow paths encourage the formation of eddies *within* the flow. Moreover, separation of the flow from the wall of the conduit creates pockets of eddying turbulence outside the flow region. In converging flow, on the other hand, there is a dampening effect on eddy formation, and the conversion from pressure energy to kinetic energy is more efficient.

We can derive an expression for the loss of head in a sudden enlargement as follows. In Fig. 7.18, section 2 corresponds to section F in Fig. 7.17, which is a section where the velocity profile has become normal again, and marks the end of the region of excess energy loss due to the turbulence created by the sudden enlargement. In Fig. 7.18 let us assume that the pressure at section 2 for ideal flow (without friction) is p_0. Then from Eq. (5.30) in this ideal case

$$\frac{p_0}{\gamma} = \frac{p_1}{\gamma} + \frac{V_1^2}{2g} - \frac{V_2^2}{2g}$$

If in the actual case, however, the pressure at section 2 is p_2, while the average pressure on the annular ring is p' then, equating the resultant force on the body of fluid between sections 1 and 2 to the time rate of change of momentum between sections 1 and 2 per Eq. (6.7a), we obtain

$$p_1 A_1 + p'(A_2 - A_1) - p_2 A_2 = \frac{\gamma}{g}(A_2 V_2^2 - A_1 V_1^2)$$

From this,

$$\frac{p_2}{\gamma} = \frac{A_1}{A_2}\frac{p_1}{\gamma} + \frac{A_2 - A_1}{A_2}\frac{p'}{\gamma} + \frac{A_1}{A_2}\frac{V_1^2}{g} - \frac{V_2^2}{g}$$

The loss of head is caused by friction, and so it is given by the difference between the ideal and actual pressure heads (or total heads) at section 2. Thus $h_x' = (p_0 - p_2)/\gamma$, and noting that

$$A_1 V_1 = A_2 V_2$$

and that $A_1 V_1^2 = A_1 V_1 V_1 = A_2 V_2 V_1$, we obtain, from substituting the preceding expressions for p_0/γ and p_2/γ into $(p_0 - p_2)/\gamma$,

$$h_x' = \frac{(V_1 - V_2)^2}{2g} + \left(1 - \frac{A_1}{A_2}\right)\left(\frac{p_1}{\gamma} - \frac{p'}{\gamma}\right)$$

We usually assume that $p' = p_1$, in which case the loss of head due to sudden enlargement is

$$h_x' = \frac{(V_1 - V_2)^2}{2g} \tag{7.76}$$

Although it is possible that, under some conditions, p' will equal p_1, it is also possible for it to be either more or less than that value, in which case the loss of head will be either less or more than that given by Eq. (7.76). The exact value of p' will depend on the manner in which the fluid eddies move around in the corner adjacent to this annular ring. However, the deviation from Eq. (7.76) is quite small and of negligible importance.

The discharge loss into still water of Sec. 7.23 is a special case where A_2 is infinite compared with A_1, or $V_2 = 0$, so that Eq. (7.76) will reduce to Eq. (7.71).

On occasion we may need to express this minor loss all in terms of one velocity (see, e.g., Sec. 7.28). If so, we can use continuity, $(\pi/4)D_1^2 V_1 = (\pi/4)D_2^2 V_2$ to obtain

$$(V_1 - V_2)^2 = \left(1 - \frac{D_1^2}{D_2^2}\right)^2 V_1^2 = \left(\frac{D_2^2}{D_1^2} - 1\right)^2 V_2^2 \tag{7.77}$$

Gradual Expansion

To minimize the loss accompanying a reduction in velocity, we can use a *diffuser* such as that shown in Fig. 7.19. The diffuser may have a curved outline, or it may be a frustum of a cone (straight sided). In Fig. 7.19 the loss of head will be some function of the angle of divergence and also of the ratio of the two areas, the length of the diffuser being determined by these two variables.

In flow through a diffuser we can consider the total loss as made up of two components. One is the ordinary pipe-friction loss, which we can represent by

$$h_f = \int \frac{f}{D} \frac{V^2}{2g} dL$$

To integrate the foregoing, we need to express the variables f, D, and V as functions of L. For our present purpose, it is sufficient, however, merely to note that the friction loss

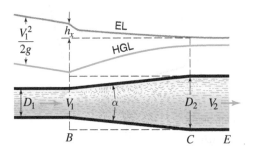

Loss due to gradual enlargement.

increases with the length of the cone. Hence, for given values of D_1 and D_2, the larger the angle of the cone, the less its length and the less the pipe friction, which we can see from the curve marked F in Fig. 7.20a. However, in flow through a diffuser there is an additional loss component, due to turbulence set up by induced currents that produce a vortex motion over and above that which normally exists. This additional turbulence loss will naturally increase with the degree of divergence, as we can see from the curve marked T in Fig. 7.20a, and if the rate of divergence is great enough then the flow may separate from the walls with eddies flowing backward along the walls. The total loss in the diverging cone consists of the sum of these two losses, marked k'. This has a minimum value at 6° for the particular case chosen, which is for a very smooth surface. If the surface were rougher, the value of the friction F would be larger. This increases the value of k', as in the dashed curve, and also shifts the angle for minimum loss to 8°. Thus the best angle of divergence increases with the roughness of the surface.

We have seen that the loss due to a sudden enlargement is very nearly equal to $(V_1 - V_2)^2/2g$. The loss due to a gradual enlargement is

$$h' = k' \frac{(V_1 - V_2)^2}{2g}$$ (7.78)

Values of k' as a function of the cone angle α are given in Fig. 7.20b, for a wider range than appears in Fig. 7.20a. Note, interestingly, that at an angle slightly above 40° the loss is the same as that for a sudden enlargement, which is 180°, and that between these two the loss is greater than for a sudden enlargement, being a maximum at about 60°. This is a result of the induced currents that develop.

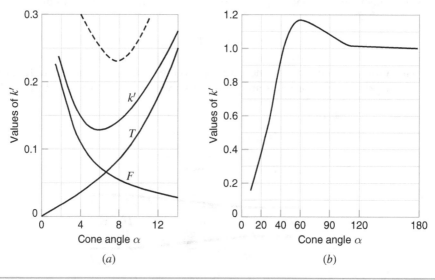

FIGURE 7.20   Loss coefficient for conical diffusers.

Sample Problem 7.10
Water flows through a 150-mm-diameter pipe at 100 L/s. (a) If the pipe suddenly
enlarges to 200 mm diameter, what is the head loss? (b) If the same enlargement is via a
diffuser with a total cone angle of 20°, what is the head loss then?

Solution

(a)
$$V_1 = \frac{Q}{A_1} = \frac{4Q}{\pi D_1^2} = \frac{4(0.100)}{\pi(0.15)^2} = 5.66 \text{ m/s}$$

$$V_2 = \frac{4(0.100)}{\pi(0.20)^2} = 3.18 \text{ m/s}$$

Eq. (7.76):
$$h'_x = \frac{(5.66 - 3.18)^2}{2(9.81)} = 0.312 \text{ m} \quad ANS$$

(b) Fig. 7.20 for $\alpha = 20°$: $k' = 0.38$

Eq. (7.78):
$$h' = 0.38(0.312) = 0.1187 \text{ m} \quad ANS$$

7.26 Loss in Pipe Fittings

For pipe fittings, we can express the loss of head as $kV^2/2g$, where V is the velocity in a
pipe of the nominal size of the fitting. Table 7.3 gives typical values of k. As an alternative, we may account for the head loss due to a fitting by increasing the pipe length by
amounts given by L/D in the table. However, we should recognize that these fittings

Fitting	k	L/D
Globe valve, wide open	10	350
Angle valve, wide open	5	175
Close-return bend	2.2	75
T, through side outlet	1.8	67
T, straight through	0.7	25
Short-radius elbow	0.9	32
Medium-radius elbow	0.75	27
Long-radius elbow	0.60	20
45° elbow	0.42	15
Gate valve, wide open	0.19	7
Gate valve, half open	2.06	72

Note: Values are approximate and will vary depending on whether a fitting is threaded or
flanged. If flanged, values may be lower than shown.

TABLE 7.3   Values of loss factors for pipe fittings

create so much turbulence that the loss caused by them is proportional to V^2, and so we should restrict this latter method to cases where the pipe friction itself is in the zone of complete turbulence (fully-rough-pipe flow). For very smooth pipes it is better to use the k values when determining the loss through fittings.

7.27   Loss in Bends and Elbows

In flow around a bend or elbow, because of centrifugal effects (Sec. 5.16), there is an increase in pressure along the outer wall and a decrease in pressure along the inner wall. In Fig. 7.21 we see the centrifugal forces acting on a number of fluid particles, each of mass m on pipe diameter CE that is normal to the pipe's plane of curvature. The centrifugal force mV^2/r on the particles near the center of the pipe, where the velocities are high, is larger than the centrifugal force on the particles near the walls of the pipe, where the velocities are low (see Fig. 7.10). Because of this unbalanced condition, a secondary flow[26] develops, as shown in the cross section of Fig. 7.21. This combines with the axial velocity to form a double spiral flow, which persists for some distance. Thus not only is there some loss of energy within the bend itself, but this distorted flow persists for some distance downstream until viscous friction dissipates it. The velocity in the pipe may not become normal again within as much as 100 pipe diameters downstream from the bend. In fact, more than half the friction loss of a bend or elbow occurs in the straight pipe following it.

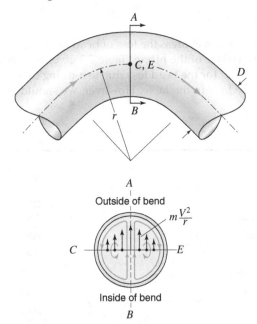

FIGURE 7.21   Secondary flow in a pipe bend.

---

[26] Secondary flow in the bends of open channels is discussed in Sec. 8.21.

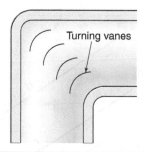

Vaned elbow.

The pressure difference that develops between A and B (Fig. 7.21), on the outside and inside of the bend, is used as the basis of the *elbow meter* (Sec. 9.16), which measures discharge.

We may eliminate most of the head loss due to a sharp bend by using a vaned elbow, such as that in Fig. 7.22. The vanes impede the formation of the secondary flows that would otherwise develop.

Note that the amount of head loss that a pipe bend or elbow causes, as given by

$$h_b = k_b \frac{V^2}{2g} \tag{7.79}$$

is the amount *in excess of the loss for an equal length of straight pipe.* It depends strongly on the ratio of the radius of curvature r (Fig. 7.21) to the diameter of the pipe D. Note also, that we cannot treat combinations of different bends placed close together by adding up the losses of each one considered separately. The total loss depends not only on the spacing between the bends, but also on the relative directions of the bends and the planes containing them. Bend loss is not proportional to the angle of the bend; for 22.5° and 45° bends the losses are respectively about 40% and 80% of the loss in a 90° bend. Figure 7.23 gives values of k_b for 90° bends, varying with bend radius and pipe roughness.

7.28 Single-Pipe Flow with Minor Losses

We have examined the fundamental fluid mechanics associated with the frictional loss of energy in single-pipe flow, caused by both the wall roughness of the pipes and by pipe fittings and the like that disturb the flow (minor losses). While the interest of the scientist extends very little beyond this, it is the task of the engineer to apply these fundamentals to various types of practical problems.

A commonly applied rule is that for pipes longer than 1000 diameters, the error from neglecting minor losses is less than that inherent in selecting a value for the friction factor (f, n, or C_{HW}). In applying this rule, one must of course use common sense and recall that a valve, for example, is a minor loss only when it is wide open; partially closed, it may be the most important loss in the system. When minor losses are negligible, as they often are, we can solve pipe-flow problems by the methods of Secs. 7.14–7.20.

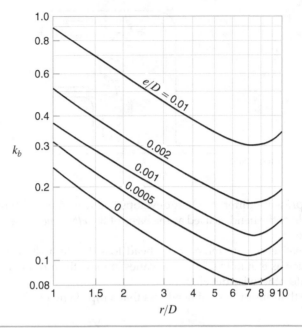

Resistance coefficients for 90° bends (resistance due to length of pipe $\pi r/2$ in the bend must be added, where r is the radius to the pipe centerline). (*From Ref. 46.*)

When we *include* minor losses, the total head h_L loss between two points is the sum of the pipe (wall) friction head loss h_f plus the minor losses, or

$$h_L = h_f + \sum h' \tag{7.80}$$

We can represent the pipe friction loss by a number of different equations, and it may depend on a number of different factors, as noted in Secs. 7.1–7.20. And, as we noted in Sec. 7.21, we can represent the minor losses as a coefficient multiplied by the velocity head, $kV^2/2g$, or as an equivalent length of pipe, expressed as a number of diameters, ND. As a result, the right-hand side of Eq. (7.80) can take on many forms. Because of the extra term added by the minor losses, such forms of this equation are often more challenging to solve than those that omit minor losses (Secs. 7.15–7.20).

Nonrigorous Equations.   In problems where a constant f value is given, Eq. (7.80) still has only one unknown, namely, h_L or V or Q or D. In most cases this equation is explicit in the unknown (Appendix B), and so is easy to solve. But for sizing problems, the resulting equation in D is of the fifth degree and implicit, requiring trial and error or an equation solver as discussed in Sec. 7.15.

The empirical equations (Secs. 7.19 and 7.20) do not necessarily continue to yield direct solutions when minor losses are included. By including $kV^2/2g$ or ND in Eq. (7.65) with Eqs. (7.67) and (7.68), we find that these equations are always explicit for Type 1 (find h_L now, not just h_f), usually explicit for Type 2 (find Q or V), and never explicit for Type 3 (find D). The exception with Type 2 occurs when using the Hazen-Williams equation and representing minor losses by $kV^2/2g$; this is because, in Eq. (7.65) with (7.67), we find that the pipe friction loss is proportional to $V^{1.852}$, so the two terms contain V to different inconvenient powers. Type 3 problems are more complex in part owing to the fact that Q is given rather than V, since this causes V to also be a function of D. We first discussed these three problem categories in Sec. 7.15.

Rigorous Equations. In practice, we may need to take into account the variation of f, which can vary by as much as a factor of five for smoother pipes (see Fig. 7.11). When including minor losses, methods involving the Darcy-Weisbach equation given by Eq. (7.13) and an unknown friction factor f remain explicit or direct for Type 1 problems (find h_L), but for the other two types of problems (find Q or V or D) they are always implicit (Appendix B). The other governing equations are the same as in Sec. 7.15.

Implicit Equations. To solve the many implicit equations that arise when we include minor losses, it is clear we must use iterative procedures, either manually or programmed. Manual iterative procedures (trial and error) for this purpose are generally like those described in Sec. 7.15. But occasionally it will be easier to make trials for D rather than f.

Automated iterative procedures are available in equation solving software packages, and in programmable calculators with an equation solving function. These unquestionably are the most convenient, and are widely accessible.

To use an equation solver it is convenient to condense the problem to one equation in one unknown. Similar to the development of Eq. (7.56), we can develop a "universal" turbulent flow equation for use in an equation solver, including minor losses, by eliminating h_f and f from Eq. (7.80) with the help of Eqs. (7.13), (7.51), and (4.7). Expressing minor losses $\Sigma h'$ in terms of $\Sigma k V^2 / 2g$, we obtain

Turbulent flow:

$$\sqrt{\frac{L/D}{\dfrac{\pi^2 g D^4 h_L}{8Q^2} - \Sigma k}} = -2\log\left(\frac{e/D}{3.7} + \frac{2.51\pi\nu D}{4Q}\sqrt{\frac{L/D}{\dfrac{\pi^2 g D^4 h_L}{8Q^2} - \Sigma k}}\right) \tag{7.81}$$

If we express minor losses in terms of an equivalent length of pipe, ND, then we obtain

Turbulent flow:

$$\frac{4Q}{\pi D^2}\sqrt{\frac{(L/D) + N}{2gh_L}} = -2\log\left(\frac{e/D}{3.7} + \frac{2.51\nu}{D}\sqrt{\frac{(L/D)+N}{2gh_L}}\right) \tag{7.82}$$

Each of these equations involves eight variables: h_L, Q, D, L, e, ν, g, and Σk or N. We note that the equations are implicit in h_L, Q, and D, but, with an equation solver we may easily solve for any one of these three quantities if the rest of the variables are known. Similar and less complex "universal" equations can also be developed using the Haaland equation given by Eq. (7.52); however, they sometimes yield false results when large diameters cause the argument of the logarithm to be very close to unity. An important reminder when using these equations is to use Eqs. (4.7) and (7.1) to check the Reynolds number and confirm that the flow is turbulent. If $\mathbf{R} < 2000$, the flow is laminar (Sec. 7.3), and we must instead solve the problem with Eq. (7.28) and Eq. (4.3) if needed.

The following sample problem illustrates the method of solution for flow through a pipeline of uniform diameter with minor losses.

If the pipe in this example discharged into a fluid that was at a pressure other than atmospheric, we would have to use the proper value of p_2/γ in the energy equation.

Sample Problem 7.11

Water at 60°F flows through the new 10-in-diameter cast-iron pipe with free discharge sketched in Fig. S7.11a. The pipe is 5000 ft long, its entrance is sharp-cornered but non-projecting, and $\Delta z = 260$ ft. Find the flow rate using (a) only the Moody diagram (Fig. 7.11) and a basic scientific calculator; (b) an equation solver.

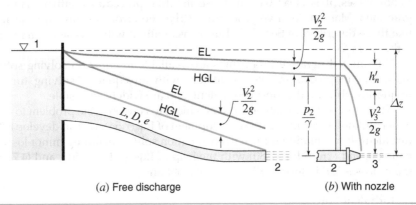

(a) Free discharge   (b) With nozzle

FIGURE S7.11   Discharge from a reservoir.

Solution

This is a Type 2 problem with minor losses.

Table 7.1 for cast iron: $e = 0.00085$ ft; $\dfrac{e}{D} = 0.00085/(10/12) = 0.001020$

Sec. 7.22 for the square-edged entrance: $k_e = 0.5$

Energy Eq. (5.28) from the water surface to the free jet at point 2:

$$260 + 0 + 0 - \left(0.5 + f\,\frac{5000}{10/12}\right)\frac{V_2^2}{2g} = 0 + 0 + \frac{V_2^2}{2g} \qquad (1)$$

(a) Rearranging: $\dfrac{V_2^2}{2g} = \dfrac{260}{1.5 + 6000f}$; $V_2 = \sqrt{\dfrac{2(32.2)260}{1.5 + 6000f}}$

Fig. 7.11, right-hand side, for $e/D = 0.001020$: $f_{min} \approx 0.0196$

$$\text{Try } f = 0.0196:\ \ V_2 = \sqrt{16{,}744/[1.5 + 6000(0.0196)]} = 11.86 \text{ fps}$$

$$D''V = 10(11.86) = 118.6$$

Fig. 7.11 for $e/D = 0.001\,020$ and $D''V = 118.6$: $f = 0.020$. Obtained and assumed f values are different, so we must try again. Tabulating all the trials:

Try f	V, ft/s	$D''V$	Obtained f	
0.0196	11.86	118.6	0.020	Try again
0.020	11.74	117.4	0.020	Converged!

Eq. (4.3):
$$Q = \frac{\pi}{4}\left(\frac{10}{12}\right)^2 (11.74) = 6.40 \text{ cfs} \qquad ANS$$

Alternatively, replace use of the chart by equations as follows:

Table A.1 for water at 60°F: $\qquad \nu = 1.217 \times 10^{-5} \text{ ft}^2/\text{sec}$

Eq. (7.54) for $e/D = 0.001\,020$: $\qquad f_{min} = 0.019\,73$

Energy equation with assumed f gives V (as before); $\mathbf{R} = DV/\nu$

Insert values of e/D and $\mathbf{R}$ into Haaland equation given by Eq. (7.52) to get a calculated f; compare with assumed f. If the difference is less than about 0.5%, we have convergence; otherwise repeat.

(b) To use an equation solver rearrange Eq. (7.81) to:

$$\frac{1}{\sqrt{\dfrac{\pi^2 g D^4 h_L}{8Q^2} - \Sigma k}} + 2\log\left(\frac{e/D}{3.7} + \frac{2.51\pi\nu D}{4Q}\sqrt{\frac{\pi^2 g D^4 h_L}{8Q^2} - \Sigma k}\right) = 0$$

Equation (7.81) does not appear to be appropriate, because it does not include for the change in velocity head from zero at point 1 to $V_2^2/(2g)$ at point 2. However, from preceding Eq. (1) we see that we *can* include for this by increasing Σk by one. Assign to the seven known variables the values (without units) $g = 32.2 \text{ ft/sec}^2$, $\nu = 1.217 \times 10^{-5} \text{ ft}^2/\text{sec}$, $h_L = \Delta z = 260 \text{ ft}$, $\Sigma k = k_e + 1 = 1.5$, $L = 5000 \text{ ft}$, $e = 0.000\,85 \text{ ft}$, and $D = 10/12 \text{ ft}$. Depending on the solver used, an estimated initial Q value may need to be assigned.

Instruct the solver to solve Eq. (7.81) for Q. The result is $Q = 6.39$ cfs. But we must check the $\mathbf{R}$. Equation (4.7): $V = 4Q/(\pi D^2) = 11.71$ fps, so $\mathbf{R} = DV/\nu = 8.02 \times 10^5$, and the turbulent flow assumption is correct. Therefore

$$Q = 6.39 \text{ cfs} \qquad ANS$$

Another example of flow from a reservoir is that of a pipeline (known as a ***penstock***) leading to an impulse turbine (Sec. 11.15). In this case the pipe does not discharge freely, but ends in a nozzle (Fig. S7.11b), which has a known or assumed loss coefficient, k_n. The head loss in the nozzle, h_n, is associated with the high issuing velocity head, $V_j^2/2g$, and is therefore not a minor loss (Sec. 7.24). We use the equation of continuity to write all losses in terms of the velocity head in the pipe. This is the logical choice for the "common unknown," because we will again build the trial-and-error solution around the pipe friction loss rather than the nozzle loss.

Sample Problem 7.12

As in Fig. S7.11b, suppose that the pipeline of Sample Problem 7.11 is now fitted with a nozzle, at the end of which discharges a 2.5-in-diameter jet and that has a loss coefficient of $k_n = 0.11$. Find the flow rate using only Fig. 7.11 and a basic scientific calculator.

Solution

Referring to Sample Problem 7.11, let point 2 now be on the pipe at the base of the nozzle and point 3 be in the jet. The head loss in the nozzle is $0.11 V_3^2/2g$ (Sec. 7.24). Writing the energy equation given by Eq. (5.28) between 1 and 3,

$$260 + 0 + 0 - \left(0.5\frac{V_2^2}{2g} + 6000 f \frac{V_2^2}{2g} + 0.11\frac{V_3^2}{2g}\right) = 0 + 0 + \frac{V_3^2}{2g}$$

Using continuity Eq. (4.17), $V_3^2/2g = (10/2.5)^4 V_2^2/2g = 256 V_2^2/2g$. Thus

$$260 = (0.5 + 1.11 \times 256 + 6000 f)\frac{V_2^2}{2g}$$

Select a trial value of f. Let $f = 0.02$ for the first assumption. Then $260 = (285 + 120) V_2^2/2g$, from which

$$\frac{V_2^2}{2g} = \frac{260}{405} = 0.643 \text{ ft}$$

and $V_2 = \sqrt{(2(32.2)0.643} = 6.43$ fps. With $D''V = 10 \times 6.43 = 64.3$ (or equivalently, $\mathbf{R} = DV/\nu = 4.40 \times 10^5$) and $e/D = 0.001$, Fig. 7.11 shows $f = 0.02$. In this case we can consider the first solution sufficiently accurate, but in general the value of f we determine from the chart may be quite different from that we assume, and we may need to make one or two more trials.

The rate of discharge is $Q = A_2 V_2 = 0.545 \times 6.43 = 3.51$ cfs   **ANS**

Then $V_3 = 16 V_2 = 16 \times 6.43 = 102.9$ fps

In addition, $H_2 = p_2/\gamma + V_2^2/2g = 260 - (0.5 + 0.02 \times 6000)0.643 = 182.58$ ft, and the pressure head $p_2/\gamma = 182.58 - 0.643 = 181.93$ ft.

This example shows that the addition of the nozzle reduces the discharge from 6.40 to 3.51 cfs, but increases the jet velocity from 11.74 to 102.9 fps. The head loss due to pipe friction is 77.1 ft and the head loss through the nozzle is 18.09 ft.

When solving Type 3 sizing problems, in general, the diameter we obtain will not be a standard pipe size, and the size we select will usually be the next largest commercially available size (Table A.9). In planning for the future, we must remember that scale deposits will increase the roughness and reduce the cross-sectional area. For pipes in water service, the absolute roughness e of old pipes (20 years and more) may increase over that of new pipes by three-fold for concrete or cement-lined steel, up to 20-fold for cast iron, and even to 40-fold for tuberculated wrought-iron and steel pipe. Substituting $V = 4Q/(\pi D^2)$ into Eq. (7.13) shows that for a constant value of f, Q varies as $D^{5/2}$. Hence for the case where minor losses are negligible and f is constant, to achieve a 100% increase in flow, we need to increase the diameter by only 32%. This amounts to a 74% increase in cross-sectional area.

7.29   Pipeline with Pump or Turbine

If a pump lifts a fluid from one reservoir to another, as in Fig. 7.24, not only does it do work in lifting the fluid the height Δz, but it also has to overcome the total friction head loss Σh_L (including minor losses) in the suction and discharge pipelines. This head loss is equivalent to some added lift, so that the effect is the same as if the pump lifted the fluid a height $\Delta z + \Sigma h_L$. Hence the power the pump delivers to the liquid is $\gamma Q(\Delta z + \Sigma h_L)$. The power required to run the pump is greater than this, depending on the efficiency of the pump (see Sec. 11.3).

The total pumping head h_p for this case is found by beginning with Eq. (5.27)

$$\left(\frac{p_1}{\gamma} + z_1 + \frac{V_1^2}{2g}\right) + h_M - h_L = \left(\frac{p_2}{\gamma} + z_2 + \frac{V_2^2}{2g}\right)$$

Since both reservoirs are open to the atmosphere, $p_1 = p_2 = 0$, and since velocities in reservoirs are negligible, $V_1 = V_2 = 0$. The only hydraulic machine in the system is a pump, so $h_M = h_p$, thus the equation simplifies to

$$h_p = \Delta z + \Sigma h_L \qquad (7.83)$$

If the pump discharges a stream through a nozzle, as in Fig. 7.25, not only does it lift the liquid a height Δz, but it also imparts a kinetic energy head of $V_2^2/2g$, where V_2 is the velocity of the jet. Thus the total pumping head is now

$$h_p = \Delta z + \frac{V_2^2}{2g} + \Sigma h_L \qquad (7.84)$$

In any case we can determine the total pumping head by writing the energy equation between any point upstream from the pump and any other point downstream, as in Eq. (5.37). For example, if the upstream reservoir were at a higher elevation than the downstream one, then the Δzs in the two foregoing equations would have negative signs.

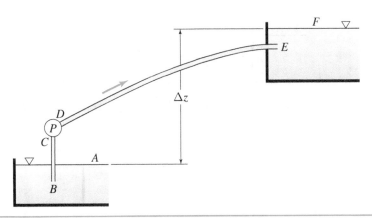

FIGURE 7.24   Pipeline with pump between two reservoirs.

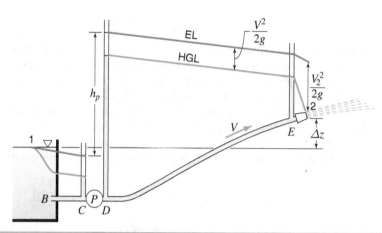

FIGURE 7.25   Pipeline with pump and nozzle.

The machine we use to convert flow energy into mechanical work we call a *turbine*. In flowing from the upper tank in Fig. 7.26 to the lower, the fluid loses potential energy head equivalent to Δz. Part of this energy is lost to hydraulic friction in the pipe, and the remainder reaches the turbine. Of that part which reaches the turbine, part is lost in hydraulic friction within the turbine, and the rest converts into mechanical work.

The power that reaches the turbine is the initial power minus the friction losses (pipe plus minor) in the pipeline, or $\gamma Q(\Delta z - \Sigma h_L)$. The power the *turbine* delivers is less than this, depending on both its hydraulic and mechanical losses (see Sec. 11.15). The head under which the turbine operates is

$$h_t = \Delta z - \Sigma h_L \tag{7.85}$$

Here Σh_L is the loss of head in the supply line plus the submerged discharge loss (pipe friction and minor losses), but it does *not* include the head loss in the *draft tube* (*DE* in Fig. 7.26), since the draft tube is considered an integral part of the turbine. A draft tube has

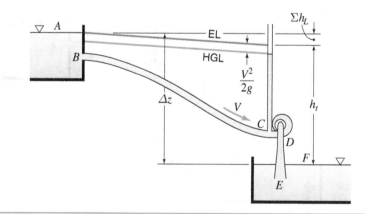

FIGURE 7.26   Pipeline with turbine.

a gradually increasing cross-sectional area, which reduces the velocity at discharge. This enhances the efficiency of the turbine because it reduces the head loss at discharge (Sec. 7.23). Note that the h_t of Eq. (7.85) represents the energy head the turbine removes from the fluid; this, of course, is the same as the energy head the fluid transfers to the turbine.

Sample Problem 7.13

The tanks, pump, and pipelines of Fig. S7.13 have the characteristics noted. The suction line entrance from the pressure tank is flush, and the discharge into the open tank is submerged. If the pump P puts 2.0 hp into the liquid, (a) determine the flow rate and (b) find the pressure in the pipe on the suction side of the pump.

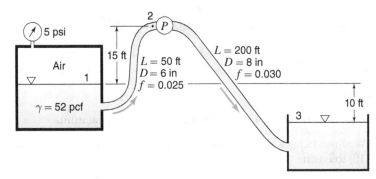

FIGURE S7.13

Solution

(a) Eq. (5.40):
$$P = \frac{\gamma Q h_p}{550} = 2.0 = \frac{52 Q h_p}{550}$$

Thus
$$h_p = \frac{21.2}{Q}$$

Figure 7.13 for flush entrance: $k_e = 0.5$

Writing energy Eq. (5.27) from 1 (datum) to 3,

$$\frac{5(144)}{52} + 0 + 0 + h_p - 0.5\frac{V_6^2}{2g} - 0.025\left(\frac{50}{6/12}\right)\frac{V_6^2}{2g} - 0.030\left(\frac{200}{8/12}\right)\frac{V_8^2}{2g} - \frac{V_8^2}{2g} = 0 - 10 + 0$$

Eq. (4.6): $V_6 = \dfrac{Q}{A_6} = \dfrac{Q}{0.1963}$ and $V_8 = \dfrac{Q}{0.349}$

Substituting for h_p, V_6, V_8, and $g = 32.2$, the energy equation reduces to

$$23.8 + \frac{21.2}{Q} - 2.48Q^2 = 0$$

which we can rewrite as the cubic expression

$$2.48Q^3 - 23.8Q - 21.2 = 0$$

Solving this by trial and error (see Sample Problems 3.5 and 5.8) or by equation or polynomial solver, $Q = 3.47$ cfs

From Eq. (B.7) the other two roots are $Q = -2.48$ and -0.987 cfs.

So $Q = 3.47$ cfs   *ANS*

(b) To obtain the pressure p_2 at the suction side of the pump,

$$V_6 = \frac{3.47}{0.1963} = 17.68 \text{ fps}$$

Writing energy Eq. (5.28) from 1 to 2,

$$\frac{5(144)}{52} + 0 + 0 - 0.5\frac{17.68^2}{2(32.2)} - 0.025\left(\frac{50}{6/12}\right)\frac{17.68^2}{2(32.2)} = \frac{p_2}{\gamma} + 15 + \frac{17.68^2}{2(32.2)}$$

from which $\dfrac{p_2}{\gamma} = -20.6 \text{ ft}; \quad p_2 = -20.6\left(\dfrac{52}{144}\right) = -7.43 \text{ psi} \quad ANS$

From appendices:

This is equivalent to $7.43(29.9/14.70) = 15.13$ inHg vacuum.

Note: We should check the absolute pressure against the vapor pressure of the liquid (Sec. 5.10) to ensure that vaporization does not occur at point 2.

7.30   Branching Pipes

For convenience, let us consider three pipes connected to three reservoirs as in Fig. 7.27 and connected together or branching at the common junction point J. Actually, we can consider that any of the pipes is connected to some other destination than a reservoir by simply replacing the reservoir with a piezometer tube in which the water level is the same as the reservoir surface. We shall suppose that all the pipes are sufficiently long that we can neglect minor losses and velocity heads, so $h_L = h_f$ which we shall designate as h.

We name the pipes and flows and corresponding friction losses as in the diagram. The continuity and energy equations require that the flow entering the junction equals

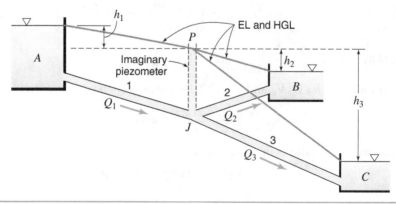

FIGURE 7.27   Branching pipes.

the flow leaving it, and that the pressure head at J (which we shall represent schematically by the imaginary open piezometer tube shown, with water at elevation P) is common to all pipes.

As there are no pumps, the elevation of P must lie between the surfaces of reservoirs A and C. If P is level with the surface of reservoir B then h_2 and Q_2 are both zero. If P is above the surface of reservoir B then water must flow into B and $Q_1 = Q_2 + Q_3$. If P is below the surface of reservoir B then the flow must be out of B and $Q_1 + Q_2 = Q_3$. So for the situation shown in Fig. 7.27 we have the following governing conditions:

1. $Q_1 = Q_2 + Q_3$.
2. Elevation of P is common to all three pipes.

The diagram suggests several different problems or cases, three of which we will discuss below using different methods of solution.

Rigorous Solutions

When we know the pipe wall material, we can estimate its e value (Table 7.1), and we know that the friction factor f varies with the e/D of the pipe and the Reynolds number of the flow. Because we are not considering minor losses, we can use the equations and methods of Secs. 7.13–7.18. In particular, using only a basic scientific calculator, we can solve pipelines for h (Type 1 problems) using the Haaland equation given by Eq. (7.52) with Eq. (7.13); we can solve for V or Q (Type 2 problems) using Eq. (7.56); and more rarely, we can solve for D (Type 3 problems) using Eqs. (7.57)–(7.62). We prefer these equations because they avoid trial and error, which can become quite confusing when combined with other trial-and-error techniques needed to solve for branching flow.

The following three cases illustrate some of the "manual" trial-and-error methods used to solve the different types of problems that can occur. In each case all the required pipe data (lengths, diameters, and materials for e values) are known.

Case 1. Find the flow to or from two reservoirs, and the surface elevation of one of these, given the other flow and two elevations and all the pipe data.

We can solve this problem directly, without trial and error. Suppose that Q_1 and the elevations of A and B are given. Then we can directly determine the head loss h_1 (Type 1), using Fig. 7.11 or Eq. (7.52) to find the proper value of f. Knowing h_1 fixes P, so now we can easily obtain h_2. Knowing h_2 enables us to directly determine the flow in pipe 2 using the Type 2 equation given by Eq. (7.56). Condition 1 (continuity at junction J) then determines Q_3, which in turn enables us to directly find h_3 (Type 1), in the same way as for line 1. Finally, P and h_3 define the required surface elevation of C.

Case 2. Find the flow to or from two reservoirs, and the surface elevation of the third, given the other flow and two elevations and all the pipe data.

Let us suppose that Q_2 and the surface elevations of A and C are given. Then we know $h_1 + h_3 = \Delta h_{13}$, say. We may use various solution approaches;[27] we shall discuss

---

[27] Other approaches include (*a*) assuming distributions of the flows Q_1 and Q_3, knowing that $Q_1 - Q_3 = Q_2$, and (*b*) by substituting for the hs in $h_1 + h_3 = \Delta h_{13}$ using Eq. (7.13) with V_3 written in terms of Q_3, and V_1 written in terms of $Q_2 + Q_3$, and successively solving the resulting quadratic equation for Q_3 while converging on f values.

a more convenient one. In this, we assume the elevation of P, which yields values for h_1 and h_3, and so Q_1 and Q_3 via Eq. (7.56). If these do not satisfy the discharge relation at J then we must adjust P until they do. To help us converge on the correct elevation of P, we can plot the results of each assumption on a graph like Fig. 7.28. For ΣQ at J, inflows to J are taken as positive and outflows as negative. Two or three points, with one fairly close to the vertical axis, determine a (near-straight) curve that intersects that axis at the equilibrium level of P, where $\Sigma Q = 0$, as required.

Last, we can determine h_2 from Q_2 and Eqs. (7.52) and (7.13), and find the required surface elevation of B.

Case 3.   Find the flow in each pipe, given the surface elevation of all three reservoirs and all the pipe data.

This is the ***classic three-reservoir problem,*** and it differs from the foregoing cases in that it is not immediately evident whether the flow is *into* or *out of* reservoir B. We can readily determine this direction by first assuming no flow in pipe 2; that is, assume the piezometer level P at the elevation of the surface of B. The head losses h_1 and h_3 then determine the flows Q_1 and Q_3 via Eq. (7.56). If $Q_1 > Q_3$, then we must raise P to satisfy continuity at J, causing water to flow *into* reservoir B, and we shall have $Q_1 = Q_2 + Q_3$; if $Q_1 < Q_3$, then we must lower P to satisfy continuity at J, causing water to flow *out of* reservoir B, and we shall have $Q_1 + Q_2 = Q_3$. From here on the solution proceeds by adjusting P as for Case 2.

Note: For any of the above three cases, we can avoid manual trial and error by setting up the governing equations and solving them simultaneously using equation solving software. There will be the usual four governing equations for each line, a flow continuity equation for the junction, and, depending on the case addressed, one or two equations relating the various head losses. With so many unknowns to solve for, the success of the procedure becomes more sensitive to the guessed values, and we may have to try different guesses. The great advantage of this approach is that it is so straightforward. It is illustrated in part (*b*) of the following sample problem.

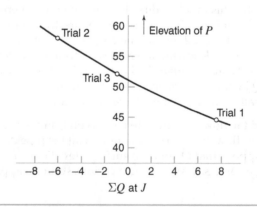

FIGURE 7.28   Convergence of solution for elevation of junction point P.

Sample Problem 7.14

Given that, in Fig. 7.27, pipe 1 is 6000 ft of 15 in diameter, pipe 2 is 1500 ft of 10 in diameter, and pipe 3 is 4500 ft of 8 in diameter, all asphalt-dipped cast iron. The elevations of the water surfaces in reservoirs A and C are 250 ft and 160 ft, respectively, and the discharge Q_2 of 60°F water into reservoir B is 3.3 cfs. Find the surface elevation of reservoir B: (a) using only a basic scientific calculator; (b) using equation solving software.

Solution

This is a Case 2 problem.

Table A.1 for water at 60°F: $\nu = 12.17 \times 10^{-6}$ ft^2/sec.

Pipe	1	2	3
L, ft	6000	1500	4500
D, ft	1.25	10/12	8/12
e, ft (Table 7.1)	0.0004	0.0004	0.0004
L/D	4800	1800	6750
$A = \pi D^2/4$, ft^2	1.227	0.545	0.349
e/D	0.00032	0.000 48	0.0006

(a) Find the elevation of P by trial and error.

Elevation of P lies between 160 and 250 ft. Calculate V from Eq. (7.56) assuming turbulent flow. Trials:

Elev. P	h_1	h_3	V_1	V_3	$\mathbf{R}_1$	$\mathbf{R}_2$	Q_1	Q_3	ΣQ	Move P?
200	50	40	6.444	4.481	662,000	245,000	7.907	1.564	+3.04	Up
230	20	70	4.013	5.984	412,000	328,000	4.925	2.088	−0.463	Down

Interpolation (Fig. 7.28): (230 − Elev. P)/(230 − 200) = 0.463/(0.463 + 3.04); Elev. $P = 226.03$.

Elev. P	h_1	h_3	V_1	V_3	$\mathbf{R}_1$	$\mathbf{R}_2$	Q_1	Q_3	ΣQ	Move P?
226	24	66	4.412	5.805	453,000	318,000	5.414	2.026	+0.088	Up

Interpolation (Fig. 7.28): (230 − Elev. P)/(230 − 226) = 0.463/(0.463 + 0.088); Elev. $P = 226.64$.

Close enough! *Note:* These adjustments are very suitable for making on a spreadsheet.

$$V_2 = \frac{Q_2}{A_2} = \frac{3.3}{0.545} = 6.055 \text{ fps}; \quad \mathbf{R}_2 = \frac{D_2 V_2}{\nu} = 416,500$$

All three $\mathbf{R}$ values are turbulent, so the use of Eq. (7.56a) and these results are valid.

Eq. (7.52): $f_2 = 0.01761$;   Eq. (7.13): $h_2 = 18.05$ ft

$$\text{Elev. } B = \text{Elev. } P − h_2 = 226.64 − 18.05 = 208.59 \text{ ft} \qquad ANS$$

(b) Using equation solving software, we note that there are 14 governing equations. Two of these are

$$h_1 + h_3 = 250 - 160 = 90, \quad Q_1 = Q_2 + Q_3$$

The remaining 12 equations are the four governing equations (Sec. 7.14) for each of the three pipes. Instruct the solver to vary the head at the junction until continuity $(Q_1 - Q_2 - Q_3 = 0)$ is satisfied.

The solver provides values for the 14 unknowns (three values each for f, V, $\mathbf{R}$, h, and two values for Q) from which it calculates

$$\text{Elev. } B = 208.53 \text{ ft} \qquad \textbf{\textit{ANS}}$$

Sample Problem 7.15

With the sizes, lengths, and material of pipes given in Sample Problem 7.14, suppose that the surface elevations of reservoirs A, B, and C are 525 ft, 500 ft, and 430 ft, respectively. (a) Does water enter or leave reservoir B? (b) Find the flow rates of 60°F water in each pipe. Use only a basic scientific calculator.

Solution
This is a Case 3 problem. Find the elevation of P by trial and error.

Table A.1 for water at 60°F: $\nu = 12.17 \times 10^{-6}$ ft²/sec.

The tabulated pipe data are the same as for Sample Problem 7.14.

(a) *Trial 1.* First, try P at elevation of reservoir surface $B = 500$ ft:

Pipe	1	2	3
h, ft	25	0	70
$\sqrt{2gDh/L}$, fps	0.579	0	0.817
V, fps [Eq. (7.56a)]	4.51	0	5.98
$Q = AV$, cfs	5.53	0	2.09

At J, $\Sigma Q =$ inflow $-$ outflow $= 5.53 - 2.09 = 3.44$ cfs. This must be zero, so P must be raised (to reduce Q_1 and increase Q_3); then water will flow *into* reservoir B.   **\textit{ANS}**

(b) *Trial 2.* Raise P. 500 ft $<$ Elev. $P <$ 525 ft. Try P at elevation 510 ft:

h, ft	15	10	80
$\sqrt{2gDh/L}$, fps	0.449	0.598	0.874
V, fps [Eq. (7.56a)]	3.46	4.49	6.41
Check $\mathbf{R}$	355,000	307,000	351,000
$Q = AV$, cfs	4.24	2.42	2.24

At J, $\Sigma Q = 4.24 - 2.42 - 2.24 = -0.42$ cfs. By interpolation (using Fig. 7.28),

$$\frac{510 - \text{Elev. } P}{510 - 500} = \frac{0.42}{0.42 + 3.44}; \quad \text{Elev. } P = 508.91 \text{ ft}$$

Trial 3. Try P at elevation 508.9 ft:

h, ft	16.1	8.9	78.9
$\sqrt{2gDh/L}$, fps	0.465	0.564	0.868
V, fps [Eq. (7.56a)]	3.59	4.19	6.36
Check $\mathbf{R}$	339,000	287,000	348,000
$Q = AV$, cfs	4.40	2.28	2.22 *ANS*

At J, $\sum Q = 4.40 - 2.28 - 2.22 = -0.10$ cfs. Close enough!
Note: These repetitive adjustments are very suitable for making on a spreadsheet.

Nonrigorous Solutions

If the value of the friction factor (constant f, C_{HW}, or Manning's n) is *given* for each pipeline, we must use the nonrigorous head-loss equations of Sec. 7.20. Then, to solve the three cases just discussed, we follow exactly the same steps as for the rigorous solutions, but instead of using the equations of Secs. 7.13–7.18, we simply use Eqs. (7.65)–(7.69). Of course, we must first determine the appropriate K and n value for each pipeline, and select accordingly from Eqs. (7.66)–(7.68). Notice, however, that we can easily solve Case 2 *directly* if $n = 2$ (Darcy-Weisbach or Manning) by writing the known elevation difference $\Delta h_{13} = h_1 + h_3 = K_1 Q_1^2 + K_3 Q_3^2 = K_1(Q_2 + Q_3)^2 + K_3 Q_3^2$, which is a quadratic equation in Q_3, the only unknown.

Sample Problem 7.16

In Fig. 7.27 pipe 1 is 300 mm diameter and 900 m long, pipe 2 is 200 mm diameter and 250 m long, and pipe 3 is 150 mm diameter and 700 m long. The Hazen-Williams coefficient for all pipes is 120. The surface elevations of reservoirs A, B, and C are 160 m, 150 m, and 120 m, respectively. (*a*) Does water enter or leave reservoir B? (*b*) Find the flow rate in each pipe. Use only a basic scientific calculator.

Solution

This is a Case 3 problem. Find the elevation of P by trial and error. Use the Hazen-Williams form given by Eq. (7.67) in SI units:

$$K = \frac{10.675\, L}{C_{HW}^{1.852} D^{4.87}}, \quad n = 1.852$$

Pipe	1	2	3
L, m	900	250	700
D, m	0.3	0.2	0.15
L/D	3 000	1 250	4 667
C_{HW}	120	120	120
n	1.852	1.852	1.852
K	480	961	10 924

(*a*) *Trial 1.* First, try *P* at elevation of reservoir surface *B* = 150 m:

h, m	10	0	30
Q, m³/s [Eq. (7.69)]	0.1236	0	0.0414

At *J*, ΣQ = inflow − outflow = 0.1236 − 0.0414 = 0.0822 m³/s. This must be zero, so we must raise *P* (to reduce Q_1 and increase Q_3), then water will flow *into* reservoir B.   **ANS**

(*b*) *Trial 2.* Raise *P*. 150 m < Elev. *P* < 160 m, so try *P* at elevation 155 m:

h, m	5	5	35
Q, m³/s [Eq. (7.69)]	0.0850	0.0584	0.0450

At *J*, ΣQ = 0.0850 − 0.0584 − 0.0450 = −0.0184 m³/s. This must be zero, so we must lower *P*. By interpolation (Fig. 7.28),

$$\frac{155 - \text{Elev. } P}{155 - 150} = \frac{0.0184}{0.0184 + 0.0822}; \quad \text{Elev. } P = 154.09 \text{ m}$$

Trial 3. Try *P* at elevation 154 m:

h, m	6	4	34	
Q, m³/s [Eq. (7.69)]	0.0938	0.0518	0.0443	**ANS**
V, m/s [Eq. (4.7)]	1.327	1.649	2.51	

At *J*, ΣQ = −0.0023 m³/s. This is close enough.
All *V*s are < 3 m/s (Sec. 7.19), so these solutions using Eq. (7.67) are valid.
Note: These adjustments are very suitable for making on a spreadsheet.

7.31 Pipes in Series

The discussion in Sec. 7.28 addressed the case of a single pipe of constant diameter. If a pipeline is made up of lengths of *different* diameters, as in Fig. 7.29, conditions must satisfy the continuity and energy equations, namely:

$$Q = Q_1 = Q_2 = Q_3 = \cdots \tag{7.86}$$

$$h_L = h_{L1} + h_{L2} + h_{L3} + \cdots \tag{7.87}$$

If we are given the rate of discharge *Q*, the problem is straightforward. We may find the head loss directly by adding the contributions from the various sections, as in Eq. (7.87). If we are given empirical coefficients or constant *f* values, we can do the same thing using Eq. (7.65) and the appropriate values of *K* and *n* selected from Eqs. (7.66)–(7.68). If, however, we are given the pipe material or *e*, the result will be more accurate, because it uses the rigorous Darcy-Weisbach approach. Then we use Eq. (7.13) to find the individual head loss contributions after finding *e/D*, *V*, **R**, and *f* for each pipe.

If we are given the total head loss h_L and want to find the flow, the problem is a little more involved.

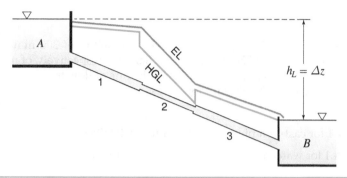

FIGURE 7.29 Pipes in series. Minor losses are not depicted.

Using the nonrigorous equations, we again substitute Eq. (7.65) into Eq. (7.87), to get

$$h_L = h_f = K_1 Q_1^n + K_2 Q_2^n + K_3 Q_3^n + \cdots$$

But since all the Qs are equal from Eq. (7.86), this becomes

$$h_L = (K_1 + K_2 + K_3 + \cdots)Q^n = (\Sigma K)Q^n \tag{7.88}$$

So, knowing the pipe information and which nonrigorous equation we must use, we can solve for Q. Last, we must check that all velocities are in the required ranges for the empirical equations to be valid (Sec. 7.19).

If we wish to use the more accurate, rigorous Darcy-Weisbach approach to find Q, we must note that in Eq. (7.88) each K has now become a function of a different f. The preferred manual method of solution is similar to that just discussed, and we call it the ***equivalent-velocity-head method.*** Substituting Eq. (7.13) into Eq. (7.87) and including minor losses if we wish (usually if $L/D < 1000$),

$$h_L = \left(f_1 \frac{L_1}{D_1} + \Sigma k_1\right)\frac{V_1^2}{2g} + \left(f_2 \frac{L_2}{D_2} + \Sigma k_2\right)\frac{V_2^2}{2g} + \cdots$$

Using continuity Eq. (4.17), we know $D_1^2 V_1 = D_2^2 V_2 = D_3^2 V_3$, etc., from which we can express all the velocities in terms of one chosen velocity. So, by assuming reasonable values for each f [e.g., from Eq. (7.54) or Fig. 7.11], for any pipeline, however complex, we can write the total head loss as

$$h_L = K\frac{V^2}{2g} \tag{7.89}$$

where V is the chosen velocity. We can solve this equation for the chosen V, and so can obtain the V and $\mathbf{R}$ and f values for each pipe. For better accuracy, we should replace the assumed values of f by the values just obtained, and obtain an improved solution. When the f values converge V is correct, and we can calculate Q.

Sample Problem 7.17

Suppose in Fig. 7.29 the pipes 1, 2, and 3 are 300 m of 300 mm diameter, 150 m of 200 mm diameter, and 250 m of 250 mm diameter, respectively, of new cast iron and are conveying 15°C water. If $\Delta z = 10$ m, find the rate of flow from A to B using only a basic scientific calculator. Neglect minor losses.

Solution

Table 7.1 for cast-iron pipe: $e = 0.25$ mm $= 0.00025$ m.

Table A.1 for water at 15°C: $\nu = 1.139 \times 10^{-6}$ m²/s.

Pipe	1	2	3
L, m	300	150	250
D, m	0.3	0.2	0.25
e/D	0.000833	0.00125	0.00100
f_{min} (Fig. 7.11)	0.019	0.021	0.020

Assuming these friction factor values,

$$\Delta z = h_L = h_f = 10 = 0.019\left(\frac{300}{0.3}\right)\frac{V_1^2}{2g} + 0.021\left(\frac{150}{0.2}\right)\frac{V_2^2}{2g} + 0.020\left(\frac{250}{0.25}\right)\frac{V_3^2}{2g}$$

From continuity,
$$\frac{V_2^2}{2g} = \frac{V_1^2}{2g}\left(\frac{D_1}{D_2}\right)^4 = \frac{V_1^2}{2g}\left(\frac{0.3}{0.2}\right)^4 = 5.06\frac{V_1^2}{2g}$$

Similarly,
$$\frac{V_3^2}{2g} = 2.07\frac{V_1^2}{2g}$$

and thus
$$10 = \frac{V_1^2}{2g}\left(0.019\frac{1000}{1} + 0.021\frac{750}{1}5.06 + 0.020\frac{1000}{1}2.07\right)$$

from which
$$\frac{V_1^2}{2g} = 0.0713 \text{ m}$$

So
$$V_1 = \sqrt{2(9.81 \text{ m/s}^2)(0.0713 \text{ m})} = 1.183 \text{ m/s}$$

The corresponding values of $\mathbf{R}$ are 0.31×10^6, 0.47×10^6, and 0.37×10^6; the corresponding friction factors are only slightly different from those we originally assumed, since the flow occurs at Reynolds numbers very close to fully-rough-pipe flow. So

$$Q = A_1V_1 = \frac{1}{4}\pi(0.30)^2 1.183 = 0.0836 \text{ m}^3/\text{s} \qquad ANS$$

Note: We would have obtained greater accuracy if we had adjusted the friction factors to match the pipe-friction chart more closely (further trials) or calculated them by Eq. (7.52), and if we had included minor losses. In that case $Q = 0.0821$ m³/s.

We can avoid manual iteration for f by solving simultaneous equations using equation solving software. There are the usual four equations for each pipe (Sec. 7.14), plus Eq. (7.87); if necessary, we may easily account for minor losses by using head loss equations with the form of Eq. (7.80). For the three pipes of Fig. 7.29, for example, there are therefore 13 simultaneous equations, which we may solve in the usual manner for 13 unknowns (see Sample Problem 7.14b). The unknowns are either the flow rate or the total head loss, and three values each of h_L, V, $\mathbf{R}$, and f.

7.32 Pipes in Parallel

In the case of flow through two or more parallel pipes, as in Fig. 7.30, conditions must satisfy the following continuity and energy equations:

$$Q = Q_1 + Q_2 + Q_3 \qquad (7.90)$$

$$h_L = h_{L1} = h_{L2} = h_{L3} \qquad (7.91)$$

because the pressures at A and B are common to all pipes. Problems may be posed in various ways.

If the head loss h_L is given, the problem is straightforward. We can find the total discharge directly by adding the contributions from the various pipes, as in Eq. (7.90). If we are given empirical roughness coefficients or constant f values, we can do the same thing using Eq. (7.69) and the appropriate values of K and n selected from Eqs. (7.66)–(7.68). If, however, we are given the pipe material or e, the result will be more accurate, because it uses the rigorous Darcy-Weisbach approach. Then we have an independent Type 2 problem for each pipe (see Secs. 7.15–7.18), which we can solve directly by Eq. (7.56) for example.

If we are given the total flow Q and want to find the head loss and individual flows, the problem is a little more involved. Using the empirical equations, and neglecting minor losses, we again substitute Eq. (7.69) into Eq. (7.90), to get

$$Q = \left(\frac{h_{f1}}{K_1}\right)^{1/n} + \left(\frac{h_{f2}}{K_2}\right)^{1/n} + \left(\frac{h_{f3}}{K_3}\right)^{1/n} + \cdots$$

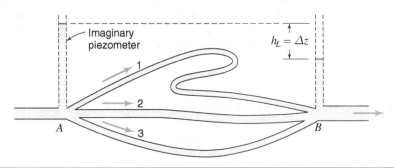

FIGURE 7.30 Pipes in parallel.

But since all the h_fs $(= h_L$s) are equal from Eq. (7.91), this becomes

$$Q = (h_f)^{1/n}\left[\left(\frac{1}{K_1}\right)^{1/n} + \left(\frac{1}{K_2}\right)^{1/n} + \left(\frac{1}{K_3}\right)^{1/n} + \cdots\right] = (h_f)^{1/n}\Sigma\left(\frac{1}{K}\right)^{1/n} \qquad (7.92)$$

So, knowing the pipe information and the empirical equation we must use, we can solve for h_f. We can then find the individual flows using Eq. (7.69). Last, we must check that all velocities are in the required ranges for the empirical equations to be valid (Sec. 7.19).

If we wish to use the rigorous Darcy-Weisbach approach to find h_L and the individual Qs, we must note that in Eq. (7.92) each K has now become a function of a different f. The preferred manual method of solution is similar to the preceding. Writing Eq. (7.13) for each line, including minor losses if we wish,

$$h_L = \left(f\frac{L}{D} + \Sigma k\right)\frac{V^2}{2g}$$

where Σk is the sum of the minor-loss coefficients, which we can usually neglect if the pipe is longer than 1000 diameters. Solving for V and then Q, we obtain the following for pipe 1:

$$Q_1 = A_1 V_1 = A_1\sqrt{\frac{2gh_L}{f_1(L_1/D_1) + \Sigma k_1}} = C_1\sqrt{h_L} \qquad (7.93)$$

where C_1 is constant for the given pipe, except for the change in f with Reynolds number. We can similarly express the flows in the other pipes, using assumed reasonable values of f from Fig. 7.11 or Eq. (7.54). Finally, Eq. (7.90) becomes

$$Q = C_1\sqrt{h_L} + C_2\sqrt{h_L} + C_3\sqrt{h_L} = (C_1 + C_2 + C_3)\sqrt{h_L} \qquad (7.94)$$

This enables us to find a first estimate of h_L and the distribution of flows and velocities in the pipes. Using these, we can next make improvements to the values of f, if indicated, and if necessary repeat them, until we finally obtain a correct determination of h_L and the distribution of flows.

If we use the turbulent-flow equation given by Eq. (7.51) or (7.52) to obtain f, we must remember to confirm that the Reynolds number is in the turbulent range. We can precheck the likelihood of laminar flow occurring in any of the pipes by calculating an "average" flow velocity from the total flow divided by the total area of all the pipes, and using this velocity to obtain an indicator **R** for each pipe.

Sample Problem 7.18

Three pipes A, B, and C are interconnected as in Fig. S7.18. The pipe characteristics are as follows:

Pipe	D (in)	L (ft)	f
A	6	2000	0.020
B	4	1600	0.032
C	8	4000	0.024

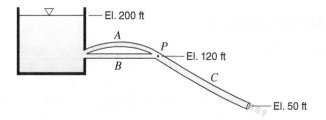

FIGURE S7.18

Find the rate at which water will flow in each pipe. Find also the pressure at point P. All pipe lengths are much greater than 1000 diameters, so neglect minor losses.

Solution

Eq. (5.28): $0 + 200 + 0 - 0.020\dfrac{2000}{6/12}\dfrac{V_A^2}{2g} - 0.024\dfrac{4000}{8/12}\dfrac{V_C^2}{2g} = 0 + 50 + \dfrac{V_C^2}{2g}$

i.e., $150 = 80\dfrac{V_A^2}{2g} + 145\dfrac{V_C^2}{2g}$ (1)

Continuity, Eq. (7.90): $6^2 V_A + 4^2 V_B = 8^2 V_C$

i.e., $36 V_A + 16 V_B = 64 V_C$ (2)

Eq. (7.91): $h_{fA} = h_{fB} = 0.020\dfrac{2000}{6/12}\dfrac{V_A^2}{2g} = 0.032\dfrac{1600}{4/12}\dfrac{V_B^2}{2g}$

i.e., $80 V_A^2 = 153.6 V_B^2, \quad V_B = 0.722 V_A$

Substituting into (2):

$$36 V_A + 16(0.722 V_A) = 64 V_C$$

$$47.5 V_A = 64 V_C, \quad V_A = 1.346 V_C$$

Substituting into (1):

$$150 = 80\dfrac{(1.346 V_C)^2}{2g} + 145\dfrac{V_C^2}{2g} = 289.9\dfrac{V_C^2}{2g}$$

$$V_C^2 = 2(32.2)150/289.9 = 33.3$$

$V_C = 5.77 \text{ fps}, \quad Q_C = A_C V_C = (0.349)5.77 = 2.01 \text{ cfs}$ *ANS*

$V_A = 1.346 V_C = 7.77 \text{ fps}, \quad Q_A = (0.1963)7.77 = 1.526 \text{ cfs}$ *ANS*

$$V_B = 0.722 V_A = 5.61 \text{ fps}$$

$Q_B = A_B V_B = (0.0873)5.61 = 0.489 \text{ cfs}$ *ANS*

As a check, note that $Q_A + Q_B = Q_C$ is satisfied.

To find the pressure at P:

Eq. (5.28): $0 + 200 + 0 - 80\dfrac{V_A^2}{2g} = \dfrac{p_P}{\gamma} + 120$

$$\dfrac{p_P}{\gamma} = 80 - 80\dfrac{(7.77)^2}{2(32.2)} = 5.01 \text{ ft}$$

Check: $120 + \dfrac{p_P}{\gamma} - 144\dfrac{V_C^2}{2g} = 50 + \dfrac{V_C^2}{2g}$

$$\dfrac{p_P}{\gamma} = 145\dfrac{(5.77)^2}{2(32.2)} - 70 = 5.01 \text{ ft}$$

So $p_P/\gamma = 5.01$ ft and $p_P = (62.4/144)5.01 = 2.17$ psi.   *ANS*

Note: In this example we were given the values of f for each pipe as known. Actually f depends on $\mathbf{R}$ [Fig. 7.11 or Eq. (7.52)]. Usually we know or assume the absolute roughness e of each pipe, and achieve an accurate solution by trial and error until the fs and $\mathbf{R}$s for each pipe have converged.

We can avoid manual iteration for f by solving simultaneous equations using equation solving software. There are the usual four equations for each pipe (Sec. 7.14), plus Eq. (7.90); if necessary, we may easily account for minor losses by using head loss equations with the form of Eq. (7.80). For the three pipes of Fig. 7.30, for example, there are therefore 13 simultaneous equations, which we may solve in the usual manner for 13 unknowns (see Sample Problem 7.14b). The unknowns are either the head loss or the total flow rate, and three values each of Q, V, $\mathbf{R}$, and f.

It is instructive to compare the solution methods for pipes in parallel with those for pipes in series. The role of the head loss in one case becomes that of the discharge rate in the other, and vice versa. Students should be already familiar with this situation from the elementary theory of dc circuits. The flow corresponds to the electrical current, the head loss to the voltage drop, and the frictional resistance to the ohmic resistance. The outstanding deficiency in this analogy occurs in the variation of potential drop with flow, which is with the first power in the electrical case ($V = IR$) and with the second power in the hydraulic case ($h_L \propto V^2 \propto Q^2$) for fully developed turbulent flow.

7.33   Pipe Networks

In municipal distribution systems, pipes are frequently interconnected so that the flow to a given outlet may come by several different paths, as in Fig. 7.31. As a result, we often cannot tell by inspection which way the flow travels, as in pipe BE. Nevertheless, the flow in any network, however complicated, must satisfy the basic relations of continuity and energy as follows:

1. The flow into any junction must equal the flow out of it.
2. The flow in each pipe must satisfy the pipe-friction laws for flow in a single pipe.
3. The algebraic sum of the head losses around any closed loop must be zero.

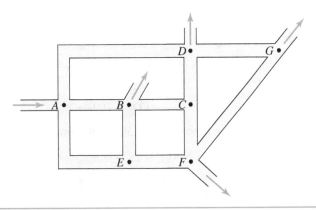

FIGURE 7.31   Pipe network.

Most pipe networks are too complicated to solve analytically by hand using rigorous (variable f) equations, as was possible in the simpler cases of parallel pipes (Sec. 7.32). They are readily solved by specially developed computer programs. However, in many cases we cannot predict the capacity requirements of water distribution systems with high precision, and flows in them vary considerably throughout the day, so high accuracy in calculating their flows is not important. As a result, the use of nonrigorous equations (Secs. 7.19–7.20) are very acceptable for this purpose. The method of successive approximations, due to Cross,[28] is such a method that was popular before the advent of computers. We will review it here to help students understand the fluid mechanics of pipe networks and evaluate computer-generated solutions. It consists of the following elements, in order:

Step 1:   By careful inspection assume the most reasonable distribution of flows that satisfies condition 1.

Step 2:   Write condition 2 for each pipe in the form

$$h_L = KQ^n \qquad (7.95)$$

where K and n are constants for each pipe as described in Sec. 7.20. If minor losses are important include them as in Eq. (7.93), which yields $K = 1/C^2$ and $n = 2$ for constant f. We may include minor losses within any pipe or loop, but must neglect them at the junction points.

Step 3:   To investigate condition 3, compute the algebraic sum of the head losses around each elementary loop, $\Sigma h_L = \Sigma K Q^n$. Consider losses from clockwise flows as positive, counterclockwise negative. Only by good luck will these add up to zero on the first trial.

Step 4:   Adjust the flow in each loop by a correction ΔQ to balance the head in that loop and give $\Sigma K Q^n = 0$. The heart of this method lies in the following determination of ΔQ. For any pipe, we may write

$$Q = Q_0 + \Delta Q$$

[28] H. Cross, Analysis of Flow in Networks of Conduits or Conductors, *Univ. Ill. Eng. Expt. Sta. Bull.* 286, 1936.

where Q is the correct discharge and Q_0 is the assumed discharge. Then, for each pipe,

$$h_L = KQ^n = K(Q_0 + \Delta Q)^n = K(Q_0^n + nQ_0^{n-1}\,\Delta Q + \cdots)$$

If ΔQ is small compared with Q_0, we may neglect the terms of the binomial series after the second one, so that

$$h_L = KQ_0^n + \Delta Q K n Q_0^{n-1}$$

For a loop, $\Sigma h_L = \Sigma KQ^n = 0$, so because ΔQ is the same for all pipes in that loop,

$$\Sigma KQ_0^n + \Delta Q \Sigma KnQ_0^{n-1} = 0$$

As we must sum the corrections of head loss in all pipes *arithmetically* (treating all terms as positive), we may solve this equation for ΔQ,

$$\Delta Q = \frac{-\Sigma KQ_0 \left| Q_0^{n-1} \right|}{n\Sigma \left| KQ_0^{n-1} \right|} = \frac{-\Sigma h_L}{n\Sigma \left| h_L / Q_0 \right|} \tag{7.96}$$

since, from Eq. (7.95), $h_L/Q = KQ^{n-1}$. We emphasize again that we must sum the numerator of Eq. (7.96) *algebraically*, with due account of each sign, while we must sum the denominator *arithmetically*. Note that the $Q_0 \left| Q_0^{n-1} \right|$ in the numerator gives this quantity the same sign as the head loss. The negative sign in Eq. (7.96) indicates that when there is an excess of head loss around a loop in the clockwise direction, we must subtract the ΔQ from clockwise Q_0 values and add it to counterclockwise ones. The reverse is true if there is a deficiency of head loss around a loop in the clockwise direction.

Step 5:   After we have given each loop a first correction, the losses will still not balance, because of the interaction of one loop upon another (pipes which are common to two loops receive two independent corrections, one for each loop). So we repeat the procedure, arriving at a second correction, and so on, until the corrections become negligible.

We may use either form of Eq. (7.96) to find ΔQ. As values of K appear in both the numerator and denominator of the first form, we can use values proportional to the actual K to find the distribution. The second form is more convenient for use with pipe-friction diagrams for water pipes.

An attractive feature of this approximation method is that errors in computation have the same effect as errors in judgment and the process eventually corrects them.

As noted earlier, varying demand rates usually make high solution accuracy unnecessary with pipe networks. However, if high manual accuracy is required for some reason, we can first solve the problem in a similar manner to the preceding example using the Darcy-Weisbach K in Eq. (7.65) and constant f values. Then we can use the resulting flows to adjust the f and K values, and repeat the process (more than once if necessary) to refine the answers. The value of such refinement is questionable, not only because of uncertainties in the demand flows, but also because of uncertainties in the e values (pipe roughness) we must use (see Sec. 7.16). Usually when we adjust f values they change by only a few percent, but we can see in Fig. 7.11 that for smoother pipe it is possible for them to change by as much as a factor of five.

Sample Problem 7.19

If the flow into and out of a two-loop pipe system are as shown in Fig. S7.19, determine the flow in each pipe using only a basic scientific calculator. The K values for each pipe were calculated from the pipe and minor loss characteristics and from an assumed value of f, and $n = 2$.

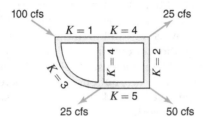

100 cfs   25 cfs

$K = 1$ $K = 4$

$K = 4$ $K = 2$

$K = 3$

$K = 5$

25 cfs   50 cfs

FIGURE S7.19

Solution

As a first step, assume a flow in each pipe such that continuity holds at all junctions. Take clockwise flows as positive. Calculate ΔQ for each loop, make corrections to the assumed Qs, and repeat several times until the ΔQs are quite small.

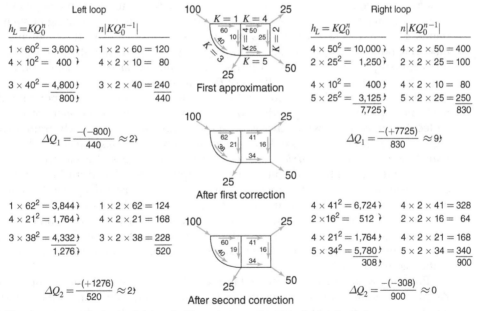

Left loop

$h_L = KQ_0^n$	$n\|KQ_0^{n-1}\|$
$1 \times 60^2 = 3{,}600 \}$	$1 \times 2 \times 60 = 120$
$4 \times 10^2 = 400 \}$	$4 \times 2 \times 10 = 80$
$3 \times 40^2 = 4{,}800 \}$	$3 \times 2 \times 40 = 240$
$800 \}$	440

$$\Delta Q_1 = \frac{-(-800)}{440} \approx 2 \}$$

$1 \times 62^2 = 3{,}844 \}$	$1 \times 2 \times 62 = 124$
$4 \times 21^2 = 1{,}764 \}$	$4 \times 2 \times 21 = 168$
$3 \times 38^2 = 4{,}332 \}$	$3 \times 2 \times 38 = 228$
$1{,}276 \}$	520

$$\Delta Q_2 = \frac{-(+1276)}{520} \approx 2 \}$$

Right loop

$h_L = KQ_0^n$	$n\|KQ_0^{n-1}\|$
$4 \times 50^2 = 10{,}000 \}$	$4 \times 2 \times 50 = 400$
$2 \times 25^2 = 1{,}250 \}$	$2 \times 2 \times 25 = 100$
$4 \times 10^2 = 400 \}$	$4 \times 2 \times 10 = 80$
$5 \times 25^2 = 3{,}125 \}$	$5 \times 2 \times 25 = 250$
$7{,}725 \}$	830

$$\Delta Q_1 = \frac{-(+7725)}{830} \approx 9 \}$$

$4 \times 41^2 = 6{,}724 \}$	$4 \times 2 \times 41 = 328$
$2 \times 16^2 = 512 \}$	$2 \times 2 \times 16 = 64$
$4 \times 21^2 = 1{,}764 \}$	$4 \times 2 \times 21 = 168$
$5 \times 34^2 = 5{,}780 \}$	$5 \times 2 \times 34 = 340$
$308 \}$	900

$$\Delta Q_2 = \frac{-(-308)}{900} \approx 0$$

Further corrections can be made if greater accuracy is desired.

We can solve simple networks without approximation and manual iteration by solving simultaneous equations using equation solving software. For networks containing i pipes, $5i$ equations are required if using the Darcy-Weisbach equation with variable f, and $2i$ equations are required if using the simplified Eq. (7.95) with constant friction factors. These required equations include (*a*) the usual (condition 2) flow equations for each pipe (four or one per pipe, depending on the equations used); (*b*) flow continuity equations (condition 1) at all but one of the j nodes (as these imply continuity at the last node); (*c*) equations for the sum of the head losses around $i - j + 1$ loops (condition 3). The unknowns we want to find for each pipe are h_L, Q, V, **R**, and f using the Darcy-Weisbach equation, or only h_L and Q using Eq. (7.95).

The pipe-network problem lends itself well to solution by use of a digital computer. Programming takes time and care, but once set up, there is great flexibility and it can save many hours of repetitive labor. Many software packages are now available to simulate water distribution networks.

7.34 Further Topics in Pipe Flow

The basics of steady incompressible flow in pressure conduits discussed in this chapter are just an introduction to many more advanced subjects involving flow in conduits. Flow through submerged culverts is a special case discussed in Sec. 8.23. Where sewers must dip to flow under obstacles like streams, they flow full and are known as *inverted siphons* or *sag pipes;* turbulence, and therefore velocities, must remain adequate when flow rates vary in order to prevent suspended solids from accumulating at such low points. Later in this text, we discuss *unsteady flow* in pipes (Chap. 10), involving both moderate rates of change and the very rapid changes associated with *water hammer.* We often investigate or simulate such flows using *numerical methods,* usually on a computer. In Chap. 9 we describe methods of *flow measurement* in pipes.

When we inject a different fluid into a pipeline, the main flow transports it (*advection*), and it also mixes and spreads through the main fluid by the processes of ionic and molecular *diffusion,* by which dissolved species move relatively slowly from areas of higher concentration to areas of lower concentration, and by *dispersion,* also known as *eddy diffusion* or *turbulent mixing.* As these latter names suggest, turbulence causes dispersion, and as a result it is usually a far more rapid process than diffusion. Injection of easily detectable fluids into pipelines, such as salts (measured by conductivity) or dyes,[29] have been used to measure flow rates and velocities. If, for example, we continuously inject dye dissolved in water into a pipeline at a known flow rate and concentration, and give it sufficient opportunity to mix completely, the dye concentration downstream will indicate the relative magnitude of the pipe flow rate to the flow rate of the injected dye solution.

Pumping with pipelines, discussed in Sec. 7.29, has many engineering applications. We provide pump stations at low areas such as at underpasses that would otherwise flood during storms. Often, we provide multiple pumps to enable a more efficient response to varying inflows. In sanitary sewer systems pump stations become necessary if the slopes needed for gravity drainage require excessive excavation; sometimes, local topography requires pumping of sewage over a hill, through a *force main.* In such

---

[29] Rhodamine and fluorescein dyes in water are detectable by a fluorometer at concentrations of one in 10^9.

cases designers must take precautions to keep solids in suspension, and to prevent accumulation of dangerous gases. Water distribution systems, discussed in the previous section, must often supply areas of uneven topography, requiring their division into subareas or *pressure zones* served by separate pump stations which boost the pressures to the appropriate levels. Advanced software exists to analyze such conditions.

To aid the dilution of sewage effluent, engineers often provide submarine and river *outfalls* with a *diffuser section.* (We discussed submarine discharge as a single jet in Sec. 5.15.) Diffusers have a number of outlets or *ports* along their length, so that the flow in that part of the main line or *manifold* decreases with distance. Manifolds also connect to navigation locks, to facilitate filling lock chambers as rapidly as possible while minimizing disturbance to moored vessels. The hydraulics of manifold flow is an advanced subject, particularly when applied to outfalls where port sizes are adjusted to maintain a uniform discharge along the length of the diffuser.

CHAPTER 8

Steady Flow in Open Channels

8.1 Open Channels

An open channel is one in which the stream is not completely enclosed by solid boundaries and therefore has a free surface subjected only to atmospheric pressure. The flow in such a channel is caused not by some external head, but rather only by the gravity component along the slope of the channel. Thus we often refer to open-channel flow as *free-surface flow* or *gravity flow.* This chapter will deal only with *steady* flow in open channels.

The principal types of open channel are natural streams and rivers; artificial canals; and sewers, tunnels, and pipelines flowing not completely full. Artificial canals may be built to convey water for purposes of water-power development, irrigation or city water supply, and drainage or flood control and for numerous other purposes. While there are examples of open channels carrying liquids other than water, few experimental data exist for such, and the numerical coefficients given here apply only to water at natural temperatures.

For convenience in dealing with large channel systems, we often divide them into reaches. A *reach* is a continuous stretch of a waterway, often chosen to have reasonably uniform properties like cross section, slope, roughness, and discharge.

The accurate solution of problems of flow in open channels is more challenging than in the case of pressure pipes. Not only are reliable experimental data more difficult to secure, but we meet a wider range of conditions than we do with pipes. Practically all pipes are round, but the cross sections of open channels can be of any shape, from circular to the irregular forms of natural streams. In pipes the degree of roughness ordinarily ranges from that of new smooth metal, on the one hand, to that of old corroded iron or steel pipes, on the other. But with open channels the surfaces vary from smooth timber or concrete to the rough or irregular beds of some rivers. Hence the friction coefficients to be used are more uncertain with open channels than with pipes.

We described uniform flow in Sec. 4.3 as it applies to hydraulic phenomena in general. In the case of open channels uniform flow means that the water cross section and depth remain constant over a certain reach of the channel as well as over time. This requires that the drop in potential energy due to the fall in elevation along the channel be exactly that consumed by the energy dissipation through boundary friction and turbulence.

247

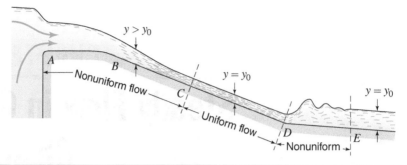

FIGURE 8.1   Steady flow down a chute or spillway.

Uniform flow is an equilibrium condition that flow tends to if the channel is sufficiently long with constant slope, cross section, and roughness. We can state this in another way: For any channel of given roughness, cross section, and slope, there exists for a given flow rate one and only one water depth, y_0, at which the flow will be uniform. Thus, in Fig. 8.1, the flow is accelerating in the reach from A to C, becomes established as uniform flow from C to D, suffers a violent deceleration due to the change of slope between D and E, and finally achieves a new depth of uniform flow somewhere beyond E. Acceleration occurs in the reach from B to C because the gravity component along the slope is greater than the boundary shear resistance. As the flow accelerates, the boundary shear increases because of the increase in velocity, until at C the boundary shear resistance becomes equal to the gravity component along the slope. Beyond C there is no acceleration, the velocity is constant, and the flow is uniform. We commonly refer to the depth in uniform flow as the ***normal depth***, y_0.

Open-channel flow is usually fully rough; that is, it occurs at high Reynolds numbers. For open channels we define the Reynolds number by $\mathbf{R} = R_h V/\nu$, where R_h is the hydraulic radius (Sec. 7.4). Since $R_h = D/4$ for pressure conduits, the critical value of Reynolds number at which the changeover occurs from laminar flow to turbulent flow in open channels is 500, whereas in pressure conduits the critical value is 2000.

In open-channel flow (Fig. 8.2) we refer to the channel ***bed slope*** S_0, the ***water surface slope*** S_w, and the slope of the energy line or the ***energy gradient*** S. It is quite evident that for open-channel flow the hydraulic grade line coincides with the water surface,

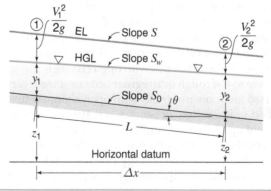

FIGURE 8.2   Open-channel flow—definition sketch.

provided there is no unusual curvature in the streamlines or stream tubes, for if we attach a piezometer tube to the side of the channel, the water will rise in it until its surface is level with the water surface in the channel at that point. We always measure water depth y vertically, and we usually define the distance between sections as the horizontal distance Δx between them. As a result, we define the three slopes just mentioned as

$$S_0 = \frac{z_1 - z_2}{\Delta x} = -\frac{\Delta z}{\Delta x} \tag{8.1}$$

$$S_w = \frac{(z_1 + y_1) - (z_2 + y_2)}{\Delta x} = -\frac{\Delta(z + y)}{\Delta x} \tag{8.2}$$

$$S = \frac{h_L}{L} = \frac{(z_1 + y_1 + V_1^2/2g) - (z_2 + y_2 + V_2^2/2g)}{L} \tag{8.3}$$

where x, y, Δx, and L are defined in Fig. 8.2. Note that the energy gradient S is defined as the head loss per length of flow path (not per horizontal distance). Also, we usually assume that $\alpha = 1.0$; this assumption is reasonable when the flow depth is less than the channel width. Furthermore, for most practical cases the angle θ between the channel bed and the horizontal is small; so L, the distance along the channel bed between the two sections, is usually almost equal to Δx, the horizontal distance between the two sections.

8.2 Uniform Flow

In uniform flow (Fig. 8.3) the cross section through which flow occurs is constant along the channel, and so also is the velocity. Thus, $y_1 = y_2$ and $V_1 = V_2$ and the channel bed, water surface, and energy line are parallel to one another. Also, $S_w = S_0 = -\Delta z/\Delta x = \tan\theta$, while $S = h_f/L = \sin\theta$, where θ is the angle the channel bed makes with the horizontal.

In most open channels (rivers, canals, and ditches) the bed slope S_0 $(= \Delta z/\Delta x)$ falls in the range of 0.01 to 0.0001. For some chutes, spillways, and sewers it may be larger, and for some large rivers it may be smaller. Because, therefore, usually $S_0 < 0.1$ so that $\theta < 5.7°$, it follows that in most cases $\sin\theta \approx \tan\theta$ (within 0.5%) and $S_0 = S_w \approx S$. In this chapter we will assume that $\theta < 5.7°$ and thus that $S \approx S_0$. In channels where $\theta > 5.7°$ we must make a distinction in the difference between S and S_0. Moreover, on such slopes air entrainment and pulsating flow (successions of traveling waves) are common.

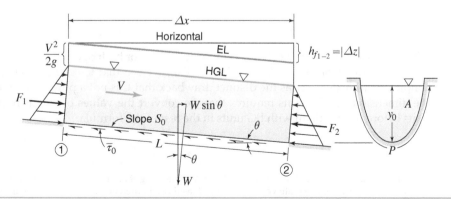

FIGURE 8.3   Uniform flow in open channel.

In Sec. 7.5 we develop a general equation for frictional resistance in a pressure conduit. The same reasoning may now be applied to uniform flow with a free surface. Consider the short reach of length L along the channel between stations 1 and 2 in uniform flow with water cross section of area A (Fig. 8.3). As the flow is neither accelerating nor decelerating, we can consider the body of water contained in the reach in static equilibrium. Summing forces along the channel, the hydrostatic-pressure forces F_1 and F_2 balance each other, since there is no change in the depth y between the stations. The only force in the direction of motion is the gravity component, and this must be resisted by the average boundary shear stress $\bar{\tau}_0$, acting over the area PL, where P is the wetted perimeter of the section. Thus

$$\gamma AL \sin\theta = \bar{\tau}_0 PL$$

But $\sin\theta = h_f/L = S$. Solving for $\bar{\tau}_0$, we have

$$\bar{\tau}_0 = \gamma\frac{A}{P}S = \gamma R_h S \tag{8.4}$$

where R_h is the hydraulic radius,[1] discussed in Sec. 7.4, and for most slopes (with $\theta < 5.7°$) S_0 may be taken as equal to S. Substituting the value of $\bar{\tau}_0$ from Eq. (7.10) and replacing S with S_0,

$$\bar{\tau}_0 = C_f \rho\frac{V^2}{2} = \gamma R_h S_0$$

We can solve this for V in terms of either the friction coefficient C_f or the conventional friction factor f [Eq. (7.15)] to give

$$V = \sqrt{\frac{2g}{C_f}R_h S_0} = \sqrt{\frac{8g}{f}R_h S_0} \tag{8.5}$$

The Chézy Formula

Antoine de Chézy (1718–1798), a French bridge engineer and hydraulics expert, proposed in 1775 that the velocity in an open channel varied as $\sqrt{R_h S_0}$. This led to the formula

$$V = C\sqrt{R_h S_0} \tag{8.6}$$

which is known by his name. It has been widely used for both open channels and pipes under pressure. Comparing Eqs. (8.5) and (8.6), we see that $C = \sqrt{8g/f}$. Despite the simplicity of Eq. (8.6), it has the distinct drawback that C is not a pure number but has the dimensions $L^{1/2}T^{-1}$. This requires that we convert the values of C which are in SI units before we use them with BG units in the rest of the formula.

---

[1] Strictly speaking, we should measure A and P in a plane at right angles to L. However, since depth is measured vertically, we calculate values of A and P as those in the vertical plane. The resulting values of R_h are identical regardless of which way we determine A and P.

As C and f are related, the same considerations that are presented in Chap. 7 regarding the determination of a value for f apply also to C. For a small open channel with smooth sides, the problem of determining f or C is the same as that in the case of a pipe. But most channels are relatively large compared with pipes, thus giving Reynolds numbers that are higher than those commonly encountered in pipes. Also, open channels are frequently rougher than pipes, especially in the case of natural streams. A study of Fig. 7.11 reveals that, as the Reynolds number and the relative roughness both increase, the value of f becomes practically independent of **R** and depends only on the relative roughness.

The Manning Formula

One of the best as well as one of the most widely used formulas for uniform flow in open channels is that published by the Irish engineer Robert Manning[2] (1816–1897). Manning had found from many tests that the value of C in the Chézy formula varied approximately as $R_h^{1/6}$, and others observed that the proportionality factor was very close to the reciprocal of n, the coefficient of roughness in the previously used, but complicated and inaccurate, Kutter formula. This led to the formula that has since spread to all parts of the world. In SI units, the Manning formula is

In SI units:
$$V(\text{m/s}) = \frac{1}{n} R_h^{2/3} S_0^{1/2} \tag{8.7a}$$

The dimensions of the friction factor n are seen to be $TL^{-1/3}$. To avoid converting the numerical value of n for use with BG units, we change the formula itself so as to leave the value of n unaffected. Thus, in BG units, the Manning formula is

In BG units:
$$V(\text{fps}) = \frac{1.486}{n} R_h^{2/3} S_0^{1/2} \tag{8.7b}$$

where 1.486 is the cube root of 3.28, the number of feet in a meter. Despite the dimensional difficulties of the Manning formula, which have long plagued those attempting to put all fluid mechanics on a rational dimensionless basis, it continues to be popular because it is simple to use and reasonably accurate. Representative values of n for various surfaces are given in Table 8.1.

In terms of flow rate, Eqs. (8.7a) and (8.7b) may be expressed as

In BG units:
$$Q(\text{cfs}) = \frac{1.486}{n} AR_h^{2/3} S_0^{1/2} \tag{8.8a}$$

In SI units:
$$Q(\text{m}^3/\text{s}) = \frac{1}{n} AR_h^{2/3} S_0^{1/2} \tag{8.8b}$$

---

[2] Robert Manning, Flow of Water in Open Channels and Pipes, *Trans. Inst. Civil Engrs. (Ireland)*, 20, 1890.

Nature of Surface	n	
	Min	**Max**
Lucite	0.008	0.010
Glass	0.009	0.013
Neat cement surface	0.010	0.013
Wood-stave pipe	0.010	0.013
Plank flumes, planed	0.010	0.014
Vitrified sewer pipe	0.010	0.017
Concrete, precast	0.011	0.013
Metal flumes, smooth	0.011	0.015
Cement mortar surfaces	0.011	0.015
Plank flumes, unplaned	0.011	0.015
Common-clay drainage tile	0.011	0.017
Concrete, monolithic	0.012	0.016
Brick with cement mortar	0.012	0.017
Cast iron, new	0.013	0.017
Riveted steel	0.017	0.020
Cement rubble surfaces	0.017	0.030
Canals and ditches, smooth earth	0.017	0.025
Corrugated metal pipe	0.021	0.030
Metal flumes, corrugated	0.022	0.030
Canals		
Dredged in earth, smooth	0.025	0.033
In rock cuts, smooth	0.025	0.035
Rough beds and weeds on sides	0.025	0.040
Rock cuts, jagged and irregular	0.035	0.045
Natural streams		
Smoothest	0.025	0.033
Roughest	0.045	0.060
Very weedy	0.075	0.150

TABLE 8.1   Values of n in Manning's formula

Some workers define $AR_h^{2/3}$ as the *conveyance* K or the *section factor* of the cross section. Then Eq. (8.8) indicates that the (uniform) flow rate depends on the surface roughness, the cross section's conveyance, and the bed slope.

Variation of n

In Sec. 7.13 we mention that e is a measure of the absolute roughness of the inside of a pipe. The question naturally arises whether e and n may be functionally related to each other.

To explore this, first note that we can combine Eq. (8.5) with Eq. (8.7), to yield

In BG units:
$$n = 1.486 R_h^{1/6} \sqrt{\frac{f}{8g}}$$
(8.9a)

In SI units:
$$n = R_h^{1/6} \sqrt{\frac{f}{8g}}$$
(8.9b)

Substituting numerical values for g gives

In BG units:
$$n = 0.0926 f^{1/2} R_h^{1/6}$$
(8.10a)

In SI units:
$$n = 0.1129 f^{1/2} R_h^{1/6}$$
(8.10b)

Thus we see that n is related to the friction factor, which depends on the relative roughness and Reynolds number and on the hydraulic radius, which is indicative of the size of the channel.

Second, since we know from Sec. 8.1 that open-channel flow is usually fully rough, we can take the fully-rough-pipe flow Eq. (7.50), substitute $4R_h$ for D per Sec. 8.1 and Eq. (7.21), and working in BG units for example, substitute for $\sqrt{f}$ from Eq. (8.9a). After simplifying and rearranging this results in

$$n = \frac{1.486 R_h^{1/6}}{4\sqrt{2g}\,\log\left(\dfrac{14.8 R_h}{e}\right)} \quad \text{or} \quad e = 14.8 R_h \exp -\left(\frac{1.486(\ln 10) R_h^{1/6}}{4\sqrt{2g}\,n}\right)$$

where for SI units we simply replace 1.486 by 1.000. These provide a relationship between e and n, which is plotted as the solid lines in Fig. 8.4 for three representative values of the hydraulic radius. The dashed line is the plot of another correlation proposed by Powell[3] that gave a better fit to experimental data for rectangular channels and small values of hydraulic radius.

One notable feature of the curves in Fig. 8.4 is that a large relative error in e causes only a small error in n. Another important observation concerns the variation of n with channel size for the same absolute roughness. For channels that are quite rough, like natural channels with $e > 0.10$ ft, we see in Fig. 8.4 that n gets smaller with increasing hydraulic radius. This, being similar to the behavior of relative roughness e/D in pipes, has long been understood. However, at lower e values the curves cross, with the result that for smoother, artificial surfaces with $e < 0.02$ ft we find that n *increases* with increasing hydraulic radius. This contrary behavior was formerly not understood, with the result that some very large concrete-lined canal systems were designed using inappropriately low n values; when put into service, designers found that the higher true n values reduced the capacity of the canals by as much as 15%.

---

[3] R. W. Powell, Resistance to Flow in Rough Channels, *Trans. Am. Geophys. Union*, 31(4):575–582, 1950.

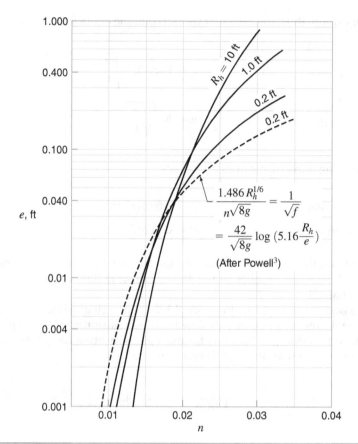

FIGURE 8.4 Correlation of n with absolute roughness e.

8.3 Solution of Uniform Flow Problems

Uniform flow problems usually involve the application of Manning's equation [Eq. (8.8)]. The selection of an appropriate value for the Manning roughness factor n is critical to the accuracy of the results of a problem. When the channel surface is concrete or some other structural material, it is possible to select a reasonably accurate value for n, but for the case of a natural channel one must rely on judgment and experience, and in many instances the selected value may be quite inaccurate.

We encounter a number of different types of problems when using Manning's equation. For example, to find the normal depth of flow for a particular flow rate in a given channel, we require a trial-and-error solution or the use of an equation solver, because y_0 is involved in A and R_h in complicated ways (see, for example, Sample Problem 8.1). On the other hand, we can solve for the expected flow in a particular channel under given conditions directly. Various types of nomographs, tables,[4] and computer

---

[4] E. F. Brater, H. W. King, J. E. Lindell, and C. Y. Wei, *Handbook of Applied Hydraulics*, 7th ed., McGraw-Hill Book Company, New York, 1996.

programs, are available to serve as an aid to the solution of open-channel problems. Some of these provide helpful visual representations of the interdependence of the various factors, but their applicability is usually limited.

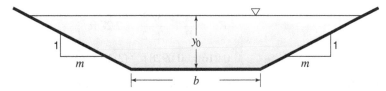

The most common shape of engineered channels is trapezoidal, for which the geometry can be combined with Manning's equation, employing Eq. (7.3) and substituting into Eq. (8.7), as

$$A = (b + my_0)y_0, \quad P = b + 2y_0\sqrt{1 + m^2}, \quad R_h = \frac{A}{P}$$

$$Q = \frac{C_1}{n} \frac{(by_0 + my_0^2)^{5/3}}{(b + 2y_0\sqrt{1 + m^2})^{2/3}} S_0^{1/2}$$

Here $C_1 = 1.486$ for BG units and $C_1 = 1.000$ for SI units. Note that side slope, m, is the horizontal run over the vertical rise. This formulation applies to triangular (with bottom width, $b = 0$) and rectangular (with side slope, $m = 0$) channel shapes as well.

Sample Problem 8.1

A trapezoidal channel has bottom width $b = 10$ ft, side slope $m = 2$, a bed slope of 0.0006 and a Manning's n of 0.016. Using only a basic scientific calculator, (a) find the uniform flow depth when the flow rate is 225 cfs and (b) compute the corresponding value of the absolute roughness e. Also, find the uniform flow depth using (c) equation solving software.

Solution

(a)
$$A = (10 + 2y_0)y_0$$

and
$$R_h = \frac{A}{P} = \frac{(10 + 2y_0)y_0}{10 + 2\sqrt{5}y_0}$$

Eq. (8.8a): $\quad Q = 225 = \dfrac{1.486}{0.016}(10 + 2y_0)y_0 \left[\dfrac{(10 + 2y_0)y_0}{10 + 2\sqrt{5}y_0}\right]^{2/3} (0.0006)^{1/2}$ \hfill (1)

By trial (see Sample Problems 3.8 and 5.8), $y_0 = 3.41$ ft, uniform flow depth **ANS**

Note: The accuracy with which flow depth can be measured can differ considerably, depending on many factors. In a laboratory, we can sometimes measure depths to 0.001 ft (0.3 mm), whereas in the field, with turbulent flow and wind, surface fluctuations (which we can average) may make it difficult to achieve an accuracy of 0.1 ft (30 mm).

Programmed computing aids could help solve problems marked with this icon.

(b)
$$A = [10 + 2(3.41)]3.41 = 57.3 \text{ ft}^2$$

$$P = 10 + 2\sqrt{5}(3.41) = 25.2 \text{ ft}$$

$$R_h = \frac{A}{P} = \frac{57.3}{25.2} = 2.27 \text{ ft}$$

Eq. (8.10a): $$0.016 = 0.0928 f^{1/2}(2.27)^{1/6}$$

from which $$f^{1/2} = 0.1503 \quad (f = 0.0226)$$

Eq. (8.9): $$\frac{1}{0.1503} = 2\log\left(\frac{14.8 \times 2.27}{e}\right)$$

from which $$e = 0.01585 \text{ ft} \qquad ANS$$

(c) Using equation solving software, set up the four governing equations,

Eq. (8.8): $$Q = \frac{C_1}{n} A R_h^{2/3} S_0^{1/2}$$

and $$A = (b + my_0)y_0, \quad P = b + 2y_0\sqrt{1 + m^2}, \quad R_h = \frac{A}{P}$$

These four equations involve 10 variables, four of which need to be found. Assign numerical values to the six known variables (Q, C_1, b, m, n, S_0) as earlier. Assign estimated initial values to the four unknown variables if needed by the software, for example $y_0 = 1.0$, $A = 10$, $P = 10$, $R_h = 1.0$.
    The results are

$$y_0 = 3.406 \text{ ft}, \quad A = 57.27 \text{ ft}^2, \quad P = 25.23 \text{ ft}, \quad R_h = 2.270 \text{ ft} \qquad ANS$$

Alternatively, the single equation used in part (a) can be rearranged to

$$\frac{1.486}{0.016}(10 + 2y_0)y_0\left[\frac{(10 + 2y_0)y_0}{10 + 2\sqrt{5}y_0}\right]^{2/3}(0.0006)^{1/2} - 225 = 0$$

which, when used with an equation solver to find the root, produces the same value for y_0 as above.

    In applying Manning's equation to channel shapes such as that in Fig. 8.5, which simulates a river with overbank flow conditions, the usual procedure is to break the section into several parts, as indicated in the figure. We assume that there is no resistance along the dashed vertical lines. Actually the flow in area A_2 tends to speed up the flow in area A_1, while the flow in A_1 tends to slow down the flow in area A_2. These two

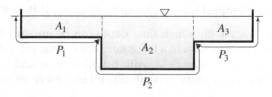

FIGURE 8.5   Manning's equation variables for a compound channel.

effects come very close to balancing one another. If A/P for the total cross section had been computed by the usual method, that is, $R_h = (A_1 + A_2 + A_3)/(P_1 + P_2 + P_3)$, it would imply that the effect of boundary resistance is uniformly distributed over the flow cross section, which, of course, is not the case.

Another advantage of breaking the total section into parts is that we can take into consideration possible variations in Manning's n. Thus, for the channel shown in Fig. 8.5

$$Q = \frac{C_1}{n_1} A_1 R_{h_1}^{2/3} S_0^{1/2} + \frac{C_1}{n_2} A_2 R_{h_2}^{2/3} S_0^{1/2} + \cdots \tag{8.11}$$

where $R_{h_1} = A_1/P_1$, $R_{h_2} = A_2/P_2$, etc. The A and P are defined in Fig. 8.5. As above, $C_1 = 1.486$ for BG units and $C_1 = 1.000$ for SI units.

8.4 Velocity Distribution in Open Channels

Due to friction around the wetted perimeter, we expect strong velocity variations across flow cross sections in open channels. A small frictional effect must also occur at the water surface, due to air drag.

Vanoni[5] has demonstrated that the universal logarithmic velocity-distribution law for pipes [Eq. (7.40)] also applies to a two-dimensional open channel, i.e., one of uniform depth that is infinitely wide (see Sec. 8.5). We can write this equation as

$$\frac{u - u_{max}}{\sqrt{gy_0 S_0}} = \frac{2.3}{K} \log \frac{y}{y_0}$$

where y_0 = depth of water in channel
u = velocity at a distance y from channel bed
K = von Kármán constant, having a value of about 0.40
S_0 = channel bed slope
Integrating this expression over the depth yields the more useful relation

$$u = V + \frac{1}{K}\sqrt{gy_0 S_0}\left(1 + 2.3\log\frac{y}{y_0}\right) \tag{8.12}$$

which expresses the distribution law in terms of the mean velocity V. This equation is plotted in Fig. 8.6, together with velocity measurements that were made on the centerline of a rectangular flume 2.77 ft (0.844 m) wide with a water depth of 0.59 ft (0.180 m). We see that the filament whose velocity u is equal to V lies at a distance of $0.632y_0$ beneath the surface.

Velocity measurements made in a trapezoidal channel result in the isovels (contours of equal velocity) shown in Fig. 8.7. The point of maximum velocity is seen to lie beneath the surface, the result of air drag. From this two-dimensional velocity distribution the values of the correction factors for kinetic energy and momentum were computed, and are given with the figure. Although they are greater than the corresponding values for pipe flow [Eq. (7.45); Secs. 5.1 and 6.3], the treatment in this chapter follows the earlier procedure of assuming the values of α and β to be unity, unless stated otherwise. Any thorough analysis would of course have to take account of their true values.

[5] V. A. Vanoni, Velocity Distribution in Open Channels, *Civil Eng.*, 11:356–357, 1941.

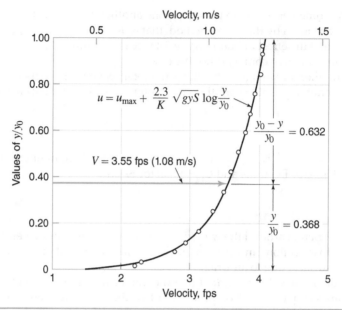

FIGURE 8.6 Velocity profile at the center of a flume 2.77 ft (0.844 m) wide for a flow 0.59 ft (0.180 m) deep (after Vanoni). The circles represent measurements.

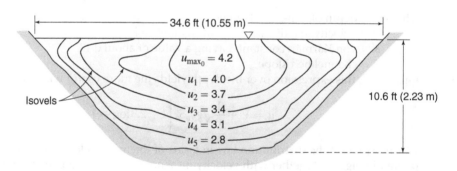

FIGURE 8.7 Velocity distribution in fps in a trapezoidal canal; $V = 3.32$ fps (1.012 m/s), $A = 230.5$ ft² (21.4 m²), $S_0 = 0.000057$, $\alpha = 1.105$, $\beta = 1.048$.

8.5 "Wide and Shallow" Flow

If the flow in a channel is sufficiently wide compared with its depth, it is reasonable to expect that the channel sides will have negligible effect on the flow in the central portion of the channel. Experiments have confirmed this to be the case provided the width-to-depth ratio $b/y > 10$. Then the central flow may be considered to be the same as in a channel of infinite width—in other words, to be two-dimensional flow.

For a small portion of such a two-dimensional flow with width Δb, the flow cross-sectional area is $y\,\Delta b$ and the wetted perimeter is Δb, so that the hydraulic radius is

Wide and shallow:
$$R_h = \frac{A}{P} = \frac{y\Delta b}{\Delta b} = y \qquad\qquad (8.13)$$

This convenient result is commonly assumed for *wide and shallow open channels.* The same result is obtained by considering the hydraulic radius of a rectangular channel,

$$R_h = \frac{A}{P} = \frac{by}{b + 2y} = \frac{y}{1 + 2y/b}$$

As the ratio $b/y \to \infty$, we see that $R_h \to y$.

Analytically, the percentage error in R_h when using y as the hydraulic radius for a rectangular channel is

$$\frac{100(y - R_h)}{R_h} = 100\left(y - \frac{by}{b + 2y}\right)\frac{b + 2y}{by}$$

i.e.,
$$\text{Error \% in } R_h = 200\frac{y}{b}$$

This results in a 20% error for $b/y = 10$, a 5% error for $b/y = 40$, and a 2% error for $b/y = 100$. If we make the assumption that a channel is "wide and shallow" and replace R_h by y in our analysis, it is important to check that the b/y ratio is sufficiently large to justify this.

Solution for the uniform flow depth is greatly simplified when a channel is wide and shallow because trial and error is no longer necessary. If we choose BG units, substituting for A and R_h into Manning's Eq. (8.8b) yields

$$Q = \frac{1.486}{n}(by_0)y_0^{2/3}S_0^{1/2}$$

from which

Wide and shallow (BG units): $\quad y_0 = \left(\frac{nQ}{1.486bS_0^{1/2}}\right)^{0.6}$

Sample Problem 8.2

A 12-m-wide stream has a reasonably rectangular cross section, and the average depth is computed to be 112 mm. What would be the percentage error in the hydraulic radius if this stream were assumed to be "wide and shallow"? Would it be reasonable to make this assumption?

Solution

$$\frac{b}{y} = \frac{12}{0.112} = 107.1$$

$$\text{Error percent} = \frac{200}{107.1} = 1.867\% \qquad ANS$$

This error is small enough that the assumption is reasonable. $\quad ANS$

8.6   Most Efficient Cross Section

Any of the open-channel formulas already given show that, for a given slope and roughness, the velocity increases with the hydraulic radius. Therefore, for a given area of water cross section, the rate of discharge will be a maximum when R_h is a maximum, i.e., when the wetted perimeter P and so the frictional resistance is a minimum. We call such a section the *most efficient cross section* or *the best hydraulic section* for the given area, and we emphasize that here we are addressing only the *hydraulic* efficiency. Or for a given rate of discharge, the cross-sectional area will be a minimum when the design is such as to make R_h a maximum (and thus P a minimum). This section would be the most efficient cross section for the given rate of discharge. These issues are important, because reducing P tends to reduce channel construction costs. However, other factors influence the *overall* cost and efficiency of an open channel. For example, the cost of deep excavation is higher than shallow excavation. Consequently most channels are designed to flow with depths less than the hydraulic optimum, resulting in R_h values a few percent smaller than $(R_h)_{max}$. In situations where right-of-way costs are very high, or where right-of-way width is limited, open channels are sometimes deeper than the hydraulic optimum.

Of all geometric figures, the circle has the least perimeter for a given area. The hydraulic radius of a semicircle is the same as that of a circle. Hence a semicircular open channel will discharge more water than one of any other shape, assuming that the area, slope, and surface roughness are the same. Semicircular open channels are often built of pressed steel and other forms of metal, but for other types of construction such a shape is impractical. For wooden flumes designers usually select a rectangular shape. Canals excavated in earth usually have a trapezoidal cross section, with side slopes less than the angle of repose of the saturated bank material. Besides hydraulic efficiency, such factors as those mentioned help determine the best cross section.

We can determine the shape of the most efficient trapezoidal cross section by using its geometric properties and a little calculus. Consider the trapezoid of Fig. 8.8, with bed width b and side slope m:1. It follows that $A = (b + my)y$ and $P = b + 2y\sqrt{1 + m^2}$. Eliminating b, we obtain

$$P = \frac{A}{y} + y\left(2\sqrt{1 + m^2} - m\right)$$

and

$$R_h = \frac{A}{P} = \frac{A}{\dfrac{A}{y} + y\left(2\sqrt{1 + m^2} - m\right)}$$

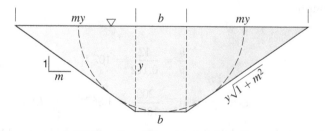

FIGURE 8.8   The most hydraulically efficient trapezoidal cross section.

Case 1   For a given A and m, we find $(R_h)_{max}$ by differentiating R_h with respect to y while holding A and m constant, and setting the result equal to zero. From this we obtain

$$y_{opt} = \sqrt{\frac{A}{2\sqrt{1+m^2} - m}}$$

$$(R_h)_{max} = \frac{y_{opt}}{2}$$

and $P_{opt} = 2y_{opt}\left(2\sqrt{1+m^2} - m\right),$ $b_{opt} = 2y_{opt}\left(\sqrt{1+m^2} - m\right)$

We may find Q_{opt} by substituting $(R_h)_{max}$ into Eq. (8.8).

Case 2   For a given Q, m, n, and S_0, as in problems where normal depth is to be determined, we rearrange Eq. (8.8) into

$$A = \frac{Qn}{C_1 R_h^{2/3} S_0^{1/2}}$$

where $C_1 = 1.486$ for BG units and $C_1 = 1.000$ for SI units. We use this to eliminate the variable A from the geometric relations. Then we find $(R_h)_{max}$ by differentiating R_h with respect to y while holding Q, m, n, and S_0 constant, and setting dR_h/dy to zero. From this we obtain

$$y_{opt} = 2^{1/4}\left[\frac{Qn}{C_1(2\sqrt{1+m^2} - m)S_0^{1/2}}\right]^{3/8}$$

$$A_{opt} = y_{opt}^2\left(2\sqrt{1+m^2} - m\right)$$

and that $(R_h)_{max}$, P_{opt}, and b_{opt} are given by the same equations as in Case 1.

Conclusions: Of these results for any trapezoid, the equation for $(R_h)_{max}$ is easy to remember. It is also interesting (see Problem 8.12) that the optimum trapezoid always envelopes a semicircle whose center is in the water surface (shown dashed in Fig. 8.8).

For the common rectangular cross section, this is a trapezoid with $m = 0$, so these results then reduce to $A = 2y^2$, $P = 4y$, $b = 2y$, and $R_h = y/2$. Thus the most efficient rectangular section has its depth equal to half its width, and this of course still envelopes a semicircle.

These results for trapezoidal cross sections are for given values of side slope m. We can find the hydraulically best shape of trapezoid (value of m) by expressing $(R_h)_{max}$ as a function of only known quantities and m, e.g., as a function of only A and m when A is given, in which case

$$(R_h)_{max} = \frac{y_{opt}}{2} = \frac{1}{2}\sqrt{\frac{A}{2\sqrt{1+m^2} - m}}$$

and then differentiating $(R_h)_{max}$ with respect to m while holding A constant. The result is that $(R_h)_{max}/dm$ is zero when $m = \infty$ or

$$\left(\frac{2m}{\sqrt{1+m^2}} - 1\right) = 0$$

We obtain the same result if instead Q, n, S_0, and m are given. Thus $(R_h)_{max}$ is maximum when $m = 1/\sqrt{3}$, i.e., the side slopes are at 60° to the horizontal and the trapezoid is a half-hexagon. We note that this is the trapezoid that most closely envelopes the semi-circle.

In a similar manner we can show that the most efficient triangular cross section is the one that has a total vertex angle of 90°.

Sample Problem 8.3
The first 95 miles of the Delta-Mendota Canal in California, designed to carry 4600 cfs, has the following dimensions (referring to Fig. 8.8): $b = 48$ ft, $m = 1.5$, $y = 16.7$ ft. Using the same cross-sectional area and side slopes, how much would the most efficient cross section increase the present hydraulic radius and flow capacity? What would be the corresponding depth and bed width?

Solution
Existing:
$$A = (48 + 1.5 \times 16.7)\,16.7 = 1220 \text{ ft}^2$$

$$P = 48 + 2\left(\frac{\sqrt{13}}{2}\right)16.7 = 108.2 \text{ ft}$$

So
$$R_h = \frac{A}{P} = \frac{1220}{108.2} = 11.27 \text{ ft}$$

which we note is considerably different from $y/2 = 8.35$ ft.

For an optimum cross section we note this is Case 1 (given A), for which

$$y_{opt} = \sqrt{\frac{1220}{2\sqrt{3.25} - 1.5}} = 24.07 \text{ ft} \qquad ANS$$

Note that this is 44% greater than the existing depth.

Also
$$(R_h)_{max} = \frac{y_{opt}}{2} = 12.04 \text{ ft}$$

This is only 6.8% greater than the existing hydraulic radius.   *ANS*

From Manning's Eq. (8.8) for the same cross-sectional area:

$$\frac{Q_{opt}}{Q} = \left[\frac{(R_h)_{max}}{R_h}\right]^{2/3} = \left(\frac{12.04}{11.27}\right)^{2/3} = 1.0450$$

Therefore Q_{opt} is 4.50% greater than the existing capacity.   *ANS*

$$b_{opt} = 2(24.07)(\sqrt{3.25} - 1.5) = 14.58 \text{ ft} \qquad ANS$$

8.7 Circular Sections Not Flowing Full

In circular pipes flow frequently occurs at partial depth, in which case they are behaving as open channels. As we can visualize from Fig. 8.9, at depths just slightly less than full depth the wetted perimeter (frictional resistance) is reduced in greater proportion than the flow area, so the hydraulic radius A/P is greater than when full. In accordance with the Manning formula, therefore, the maximum rate of discharge in such a section occurs at slightly less than full depth.

We can explore such maximum condition analytically. Referring to Fig. 8.9,

$$y = \frac{D}{2}(1 - \cos(\theta/2)) \quad \text{or} \quad \frac{y}{D} = \frac{1}{2}(1 - \cos(\theta/2))$$

$$A = \frac{D^2}{8}(\theta - \sin\theta) \quad \text{and} \quad P = D\frac{\theta}{2}$$

where θ is expressed in radians. This gives

$$R_h = \frac{A}{P} = \frac{D}{4}\left(1 - \frac{\sin\theta}{\theta}\right)$$

For the maximum rate of *discharge,* the Manning formula indicates that $AR_h^{2/3}$ must be a maximum. Substituting the preceding expressions for A and R_h into $AR_h^{2/3}$ and differentiating with respect to θ, setting equal to zero and solving for θ gives $\theta = 302.4°$, which corresponds to $y = 0.938D$ for the condition of maximum discharge. By differentiating $R_h^{2/3}$ the maximum *velocity* is found to occur at $0.813D$.

These analyses of optimum conditions *are theoretical only,* and not recommended for general use, because they assume a constant roughness coefficient in Manning's equation. In Sec. 8.2 we saw how, for constant e, the friction factor n varied with R_h For circular pipes investigators have found that the value of n increases by as much as 28% from the full depth to about one-quarter full depth, where it appears to be a maximum (Fig. 8.10). This effect causes the actual maximum discharge and velocity to occur at water depths of about 0.97 and 0.94 full depth, respectively, as indicated in Fig. 8.10.[6] The interesting result is that in the small topmost depth range where the discharge is equal to or greater than Q_{full}, for a given discharge two different depths may occur.

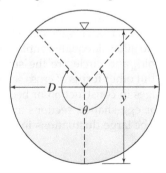

FIGURE 8.9 Circular section not full.

---

[6] See Design and Construction of Urban Stormwater Management Systems, *ASCE Manuals and Reports on Engineering Practice No. 77,* 146, 1992.

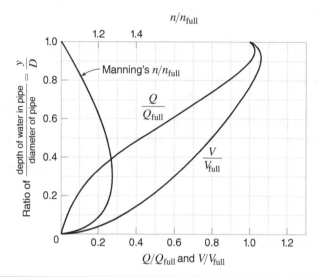

FIGURE 8.10 Hydraulic characteristics of circular pipe flowing partly full (including for the effects of *n* varying with depth).

The simplest way to handle the problem of a partially full circular section is to compute the velocity or flow rate for the pipe-full condition and then adjust to partly full conditions by using a chart such as Fig. 8.10. On this chart the effect of the variation of Manning's *n* with depth is already taken into consideration. Manning's equation can be conveniently expressed in terms of θ[7] (in radians)

$$\theta^{-\frac{2}{3}}(\theta - \sin\theta)^{\frac{5}{3}} - C_2 nQD^{-\frac{8}{3}}S_0^{-\frac{1}{2}} = 0 \quad [C_2 = 20.16 \text{ (SI units)}; C_2 = 13.53 \text{ (BG units)}]$$

which can be solved by trial and error or using an equation solver. This avoids the uncertainties of interpolating from Fig. 8.10, though the solution assumes *n* does not vary with depth. Explicit approximations have been developed to apply Manning's equation for partially full pipes including the effect of varying *n* with depth.[8]

Despite the analysis just made of optimum conditions, circular sections in some installations are designed to carry the design capacity when flowing full, since the conditions producing maximum flow frequently include sufficient backwater to place the conduit under slight pressure. For sanitary sewers the design is often at some level less than full (e.g., 80% full or $y/D = 0.8$) to allow adequate ventilation of corrosive or odorous gases.[9]

The rectangle, trapezoid, and circle are the simplest geometric shapes from the standpoint of hydraulics, but other forms of cross section are often used, either because they have certain advantages in construction or because they are desirable from other standpoints. Thus oval- or egg-shaped sections are common for sewers and similar channels where there may be large fluctuations in the rate of discharge. They are used

[7] D.A. Chin, *Water-Resources Engineering*, 3rd ed., Pearson, 2012.
[8] Ö Akgiray, Explicit Solutions of the Manning Equation for Partially Filled Circular Pipes. *Canadian J. Civil Engineering*, 32(3):490–499, 2005.
[9] Gravity Sanitary Sewer Design and Construction, *ASCE Manuals and Reports on Engineering Practice No. 60*, 2007.

because it is desirable that the velocity, when a small quantity is flowing, be kept high enough to prevent the deposit of sediment, and when the conduit is full, the velocity should not be so high as to cause excessive wear of the pipe wall. For example, in sanitary sewers we usually maintain a velocity of at least 2 fps to prevent deposition of sediment, while in concrete channels conveying storm water, velocities in excess of 10 or 12 fps may result in excessive abrasion of the channel sides and bottom.

Sample Problem 8.4
A 36-in-diameter pipe flows just full when it carries 30 cfs. What will be the discharge and depth when the velocity is 2 fps?

Solution

$$V_{full} = \frac{Q_{full}}{A} = \frac{30}{\pi 3^2/4} = 4.24 \text{ fps}$$

At 2 fps: $\quad V/V_{full} = 2/4.24 = 0.471$

Fig. 8.10 for $V/V_{full} = 0.471$: $\quad y/D \approx 0.20$

So $\quad y \approx 0.20(36) = 7.20 \text{ in}$ **ANS**

Fig. 8.10 for $y/D = 0.20$: $\quad Q/Q_{full} \approx 0.077$

So $\quad Q \approx 0.077(30) = 2.31 \text{ cfs}$ **ANS**

Sample Problem 8.5
A new 30-in-diameter precast concrete pipe is installed with a slope of $S_0 = 0.005$ (0.5%). It is to carry its peak flow when flowing 80% full ($y/D = 0.8$). Determine what flow rate, Q, this can accommodate using (a) Fig. 8.10, and (b) the geometric equations for hydraulic geometry. Also, (c) use equations to find the flow depth for a 36-in-diameter pipe at $S_0 = 0.005$ carrying $Q = 30$ ft³/s, assuming constant $n = 0.013$.

Solution
Table 8.1: use $n = 0.013$ for precast concrete pipe.
(a) For full flow

$$A_{full} = \frac{\pi}{4}\left(\frac{30}{12}\right)^2 = 4.91 \text{ ft}^2$$

$$R_{full} = \frac{A}{P} = \frac{D}{4} = 0.625 \text{ ft}$$

Eq. (8.8a): $Q_{full} = 1.486\frac{1}{0.013}(4.91)(0.625)^{2/3}(0.005)^{1/2} = 29.0 \text{ ft}^3/s$

Fig. 8.10: for $y/D = 0.8$, $Q/Q_{full} \approx 0.87$

$$Q = 0.87 \times Q_{full} = 0.87 \times 29.0 = 25.2 \text{ ft}^3/s \quad \textbf{ANS}$$

(b) First find θ for a pipe 80% full

$$\theta = 2\cos^{-1}\left(1 - 2\frac{y}{D}\right) = 2\cos^{-1}(1 - 2 \cdot 0.8) = 253.7°\left(\frac{\pi \text{ radian}}{180°}\right) = 4.43 \text{ radians}$$

$$A = \frac{(30/12)^2}{8}(4.43 - \sin 4.43) = 4.21 \text{ ft}^2$$

$$P = \frac{4.43(30/12)}{2} = 5.54 \text{ ft}$$

$$R = \frac{A}{P} = \frac{4.21}{5.54} = 0.76 \text{ ft}$$

$$Q = 1.486\frac{1}{0.013}(4.21)(0.76)^{2/3}(0.005)^{1/2} = 28.3 \text{ ft}^3/\text{s} \qquad ANS$$

The result for (b) is greater than for part (a) because the equation assumes a constant n with depth, while Fig. 8.10 accounts for varying n with depth. To include the effects of varying n, use the n/n_{full} curve in Fig. 8.10. For $y/D = 0.8$, $n/n_{full} = 1.12$; use $n = 1.12(0.013) = 0.0146$ in the Manning equation:

$$Q = 1.486\frac{1}{0.0146}(4.21)(0.76)^{2/3}(0.005)^{1/2} = 25.2 \text{ ft}^3/\text{s}$$

While it is known that n varies with depth, design guidance often specifies the use of constant n values,[10] in which case the adjustment could be applied to the answer from part (a) to match that of part (b).

(c) Substitute known values for D, S_0, Q, n, and C_2 ($C_2 = 13.53$ since BG units are used) into the Manning formulation for circular pipes flowing partially full:

$$\theta^{-2/3}(\theta - \sin\theta)^{5/3} - C_2 nQD^{-8/3}S_0^{-1/2} = 0$$

Use an equation solver to find the solution of $\theta = 3.456$ radians. Then

$$y = \frac{D}{2}\left(1 - \cos\frac{\theta}{2}\right) = \frac{3.0}{2}\left(1 - \cos\frac{3.456}{2}\right) = 1.735 \text{ ft} \qquad ANS$$

8.8   Laminar Flow in Open Channels

Laminar flow in open channels is sometimes encountered in industrial processes where a very viscous liquid is flowing in a trough or similar conveyance structure. More commonly, however, laminar flow occurs as *sheet flow,* a thin sheet of flowing liquid, such as that in drainage from sidewalks, streets, and airport runways.

Sheet flow is essentially two-dimensional and can be analyzed in that fashion. Consider two-dimensional uniform laminar flow at depth y_0 as shown in Fig. 8.11. At a

---

[10] Federal Highway Administration, *Introduction to Highway Hydraulics June 2008 Hydraulic Design Series Number 4*, 4th ed., 2008.

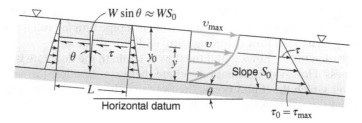

FIGURE 8.11 Laminar flow in open channel of infinite width and uniform depth, showing the velocity profile and the shear-stress distribution.

distance y above the channel bed the shear stress in the flow is τ and the velocity is v. The forces acting on a unit width of the liquid represented by the shaded area include the hydrostatic forces which cancel out, the gravity force[11] component parallel to the slope $\gamma(y_0 - y)LS_0$ and the shear force τL exerted along the lower boundary of the shaded liquid by the liquid below it. Since the flow is uniform, there is no acceleration and thus these two forces must balance one another. Hence

$$\gamma(y_0 - y)LS_0 = \tau L$$

From this expression, it can be seen that the shear stress must vary linearly from zero at the liquid surface to a maximum value at the channel bed.

Since the flow is laminar, we can use Eq. (2.9) to replace τ with $\mu \, dv/dy$. Making this substitution, we get

$$\gamma(y_0 - y)S_0 = \mu \frac{dv}{dy}$$

Separating variables and integrating, noting that $v = 0$ when $y = 0$, gives

$$dv = \frac{\gamma}{\mu}(y_0 - y)\, S_0 dy$$

and
$$v = \frac{\gamma}{\mu}\left(y_0 y - \frac{y^2}{2}\right)S_0 = \frac{g}{\nu}\left(y_0 y - \frac{y^2}{2}\right)S_0 \qquad (8.14)$$

This is the equation of a parabola; thus the velocity profile is parabolic, as it was for laminar flow in a pipe (Sec. 7.8).

We can now integrate Eq. (8.14) over the depth from $y = 0$ to $y = y_0$ to obtain an expression for q, the flow rate per unit width:

$$dq = \int_0^{y_0} v \, dy$$

$$q = \frac{g}{\nu}\frac{y_0^3}{3}S_0 \qquad (8.15)$$

---

[11] We always measure depth y vertically. Thus, $\gamma(y_0 - y)L$ is not precisely the weight of the shaded volume of liquid. However, for the small slopes usually encountered it is an excellent approximation.

From Eqs. (8.14) and (8.15), it can be shown that the average velocity V for this case is equal to $\frac{2}{3}v_{max}$. In contrast, for laminar flow in a pipe flowing full it was shown (Sec. 7.7) that $\frac{2}{3}v_{max}$. The total flow Q through any width B will be qB.

Equation (8.15) shows that if the flow is laminar, the flow rate is independent of the roughness. However, in situations where there are significant irregularities in the surface over which the liquid flows, the flow may be disturbed such that it is not everywhere laminar. Consequently, we must apply Eqs. (8.14) and (8.15) with caution.

Sample Problem 8.6
Water is observed to flow uniformly across a glass roof at a depth of 1 mm. (a) If the temperature is 15°C and the slope of the roof is 0.08, what is the discharge rate from the edge of the roof per meter length? (b) How fast should we expect an oil mark on the water to move?

Solution

Table A.1 for 15°C: $\nu = 1.139 \times 10^{-6} \text{ m}^2/\text{s}$

(a) Assuming laminar flow:

Eq. (8.15): $q = \dfrac{9.81}{1.139 \times 10^{-6}} \dfrac{(0.001)^3}{3} 0.08 = 0.000230 \text{ m}^3/\text{s per meter}$

Since $b/y > 1/0.001 = 1000$, per Eq. (8.13): $R_h = y_0 = 0.001$ m

$$V = \frac{Q}{A} = \frac{q}{y_0} = \frac{0.000230}{0.001} = 0.230 \text{ m/s}$$

Sec. 8.1: $\mathbf{R} = \dfrac{R_h V}{\nu} = \dfrac{(0.001)0.230}{1.139 \times 10^{-6}} = 201$

As $\mathbf{R} < \mathbf{R}_c = 500$, flow is laminar and Eqs. (8.14) and (8.15) apply. Therefore $q = 0.000230 \text{ m}^3/\text{s per meter}$ *ANS*

(b) Laminar flow Eq. (8.14) at the surface, where $y = y_0$:

$$v = \frac{g}{\nu}\frac{y_0^2}{2}S_0 = \frac{9.81}{1.139 \times 10^{-6}}\frac{(0.001)^2}{2}0.08 = 0.345 \text{ m/s} \quad ANS$$

8.9 Specific Energy and Alternate Depths of Flow in Rectangular Channels

For *any* cross-section shape, the *specific energy* E at a particular section is defined as the energy head referred to the channel bed as datum. Thus, if the channel bed rises abruptly E decreases and if the channel bed drops E increases (ignoring head losses). Thus

$$E = y + \alpha \frac{V^2}{2g} \tag{8.16}$$

where α is the kinetic energy correction factor (Sec. 5.1), which accounts for velocity variations across the section. Friction at the channel walls reduces velocities near the wetted perimeter, as indicated in Fig. 8.7 for which $\alpha = 1.105$. As noted in Sec. 8.4, the value of α is usually assumed to be 1.0; for typical velocities this results in only a small error in E.

For *rectangular* channels that we discuss in this section, provided they are not unusually narrow so that α is large (Sec. 5.1), a representative average value of the flow q per unit width can be expressed as $q = Q/b$. The average velocity $V = Q/A = qb/by = q/y$ and so Eq. (8.16) with $\alpha = 1$ can be expressed as

$$E = y + \frac{1}{2g}\frac{q^2}{y^2} \tag{8.17}$$

Let us consider how E will vary with y if q remains constant. Physically, this situation would occur if the slope of a rectangular channel could be changed with the flow rate remaining constant. Manning's equation [Eq. (8.8)] indicates that on a steep slope, with a given flow rate, the normal depth of flow y_0 will be relatively small in contrast to a larger depth on a flatter slope. That this is so can be seen more easily for a wide and shallow channel by writing the Manning equation in terms of the flow q per unit width, noting that $A = by$ and $R_h \approx y$ for $b >> y$. Thus, for uniform flow ($y = y_0$), in BG units

Wide and shallow, uniform flow:

$$q = \frac{Q}{b} = \frac{1.486}{nb}AR_h^{2/3}S_0^{1/2} = \frac{1.486}{n}\frac{by_0}{h}y_0^{2/3}S_0^{1/2} = \frac{1.486}{n}y_0^{5/3}S_0^{1/2} \tag{8.18}$$

For the case of constant q, we can restate Eq. (8.17) as

$$(E - y)y^2 = \frac{q^2}{2g} = \text{constant} \tag{8.19}$$

A plot of E versus y is hyperbola-like with asymptotes $(E - y) = 0$ (that is, $E = y$) and $y = 0$. Such a curve, shown in Fig. 8.12, is known as the *specific-energy diagram*.

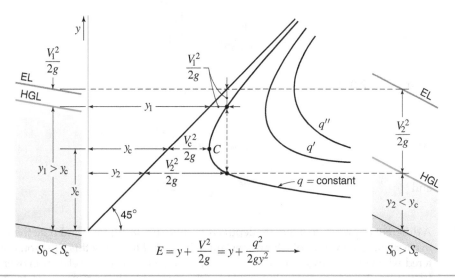

Figure 8.12 Specific-energy diagram for three constant rates of discharge in a rectangular channel. (Bed slopes are greatly exaggerated.)

Actually, each different value of q will give a different curve, as shown in Fig. 8.12. For a particular q, we see there are two possible values of y for a given value of E. These are known as **alternate depths.** Equation (8.19) is a cubic equation with three roots, the third root being negative has no physical meaning. The two alternate depths represent two totally different flow regimes—slow and deep on the upper limb of the curve and fast and shallow on the lower limb of the curve. Point C represents the dividing point between the two regimes of flow. At C, for a given q, the value of E is a *minimum* and we refer to the flow at this point as **critical flow.** The depth of flow at that point is the **critical depth** y_c and the velocity is the **critical velocity** V_c. A relation for critical depth in a wide rectangular channel can be found by differentiating E of Eq. (8.17) with respect to y to find the value of y for which E is a minimum. Thus

$$\frac{dE}{dy} = 1 - \frac{q^2}{gy^3} \tag{8.20}$$

and when E is a minimum, $dE/dy = 0$, and letting $y = y_c$, then

$$0 = 1 - \frac{q^2}{gy_c^3}, \quad \text{or} \quad q^2 = gy_c^3 \tag{8.21}$$

Substituting $q = Vy = V_c y_c$ gives

$$V_c^2 = gy_c \quad \text{and} \quad V_c = \sqrt{gy_c} = \frac{q}{y_c} \tag{8.22}$$

where the subscript c indicates critical flow conditions (minimum specific energy for a given q).

We can also express Eq. (8.22) as

$$y_c = \frac{V_c^2}{g} = \left(\frac{q^2}{g}\right)^{1/3} \tag{8.23}$$

From Eq. (8.22),

$$\frac{V_c^2}{2g} = \frac{1}{2}y_c \tag{8.24}$$

So

$$E_c = E_{min} = y_c + \frac{V_c^2}{2g} = \frac{3}{2}y_c \tag{8.25}$$

and

$$y_c = \frac{2}{3}E_c = \frac{2}{3}E_{min} \tag{8.26}$$

A different approach to alternate depths is to solve Eq. (8.19) for q and note the variation in q for changing values of y for a constant value of E. Physically this situation is encountered when water flows from a larger reservoir of constant surface elevation over a high, frictionless, broad-crested weir provided with a moveable sluice gate near its downstream end (Fig. 8.13a). As the gate is opened the flow rate increases until the opening becomes

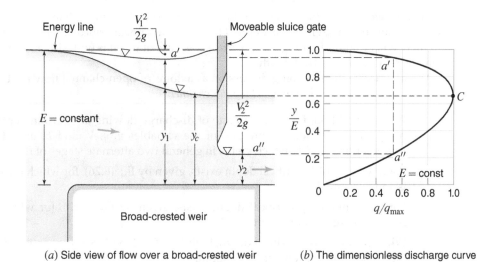

FIGURE 8.13 Variation of depth y and discharge per unit width q for constant specific energy.

just large enough for critical depth to occur. With further opening of the gate, there is no increase in flow rate. As long as the water impinges on the gate (Fig. 8.13a), the flow is subcritical (a') upstream of the gate and supercritical (a'') downstream of the gate.

Rewriting Eq. (8.19) gives

$$q = y\sqrt{2g(E - y)} \tag{8.27}$$

This is the equation of the curve that is shown in dimensionless form in Fig. 8.13b. It is seen that maximum discharge for a given specific energy occurs when the depth is between $0.6E$ and $0.7E$. This may be established more exactly by differentiating Eq. (8.27) with respect to y and equating to zero. Thus

$$\frac{dq}{dy} = \sqrt{2g}\left(\sqrt{E - y} - \frac{1}{2}\frac{y}{\sqrt{E - y}}\right) = 0$$

from which

$$y_c = \frac{2}{3}E \tag{8.28}$$

where y_c is called the **critical depth** *for the given specific energy*. This equation is identical to Eq. (8.26). Thus there is a maximum value of q for a given E as indicated by point C of Fig. 8.13b. We often refer to this curve as the **discharge curve.** The flow depicted by the upper limb of the curve has characteristics similar to those of the upper limb of that in Fig. 8.12. Similarly, the flow depicted by the lower limb of the curve has characteristics similar to the lower limb of that of Fig. 8.12. The significance of these two regimes of flow is discussed in Sec. 8.10. Point C of Fig. 8.13b represents critical flow conditions.

The discharge is a maximum when $y = y_c$, as indicated on the discharge curve of Fig. 8.13b. We can obtain an expression for q_{max} by substituting $E = 1.5y_c$ from Eq. (8.28) into Eq. (8.27), to yield

$$q_{max} = \sqrt{gy_c^3} \tag{8.29}$$

In Fig. 8.12 we can see that when the flow is near critical, a small change in specific energy results in a large change in depth. Hence flow at or near critical depth is unstable and there will be an undulating water surface. Because of this it is undesirable to design channels to flow near critical depth.

We summarize the foregoing discussion as axioms of open-channel flow related to conditions at a given section in a *rectangular channel:*

1. A flow condition, i.e., a certain rate of discharge flowing at a certain depth, is completely specified by any two of the variables y, q, V, and E, except the combination q and E, which yields in general two alternate stages of flow.

2. For any value of E a critical depth exists, given by Eq. (8.26), for which the flow is a maximum.

3. For any value of q a critical depth exists, given by Eq. (8.23), for which the specific energy is a minimum.

4. When flow occurs at critical depth, both Eqs. (8.22) and (8.26) are satisfied and the velocity head is one-half the depth.

5. For any flow condition other than critical, an alternate stage exists at which the same rate of discharge is carried by the same specific energy. We can find the alternate depth from either the specific-energy diagram (Fig. 8.12) or the discharge curve (Fig. 8.13b), by extending a vertical line to the alternate limb of the curve. Analytically, the alternate depth is found by solving Eq. (8.17).

Sample Problem 8.7

Water in a rectangular channel flows 0.5 m deep at a velocity of 6 m/s. Find (a) the alternate depth, (b) the critical depth for this discharge, and (c) the percentage by which the specific energy exceeds the minimum.

Solution

(a) Eq. (8.16):
$$E = 0.5 + \frac{6^2}{2(9.81)} = 2.33 \text{ m}$$

$$q = Vy = 6(0.5) = 3 \text{ m}^2/\text{s per m}$$

Eq. (8.17):
$$2.33 = y + \frac{3^2}{2(9.81)y^2} \tag{1}$$

i.e.,
$$y^3 - 2.33y^2 + 0.459 = 0$$

We can solve this cubic equation with a polynomial solver (Appendix B). Or, knowing that 0.5 m is one root of this equation, we can in effect divide through by $(y - 0.5)$ and solve the resulting quadratic by obtaining the other two roots from Eq. (B.9):

$$y = -0.409 \text{ m (meaningless)} \quad \text{or} \quad y = 2.24 \text{ m} \qquad ANS$$

(b) Eq. (8.23): $\qquad y_c = \left(\dfrac{3^2}{9.81}\right)^{1/3} = 0.972 \text{ m} \qquad ANS$

(c) Eq. (8.25): $\qquad E_c = E_{min} = 1.5y_c = 1.5(0.972) = 1.458 \text{ m}$

$$\frac{E}{E_{min}} = \frac{2.33}{1.458} = 1.602, \quad \text{so exceedance} = 60.2\% \qquad ANS$$

Note: A popular alternative manual approach is to solve (1) by trial and error (see Sample Problems 3.5 and 5.8); values of y are tried until the right side equals the left. As there are two meaningful roots, helpful guidance will be provided by first finding y_c and E_{min} and making a sketch of the specific energy diagram. Some programmable scientific calculators have equation solvers, which in effect automate the trial and error. Equation solving software packages can also conveniently solve such equations.

8.10   Subcritical and Supercritical Flow

In Sec. 8.9 we referred to the upper and lower portions of the specific-energy diagram (Fig. 8.12) and the discharge curve (Fig. 8.13). The upper limb of those curves, where velocities are less than critical, represent *subcritical* (also known as tranquil or upper-stage) flow, while the lower limb of the curves where velocities are greater than critical represent *supercritical* (also known as rapid or lower-stage) flow. We discussed how to identify the critical point separating these two portions in Sec. 8.9.

The bed slope required to give uniform flow at critical depth ($y_0 = y_c$) for a given discharge is known as the *critical slope* S_c. Note that S_c varies with discharge. We obtain an expression for the critical slope of a *wide and shallow* rectangular channel ($R_h = y$) when we combine Eq. (8.23) for critical flow with Eq. (8.18) for uniform flow, eliminating q as

Wide and shallow: $\qquad S_c = \left(\dfrac{n}{C_1}\right)^2 \dfrac{g}{y_c^{1/3}}$ (8.30)

Substituting for y_c from Eq. (8.23), we can also write this as

Wide and shallow: $\qquad S_c = \left(\dfrac{n}{C_1}\right)^2 \dfrac{g^{10/9}}{q^{2/9}}$ (8.31)

As above, $C_1 = 1.486$ for BG units and $C_1 = 1.000$ for SI units.

For channels with cross sections other than wide and shallow, an expression for the critical slope is more complex than the above. However, we seldom need to calculate S_c because, as we shall see, we can more easily determine how S_0 compares with S_c by comparing y_0 with y_c.

If the bed slope $S_0 > S_c$, the slope is known as a *steep slope* for the given discharge. Normal depth for uniform flow on such a slope will be less than critical depth and hence normal flow will be supercritical. In contrast, if $S_0 < S_c$, the normal depth will be greater than critical and normal flow is subcritical. Such a slope is referred to as a *mild slope.* By referring to Eq. (8.30), we see that the hydraulic steepness of a channel slope is determined by more than its elevation gradient. A steep slope for a channel with a

Characteristic	Subcritical	Critical	Supercritical
Depth of flow, y	$y > y_c$	$y = y_c = \left(\dfrac{q^2}{g}\right)^{1/3}$	$y < y_c$
Velocity of flow, V	$V < V_c$	$V = V_c = \sqrt{gy}$	$V > V_c$
Slope for uniform flow, S_0	Mild slope $S_0 < S_c$	Critical slope $S_0 = S_c$ [Eq. (8.30) if wide and shallow]	Steep slope $S_0 > S_c$
Froude number, $\mathbf{F} = \dfrac{V}{\sqrt{gy}} = \dfrac{q}{\sqrt{gy^3}}$	$\mathbf{F} < 1.0$	$\mathbf{F} = 1.0$	$\mathbf{F} > 1.0$
Disturbance waves (Sec. 8.20)	Will propagate in all directions	Will hold fast, not propagate upstream	Will form standing wave with $\sin \beta = c/V$ downstream only
Velocity head compared with half-depth	$\dfrac{V^2}{2g} < \dfrac{y}{2}$	$\dfrac{V^2}{2g} = \dfrac{y}{2}$	$\dfrac{V^2}{2g} > \dfrac{y}{2}$
Can be followed by a hydraulic jump? (Sec. 8.16)	No	No	Yes

Tᴀʙʟᴇ 8.2   Characteristics of subcritical, critical, and supercritical flow in rectangular channels

smooth lining could be a mild slope for the same flow with a rough lining. Even for a given channel with a given boundary roughness, the slope may be mild for a low rate of discharge and steep for a higher one.

The ratio of inertial to gravity forces in flowing water is expressed by the *Froude number*, defined as $\mathbf{F} = V/\sqrt{gL}$. If for rectangular channels the depth of flow is used to represent the significant length parameter in the Froude number (that is, $\mathbf{F} = V/\sqrt{gy}$), we find by comparing this with Eq. (8.22) that the flow is critical if $\mathbf{F} = 1.0$, the flow is subcritical if $\mathbf{F} < 1.0$, and the flow is supercritical if $\mathbf{F} > 1.0$.

Another convenient way to determine the type of flow in rectangular channels is to compare the velocity head with half the depth. Figure 8.12 clearly indicates that if $V^2/2g > y/2$ the flow is supercritical, and if $V^2/2g < y/2$ the flow is subcritical. Important characteristics of subcritical, critical, and supercritical flow in rectangular channels are summarized in Table 8.2.

8.11   Critical Depth in Nonrectangular Channels

For nonrectangular channels, the definition of the Froude number is

$$\mathbf{F} = \frac{V}{\sqrt{gy_h}}$$

where we define $y_h = A/B$ as the *hydraulic depth* (or *hydraulic mean depth*), where A is the cross-sectional flow area and B is the width of the flow area at the water surface. The expression just given for Froude number is the same as $\mathbf{F} = Q\sqrt{B/(gA^3)}$.

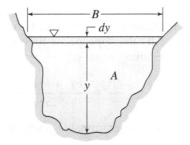

F<small>IGURE</small> 8.14 A cross section of an irregularly shaped channel.

Let us now consider an irregularly shaped flow section (Fig. 8.14) of area A carrying a flow Q. Thus, Eq. (8.17) becomes

$$E = y + \frac{Q^2}{2gA^2} \tag{8.32}$$

Differentiating with respect to y and setting to zero to find y_c, that is, the value of y for which E is a minimum,

$$\frac{dE}{dy} = 1 - \frac{Q^2}{2g}\left(\frac{2}{A^3}\frac{dA}{dy}\right) = 0$$

As A may or may not be a reasonable function of y, it is helpful to observe that $dA = B\,dy$, and thus $dA/dy = B$, *the width of the water surface.* Substituting this in the preceding expression results in

$$\frac{Q^2}{g} = \left(\frac{A^3}{B}\right)_{y=y_c} \tag{8.33}$$

as the equation that must be satisfied for critical flow. For a given cross section, the right-hand side is a function of y only. We generally need a trial-and-error solution to find y_c, which is the value of y that causes A and B to satisfy Eq. (8.33). Once we find this, we call A and B A_c and B_c. It is easy to confirm that Eq. (8.33) causes $\mathbf{F} = 1$.

We can next solve for V_c, the critical velocity in the irregular channel, by observing that $Q = A_c V_c$. Substituting this in Eq. (8.33) yields

$$\frac{V_c^2}{g} = \frac{A_c}{B_c}, \quad \text{or} \quad V_c = \sqrt{\frac{gA_c}{B_c}} = \sqrt{g(y_h)_c} \tag{8.34}$$

If the channel is rectangular, $A_c = B_c y_c$, and Eq. (8.34) reduces to Eq. (8.22); also, then $Q = B_c q$, and Eq. (8.33) reduces to Eq. (8.23).

The cross sections most commonly encountered in open-channel hydraulics are not rectangular but are *trapezoidal* or part-full circular sections. We discussed the latter in Sec. 8.17. For the very common trapezoidal cross section, practicing hydraulic engineers in the past made use of numerous tables, curves, and computer programs to avoid tedious repeated trial-and-error solutions of Eq. (8.33) for y_c. Now it is more convenient to replace A and B by the appropriate functions of y_c, b, and m (defined as in Fig. 8.8), and to solve the resulting equation

$$\frac{Q^2}{g} = \frac{(by_c + my_c^2)^3}{b + 2my_c}$$

using an equation solver in a programmable scientific calculator or in a mathematics software package.

Sample Problem 8.8

In Fig. S8.8 water flows uniformly at a steady rate of 14.0 cfs in a very long triangular flume that has side slopes 1:1. The flume is on a slope of 0.006, and $n = 0.012$. (*a*) Is the flow subcritical or supercritical? (*b*) Find the relation between $V_c^2/2g$ and y_c for this channel.

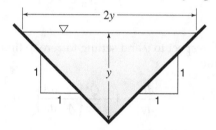

Solution

(*a*)
$$A = \frac{1}{2}(2y)y = y^2$$

$$P = 2\sqrt{2}y = 2.83y$$

$$R_h = \frac{A}{P} = 0.354y$$

Eq. (8.8*a*) for uniform flow: $14 = \dfrac{1.486}{0.012} y_0^2 (0.354y_0)^{2/3}(0.006)^{1/2}$

from which $y_0 = 1.494$ ft. When $y = y_c$, from Eq. (8.33):

$$\frac{(14)^2}{32.2} = \frac{(y_c^2)^3}{2y_c}$$

and $y_c = 1.648$ ft

Since 1.648 ft $= y_c > y_0 = 1.494$ ft, the flow is supercritical.   ***ANS***

Note: If the data in this problem had been given in SI units rather than in BG units, the procedure for solution would have been the same except that Eq. (8.8*b*) would have been used instead of Eq. (8.8*a*).

(b) Eq. (8.34):

$$\frac{V_c^2}{g} = \frac{A_c}{B_c} = \frac{y^2}{2y} = \frac{y}{2}$$

So

$$\frac{V_c^2}{2g} = \frac{y_c}{4} \qquad ANS$$

Consequently, we see that the relation between $V^2/2g$ and y for critical flow conditions depends on the geometry of the flow section. If the vertex angle of the triangle had been different from 90°, the relation would have been different.

8.12 Occurrence of Critical Depth

When flow changes from subcritical to supercritical or vice versa, the depth must pass through critical depth. In the former, the phenomenon gives rise to what is known as a *control section.* In the latter a hydraulic jump (Sec. 8.18) usually occurs. In Fig. 8.15 is depicted a situation where the flow changes from subcritical to supercritical. Upstream of the break in slope there is a *mild* slope, the flow is subcritical, and $y_{0_1} > y_c$. Downstream of the break there is a *steep* slope, the flow is supercritical, and $y_{0_2} < y_c$. At the break in slope the depth passes through critical depth. This point in the stream is referred to as a control section since the depth at the break controls the depth upstream. A similar situation occurs when water from a reservoir enters a canal in which the uniform depth is less than critical. In such an instance (Fig. 8.16), the depth passes through

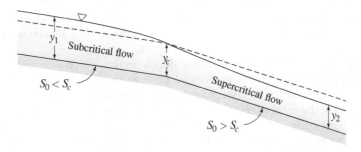

FIGURE 8.15   Change in flow from subcritical to supercritical at a break in slope.

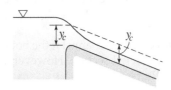

FIGURE 8.16   Hydraulic drop entering a steep slope.

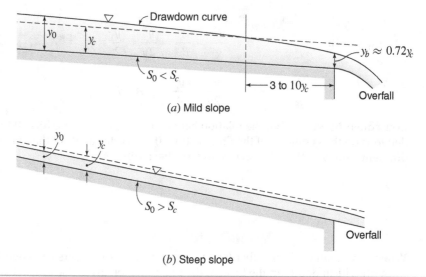

FIGURE 8.17   Free overfall.

critical depth in the vicinity of the entrance to the canal. Therefore, this section is also known as a control section. By measuring the depth at a control section, we can compute a reasonably accurate value of Q by applying Eq. (8.23) for rectangular channels or Eq. (8.33) for nonrectangular channels.

At a control section there is a specific relationship between the stage and the discharge, regardless of the channel roughness and slope. This indicates that such locations are desirable sites for gaging stations. It also means that when the upstream or downstream flow conditions change, the transmission of such changes through a control section are restricted. Note that not only critical depth creates a control section; control structures, such as sluice gates and orifices, may also be used to create control sections.

Another instance where critical depth occurs is that of a *free overfall* (Fig. 8.17a) with subcritical flow in the channel prior to the overfall. Since friction produces a constant diminution in energy in the direction of flow, it is obvious that at the point of overfall the total energy must be less than at any point upstream. As critical depth is the value for which the specific energy is a minimum, one would expect critical depth to occur at the overfall. However, the value for the critical depth is derived on the assumption that the water is flowing in straight lines. In the free overfall, gravity creates a curvature of the streamlines, with the result that the depth at the brink is less than critical depth. Experimenters have found that the depth at the brink $y_b \approx 0.72y_c$. Also, critical depth generally occurs upstream of the brink a distance of somewhere between $3y_c$ and $10y_c$. If the flow is supercritical, there is no drawdown curve (Fig. 8.17b).

Critical depth *may* occur in a channel if the bottom is humped or if the sidewalls are moved in to form a contracted section. However, in such cases critical depth will not always occur (Sample Problem 8.9).

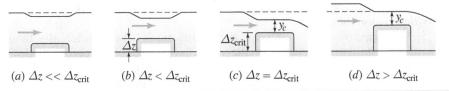

(a) $\Delta z \ll \Delta z_{crit}$ (b) $\Delta z < \Delta z_{crit}$ (c) $\Delta z = \Delta z_{crit}$ (d) $\Delta z > \Delta z_{crit}$

FIGURE 8.18   Subcritical flow over a hump in a rectangular channel. Flow rate is the same in all four cases. Δz = hump height; Δz_{crit} = critical hump height.

8.13   Humps and Contractions

Let us examine the case of a hump or weir in a rectangular channel (Fig. 8.18). We shall assume the hump is streamlined and neglect head loss. In Fig. 8.18*a* we see a small hump. If the flow upstream of the hump is *subcritical,* there will be a slight depression in the water surface over the hump. Inspection of the subcritical (upper) portion of the specific-energy diagram (Fig. 8.12) confirms this statement. The hump (locally raised bed) causes a drop in specific energy E, since the energy line is unchanged. A decrease in E with unchanged q is accompanied by a decrease in y that exceeds the decrease in E, so there must be a depression in the water surface over the hump. If the height of the hump Δz is increased with no change in q, there will be a further drop in the water surface over the hump (Fig. 8.18*b*) for the same reason. Further increases in hump height create further depression of the water surface over the hump until finally the depth on the hump becomes critical (Fig. 8.18*c*). We shall call the minimum height of the hump that causes critical depth the ***critical hump height,*** Δz_{crit}. If the hump is made still higher (Fig. 8.18*d*), critical depth remains on the hump and the depth upstream of the hump increases (tapering off with distance as in Fig. 8.21) until it gains sufficient energy to be able to flow over at the same rate. The more the hump is raised, the greater will the upstream depths become, and the further upstream will these effects extend. We refer to this phenomenon as ***damming action.*** With damming action, immediately downstream of y_c the flow must be supercritical (to comply with Fig. 8.20). In such a situation, the hump, if it has a flat top, becomes a ***broad-crested weir*** (Sec. 9.12).

Let us next consider contractions in rectangular channels. We shall again assume them to be streamlined so that we can neglect any head loss they may cause. Contractions change q, the flow per unit width, and we note that curves for increasing q move to the right on Fig. 8.12. It follows from Fig. 8.12 that a contraction, which causes q to increase, has a similar effect on y as does a hump, which causes E to decrease. If the approaching flow is subcritical, a small contraction will cause a slight depression in the water surface through the contraction. Further contraction creates further depression in the water surface until critical depth occurs in the contraction. This happens when point C on Fig. 8.12 has energy E. Upon further contraction, the depth remains critical in the contracted section and upstream depths increase. However, unlike for humps, the depth in the contracted section continues to increase. This is because the critical depth there depends upon q via Eq. (8.23), and q increases with contraction. Figure 8.12 indicates how contraction beyond the onset of critical depth requires the flow to have extra energy, and this again comes from the damming action, which causes the upstream water surface to rise for a considerable distance. We shall here call the maximum width (i.e., minimum contraction) at which critical depth occurs in the contracted section the *critical contracted width.*

With *supercritical* flow, humps and contractions behave differently than they do with subcritical flow according to the trends indicated by the supercritical (lower) limbs of Figs. 8.12 and 8.13. Thus, when the flow is supercritical, the water depth at the hump or contraction must be slightly *greater* than the depth immediately upstream. And as the hump height or amount of contraction is made larger, the water depth at the hump or contraction *increases,* until it reaches critical depth. Beyond that, further enlargement of the hump height or amount of contraction causes damming action. However, as we shall see in Secs. 8.16–8.19, the damming action effects on the upstream water surface will be much less extended and very different from those for subcritical flow.

A hump may be combined with a contraction to give a section not unlike a venturi meter (Sec. 9.7). We may apply the same principles discussed above for the hump and the contraction to such a situation. It is important to note that the hump changes E and the contraction changes q in the *single* equation [Eq. (8.17)] to be solved; using a separate equation for each effect would be incorrect. Accompanying such calculations by a labeled sketch of the specific-energy diagram is extremely helpful.

Sample Problem 8.9

In Fig. S8.9 uniform flow of water occurs at 135 cfs in a 20-ft-wide rectangular channel at a depth of 2.00 ft. (*a*) Is the flow subcritical or supercritical? (*b*) If a hump of height $\Delta z = 0.30$ ft is placed across the bottom of the channel, calculate the water depth on the hump, and the change in the water surface level at the hump. (*c*) If the hump height is raised to $\Delta z = 0.60$ ft, what then are the water depths immediately upstream and downstream of the hump? (*d*) If the 0.30-ft hump is accompanied by a local contraction to 18 ft, find the water depth on the hump. In all cases neglect head losses over the hump and through the contraction.

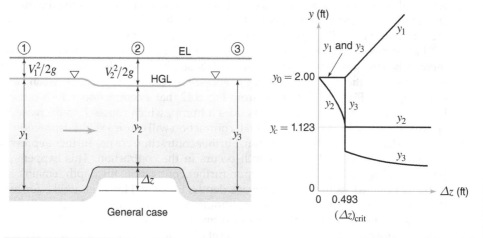

General case

Figure S8.9

Solution

(a)
$$q = 135/20 = 6.75 \text{ cfs/ft}$$

Eq. (8.23):
$$y_c = \left(\frac{q^2}{g}\right)^{1/3} = \left[\frac{(6.75)^2}{32.2}\right]^{1/3} = 1.123 \text{ ft}$$

Since the normal depth (2.00 ft) is greater than the critical depth, the flow is subcritical and the channel slope is mild.   *ANS*

(b) First find the critical hump height. Write the energy equation between sections 1 and 2, assuming critical flow on the hump.

Eq. (8.17):
$$E_1 = 2.00 + \frac{1}{2(32.2)}\left(\frac{6.75}{2.00}\right)^2 = 2.18 \text{ ft}$$

Energy:
$$E_1 = \Delta z_{crit} + (E_{min})_2$$

Using Eq. (8.25):
$$2.18 = \Delta z_{crit} + 1.5(1.123)$$

$$\Delta z_{crit} = 0.493 \text{ ft}$$

Thus the minimum-height hump that will produce critical depth on the hump is 0.493 ft.
    Since the actual hump height, $\Delta z = 0.30$ ft, is less than the critical hump height, 0.493 ft, critical flow does not occur on the hump and there is no damming action.
    Let us now find the depth y_2 on the hump.
Energy in ft of head from section 1 to section 2:

$$2.18 = 0.30 + y_2 + \frac{1}{2(32.2)}\left(\frac{6.75}{y_2}\right)^2$$

We can solve this with an equation solver in a programmable scientific calculator or in computer software. Alternatively, we can rearrange it into a cubic equation as follows,

$$y^3 - 1.877y^2 + 0.707 = 0$$

We can solve this cubic equation by trial and error (see Sample Problems 3.5 and 5.8) followed by Eq. (B.9), or with a polynomial solver (Appendix B). We obtain three roots, 1.601 ft, 0.817 ft, and a negative answer (−0.541 ft) that has no physical meaning. Since $\Delta z < \Delta z_{crit}$, the flow on the hump must be subcritical; that is, $y_2 > y_c$. Hence $y_2 = 1.601$ ft.   *ANS*
    The change in the water surface over the hump $= 2.00 - (0.30 + 1.60) = 0.10$ ft drop.   *ANS*

(c) In this case the hump height $\Delta z = 0.60$ ft, which is greater than the critical hump height. Hence there will be damming action and critical depth ($y_c = 1.123$ ft) will occur on the hump. Writing the energy equation for this case, from section 1 to section 2:

$$y_1 + \frac{1}{2(32.2)}\left(\frac{6.75}{y_1}\right)^2 = 0.60 + 1.5(1.123) = 2.28 \text{ ft}$$

Solving the resulting cubic for y_1 gives three roots, 2.13 ft, 0.660 ft, and a negative answer (−0.504 ft) that has no physical meaning. In this case damming action occurs and the depth y upstream of the hump is 2.13 ft. On the hump, the depth passes through critical depth of 1.123 ft and just downstream of the hump the depth will be 0.660 ft.   *ANS*

The depth will then increase in the downstream direction following an M_3 water surface profile (Fig. 8.20) until a hydraulic jump (Sec. 8.18) occurs to return the depth to the normal uniform flow depth of 2.00 ft. The depth just downstream of the hump cannot be 2.13 ft, because then there would be no way for the flow to dissipate its extra energy and return to normal depth; this is explained in Sec. 8.16 and Fig. 8.20.

This higher type of hump is commonly built of concrete and is used for flow measurement purposes. We discuss such structures under broad-crested weirs, in Sec. 9.12. The right-hand sketch of Fig. S8.9 shows the relation between the hump height Δz, and the water depths y_1, y_2, and y_3 for the condition where $Q = 135$ cfs.

(*d*) At section 2: $q = 135/18 = 7.50$ cfs/ft

$$y_{c2} = \left(\frac{7.50^2}{32.2}\right)^{1/3} = 1.204 \text{ ft}$$

$$(E_{min})_2 = 1.5(1.204) = 1.807 \text{ ft}$$

$$\Delta z_{crit} = E_1 - (E_{min})_2 = 2.18 - 1.807 = 0.370 \text{ ft}$$

So Δz is less than Δz_{crit}; therefore the combination of hump and contraction will not cause critical depth or damming action.

Energy from section 1 to section 2:

$$E_2 = E_1 - \Delta z = 2.18 - 0.300 = 1.877 \text{ ft}$$

Eq. (8.17): $$1.877 = y_2 + \frac{1}{2(32.2)}\left(\frac{7.50}{y_2}\right)^2$$

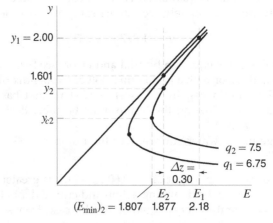

From the sketch of the specific-energy diagram, we expect $1.204 < y_2 < 1.601$ ft. The sketch helps direct manual trials.

By trials, by polynomial solver, or by equation solver, $y_2 = 1.476$ ft.   *ANS*

Sample Problem 8.10

Suppose that the depth of uniform flow in a 1.8-m-wide rectangular channel is 55 cm. Find the change in water-surface elevation caused by a 30-cm-wide bridge pier placed in the middle of the channel. The flow rate is 1.1 m³/s.

Solution
Find upstream flow conditions:

$$V = \frac{Q}{by_0} = \frac{1.1}{1.8(0.55)} = 1.11 \text{ m/s}$$

Eq. (8.17): $$E = 0.55 + \frac{1.11^2}{2(9.81)} = 0.613 \text{ m}$$

Eq. (8.23): $$y_c = \left(\frac{q^2}{g}\right)^{1/3} = \left[\frac{(1.1/1.8)^2}{9.81}\right]^{1/3} = 0.336 \text{ m}$$

Since the upstream depth of 0.55 m is greater than the critical depth, the flow is subcritical and the channel slope is mild. The flow level will therefore drop as the flow encounters a restriction, as long as the width available for flow remains above the minimum value of the opening, b_{crit}, at which critical flow will occur as water flows past the pier. When this condition occurs, at the critical constriction $E = E_{min}$.

Eq. (8.28): $$E_{min} = 0.613 = \frac{3}{2}y_c \quad \text{so} \quad y_c = 0.409 \text{ m}$$

At this critical constriction, the flow per unit width is

Eq. (8.23): $$q = \sqrt{gy_c^3} = \sqrt{9.81(0.409^3)} = 0.818 \text{ m}^2/\text{s}$$

Then: $$b_{crit} = \frac{Q}{q} = \frac{1.1}{0.818} = 1.345 \text{ m}$$

The maximum obstruction width is 1.8 − 1.345 = 0.455 m. Obstructions greater than 0.455 m in width will cause damming at the bridge pier and result in supercritical flow depth downstream of the pier. Since the 30-cm obstruction is less than the maximum, the flow depth will drop as it passes around the pier and remain subcritical. The depth is determined by finding the roots of Eq. (8.17).

$$y + \frac{[1.1/((1.8 - 0.3)y)]^2}{2(9.81)} - 0.613 = 0$$

which can be rearranged to $\quad y^3 - 0.613y^2 + 0.0274 = 0$

The roots of the cubic equation can be determined by trial and error or using an equation solver, as $y = [-0.185; 0.292; 0.506]$. The two positive roots are the supercritical and subcritical flow depths, respectively. Since we have determined that the flow remains subcritical past the pier, the depth $y = 0.506$ m.   *ANS*

Cause	Damming Action Occurs When:
Hump only (constant q)	$\Delta z > \Delta z_{crit} = E_u - E_{min} = E_u - \frac{3}{2}y_c$
Contraction only (constant E)	$b_2 < b_{crit}$, so $E_u < (E_{min})_2 = \frac{3}{2}y_{c2}$
Hump with contraction	Δz and b_2 are such that $E_u - \Delta z < (E_{min})_2 = \frac{3}{2}y_{c2}$

Subscript u represents upstream conditions with no constrictions.
Subscript 2 represents a station *at* the constriction.

TABLE 8.3   Occurrence of damming action

An unnecessary difficulty sometimes baffles students trying to solve problems like part (c) of Sample Problem 8.9 or Sample Problem 8.10. It happens if they forget to first calculate Δz_{crit} or b_{crit} and forget that damming action may have occurred. When they try to solve for the water depth on the hump or in the constriction, they cannot get a solution, or get only negative and imaginary roots. This is because without damming action E on the hump is less than E_{min} (see Fig. 8.12), and they are trying to solve an impossible situation. Similar logic can be applied to solve problems involving depressions in the channel bed (negative humps) and channel expansions (negative contractions). Of course, opposite effects are produced to those that occur with positive humps and contractions.

Because identifying damming action is an important key to solving flows past constrictions, and the relevant relationships can be confusing, we summarize situations when damming action will occur in Table 8.3; with the contrary inequalities it will not occur. Students are strongly urged to sketch appropriate specific-energy diagrams, like the one at the end of Sample Problem 8.9, to help clarify situations.

8.14   Nonuniform, or Varied, Flow

As a rule, uniform flow is found only in artificial channels of constant shape and slope, although even under these conditions the flow for some distance may be nonuniform, as shown in Fig. 8.1. But with a natural stream, the slope of the bed and the shape and size of the cross section usually vary to such an extent that true uniform flow is rare. So we can expect the application of the equations given in Sec. 8.2 to natural streams to yield results that are only approximations to the truth. To apply these equations at all, we must divide the stream into lengths (reaches) within which the conditions are approximately the same.

In the case of artificial channels that are free from the irregularities found in natural streams, it is possible to apply analytical methods to the various problems of nonuniform flow. In many instances, however, particularly when dealing with natural channels, we must resort to trial solutions and graphical methods.[12]

[12] For details on various methods of solving gradually varied open-channel flow, see Richard H. French, *Open-Channel Hydraulics*, McGraw-Hill Book Company, New York, 201–247, 1985.

In the case of pressure conduits, we have dealt with uniform and nonuniform flow without drawing much distinction between them. We can do this because in a closed pipe the area of the water section, and hence the mean velocity, is fixed at every point. But in an open channel these conditions are not fixed, and the stream adjusts itself to the size of cross section that the energy gradient (i.e., slope of the energy line) requires.

In an open stream on a falling grade, without friction, the effect of gravity is to tend to produce a flow with a continually increasing velocity along the path, as in the case of a freely falling body. However, the gravity force is opposed by frictional resistance. The frictional force increases with velocity, while gravity is constant; so eventually the two will be in balance, and uniform flow will occur. When the two forces are not in balance, the flow is nonuniform.

There are two types of nonuniform flow. In one the changing conditions extend over a long distance, and we call this *gradually varied flow*. In the other the change may take place very abruptly, and the transition is thus confined to a short distance. We designate this as a *local nonuniform phenomenon* or *rapidly varied flow*. Gradually varied flow can occur with either subcritical or supercritical flow, but the transition from one condition to the other is ordinarily abrupt, as between D and E in Fig. 8.1. Other cases of local nonuniform flow occur at the entrance and exit of a channel, at changes in cross section, at bends, and at obstructions such as dams, weirs, or bridge piers.

8.15 Energy Equation for Gradually Varied Flow

The principal forces involved in flow in an open channel are inertia, gravity, net hydrostatic force due to change in depth, and friction. The first three represent the useful kinetic and potential energies of the liquid, while the fourth dissipates useful energy into the useless kinetic energy of turbulence and eventually into heat because of the action of viscosity. Referring to Fig. 8.19, the total energy of the elementary volume of liquid shown is proportional to

$$H = z + y + \alpha \frac{V^2}{2g} \tag{8.35}$$

where $z + y$ is the potential energy head above the arbitrary datum, and $\alpha V^2/2g$ is the kinetic energy head, V being the mean velocity in the section. Each term of the equation

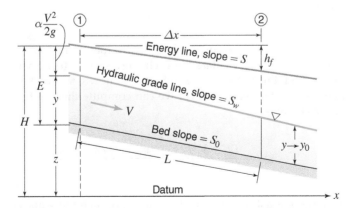

FIGURE 8.19   Energy relations for nonuniform flow.

represents energy in foot-pounds per pound of fluid (or newton-meters per newton of fluid in SI units). Once again, as in Secs. 8.1 and 8.2, we define L as the distance along the channel bed between any two sections and Δx as the horizontal distance between the sections. For all practical purposes, these two distances can be considered as being equal to one another.

We will generally find the value of α to be higher in open channels than in pipes, as was explained in Sec. 8.4. It may range from 1.05 to 1.40, and in the case of a channel with an obstruction the value of α just upstream may be as high as 2.00 or even more. As the value of α is not known unless the velocity distribution is determined, it is often omitted from the equations, but an effort should be made to employ it if a high degree of accuracy is necessary. In the numerical problems in this chapter, α is assumed to be unity.

The energy equation for steady flow between sections 1 and 2 of Fig. 8.19 a distance Δx apart is

$$z_1 + y_1 + \alpha_1 \frac{V_1^2}{2g} = z_2 + y_2 + \alpha_2 \frac{V_2^2}{2g} + h_f \tag{8.36}$$

As $z_1 - z_2 = S_0 \Delta x$ and $h_f \approx S \Delta x$, the energy equation may also be written (with $\alpha_1 = \alpha_2 = 1$) in the form

$$y_1 + \frac{V_1^2}{2g} = y_2 + \frac{V_2^2}{2g} + (S - S_0)\, \Delta x \tag{8.37}$$

We can make an approximate analysis of gradually varied, nonuniform flow by considering a length of stream to consist of a number of successive reaches, in each of which uniform flow occurs. Greater accuracy results from smaller depth variations in each reach. The Manning formula [Eq. (8.7)] is rearranged and applied to average conditions in each reach to provide an estimate of the value of the energy gradient S for that reach as follows:

In BG units:
$$S = \left(\frac{n\bar{V}}{1.486\bar{R}_h^{2/3}} \right)^2 \tag{8.38a}$$

In SI units:
$$S = \left(\frac{n\bar{V}}{\bar{R}_h^{2/3}} \right)^2 \tag{8.38b}$$

where $\bar{V}$ and $\bar{R}_h$ are the means of the respective values at the two ends of the reach. If n is not constant, S can be calculated at each cross section then then averaged for each stream reach. With this value of S, with S_0 and n known, and with the depth and velocity at both ends of the reach known, we can compute the length Δx from Eq. (8.37), rearranged as follows:

$$\Delta x = \frac{(y_1 + V_1^2/2g) - (y_2 + V_2^2/2g)}{S - S_0} = \frac{E_1 - E_2}{S - S_0} \tag{8.39}$$

Only the depth y must be prescribed at each end; then the values of V and R_h follow from the discharge and cross-section shape. As the calculation for Δx is direct, this approach is much easier than prescribing Δx and trying to solve for y_2, which would require trial and error because V_2, R_{h2}, and therefore S are all functions of y_2.

In practice, therefore, we divide the depth range of interest into small increments, usually equal, which define reaches whose lengths can be found by using Eq. (8.39). We must begin the discretization process from a section where y is known, i.e., from calculations for a control section (Sec. 8.12) or from a field measurement at any section. If the resulting value of Δx is negative, from Eq. (8.39) and Fig. 8.19 we can deduce this means that section 2 is *upstream* of section 1.

Sample Problem 8.11

At a certain section in a very smooth rectangular channel 6 ft wide the depth is 3.00 ft when the flow rate is 160 cfs. Compute the distance to the section where the depth is 3.20 ft if $S_0 = 0.0020$ and $n = 0.012$.

Solution

The calculations are shown in the following table, where $\bar{V} = (V_1 + V_2)/2$ and $\bar{R}_h = (R_{h1} + R_{h2})/2$. The total distance is calculated to be 71.3 ft.

y, ft	$A =$ $6y$, ft^2	$P =$ $6+2y$, ft	$R_h =$ A/P, ft	$V =$ Q/A, fps	$V^2/2g$, ft	$E =$ $y + \dfrac{V^2}{2g}$, ft	Numerator $E_1 - E_2$, ft	$\bar{V}$, fps	$\bar{R}_h$, ft	S Eq. (8.38)	Denominator $S - S_0$	$\Delta x =$ $\dfrac{E_1 - E_2}{S - S_0}$, ft
3.00	18.00	12.00	1.500	8.889	1.2269	4.2269						
							−0.0221	8.746	1.512	0.002 873	0.000 873	−25.3
3.10	18.60	12.20	1.525	8.602	1.1490	4.2490						
							−0.0293	8.468	1.536	0.002 637	0.000 637	−46.0
3.20	19.20	12.40	1.548	8.333	1.0783	4.2783						

ANS: $\Sigma(\Delta x) = -71.3$

Notes:
1. The accuracy of the calculated total distance (71.3 ft) could be improved by taking more steps, i.e., smaller Δy.
2. As the resulting Δx are negative, the sections progress *upstream*, and the water depth decreases downstream.
3. The *magnitude* (absolute value) of Δx increases with depth (sketch the profile by reaches to see this), so the surface curvature is concave down.
4. $y_c = 2.81$ ft [from Eq. (8.23)] and $y_0 = 3.50$ ft [from Eq. (8.8a)], so $y_c < y < y_0$. In Sec. 8.16 we shall see that this tells us the water surface must have the shape of an M_2 profile, with depth decreasing downstream and curved concave down.

An important issue in calculations of this type (as illustrated in Sample Problem 8.11) is working with sufficient precision, i.e., carrying enough significant figures. In two instances we must deal with the small difference between two numbers whose values are close to one another. In the example just presented, we see that we lost two significant figures in calculating $E_1 - E_2$ and we lost one in calculating $S - S_0$. Therefore the calculations for these quantities, and for V and R_h on which they are based, need to be carried to more significant figures so that the final results are always reliable to three significant figures.

If the data for the preceding example had been given in SI rather than BG units, the procedure for solution would have been the same except that Eqs. (8.38b) and (8.7a) would have been used rather than Eqs. (8.38a) and (8.7b). Note that $\bar{V}$ is *not* calculated from $Q/\bar{A}$ nor from $\bar{y} = (y_1 + y_2)/2$; these would give different results. Because of the extensive calculations required for many reaches, we commonly solve this type of problem by computer, either by using applications software, or using a spreadsheet or other computational tool.

Section 8.16 will explain that depth ranges for this procedure must not cross the normal or critical depths. Therefore, to check this, first compute y_0 from Eq. (8.8) and y_c from Eq. (8.23) for rectangular channels or from Eq. (8.33) for other channels.

8.16 Water-Surface Profiles in Gradually Varied Flow (Rectangular Channels)

As there are some 12 different circumstances giving rise to as many different fundamental types of gradually varied flow, having a logical scheme of type classification is helpful. In general, any problem of varied flow, no matter how complex it may appear, with the stream passing over dams, under sluice gates, down steep chutes, on the level, or even on an upgrade, can be broken down into reaches such that the flow within any reach is either uniform or falls within one of the given nonuniform classifications. We then analyze the stream one reach at a time, proceeding from one reach to the next until the desired result is obtained.

The following treatment is based, for simplicity, on *channels of rectangular section*. The section will be considered sufficiently wide and shallow so that we may confine our attention to a section 1 ft (or 1 m) wide through which the velocity is essentially uniform. It is important to bear in mind that this development is based on a constant value of the discharge per unit width q, which can only be used with rectangular channels, and on one value of n, the roughness coefficient.

Differentiating Eq. (8.35) with respect to x, the horizontal distance along the channel, the rate of energy dissipation is found (with $\alpha = 1$) to be

$$\frac{dH}{dx} = \frac{dz}{dx} + \frac{dy}{dx} + \frac{1}{2g}\frac{d(V^2)}{dx} \tag{8.40}$$

The energy gradient $S = -dH/dL \approx -dH/dx$, while the slope of the channel bed is $S_0 = -dz/dx$, and the slope of the hydraulic grade line or water surface is given by $S_w = -dz/dx - dy/dx$.

Commencing with the last term of Eq. (8.40), we may observe that since $V = q/y$,

$$\frac{1}{2g}\frac{d(V^2)}{dx} = \frac{1}{2g}\frac{d}{dx}\left(\frac{q^2}{y^2}\right) = -\frac{q^2}{g}\frac{1}{y^3}\frac{dy}{dx}$$

Substituting this, plus the S and S_0 terms, into Eq. (8.40), yields

$$-S = -S_0 + \frac{dy}{dx}\left(1 - \frac{q^2}{gy^3}\right)$$

or $\qquad$ $$\frac{dy}{dx} = \frac{S_0 - S}{1 - q^2/gy^3} = \frac{S_0 - S}{1 - V^2/gy} = \frac{S_0 - S}{1 - \mathbf{F}^2} \tag{8.41}$$

Evidently, if the value of dy/dx as determined by Eq. (8.41) is positive, the water depth will be increasing along the channel; if negative, it will be decreasing. Let us examine the numerator and denominator of Eq. (8.41).

Looking first at the numerator, we can consider S as the energy gradient [such as would be obtained from Eq. (8.38)] that would carry the given discharge at depth y with uniform flow. For a wide and shallow rectangular channel, substituting $V = q/y$ and $R_h = y$ into Eq. (8.38b), we obtain

Wide and shallow:
$$S = \left(\frac{nq}{y^{5/3}}\right)^2$$

Similarly substituting these into Eq. (8.7a), the slope, S_0, for normal depth, y_0, is

Wide and shallow:
$$S_0 = \left(\frac{nq}{y_0^{5/3}}\right)^2$$

Comparing these two expressions, we see that $S/S_0 = (y_0/y)^{10/3}$. Consequently, for constant q and n, when $y > y_0$, $S < S_0$, and the numerator $(S_0 - S)$ is positive. Conversely, when $y < y_0$, $S > S_0$, and the numerator is negative.[13]

To investigate the denominator of Eq. (8.41), we observe that if $\mathbf{F} = 1$ (critical flow, Sec. 8.10, $y = y_c$), the denominator is zero and $dy/dx = \infty$; if $\mathbf{F} > 1$ (supercritical flow, $y < y_c$), the denominator is negative; and if $\mathbf{F} < 1$ (subcritical flow, $y > y_c$), the denominator is positive.

In this manner the signs of the numerator and the denominator of Eq. (8.41) can be found for any depth by comparing it with y_0 and y_c. These two signs together give the sign of dy/dx, which in turn defines the slope of the water surface.

The foregoing analyses have been combined graphically into a series of water-surface profiles in Fig. 8.20. The surface profiles are classified according to slope and depth as follows. If S_0 is positive, the bed slope is termed *mild* (M) when $y_0 > y_c$, *critical* (C) when $y_0 = y_c$, and *steep* (S) when $y_0 < y_c$; if $S_0 = 0$, the channel is *horizontal* (H); and if S_0 is negative, the bed slope is called *adverse* (A). If the stream surface lies above both the normal (uniform flow) and critical depth lines, it is of type 1; if between these lines, it is of type 2; and if below both lines, it is of type 3. The 12 forms of surface curvature are labeled accordingly in Fig. 8.20. Each is discussed in more detail in the next section.

Note that the scale of the drawings in Fig. 8.20 is greatly compressed in the horizontal direction. The problems at the end of this chapter demonstrate that gradually varied flow generally extends over many hundreds of feet, and if plotted to an undistorted scale, the rate of change in depth would be scarcely discernible. Note also that, since even a hydraulically steep slope varies only a few degrees from the horizontal, it makes little difference whether the depth y is measured vertically (as shown) or perpendicular to the bed.

We see that some of the curves of Fig. 8.20 are concave upward while others are concave downward. Although the mathematical proof for this is not given, the physical explanation is not hard to find. In the case of the type 1 curves, the surface must approach a horizontal asymptote as the velocity is progressively slowed down because

---

[13] Combining these deductions with the fact that E must decrease downstream if $S > S_0$, and vice versa, yields the result that E must decrease downstream when $y < y_0$ and vice versa.

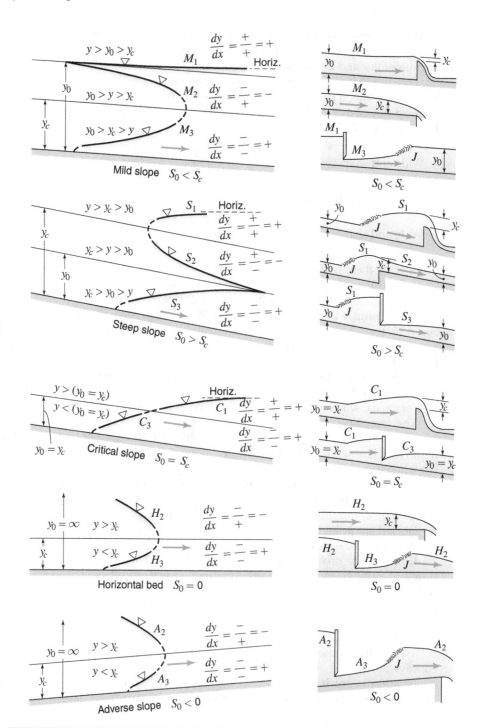

FIGURE 8.20 Various types of nonuniform flow, with flow from left to right. Signs for dy/dx pertain to Eq. (8.41).

of the increasing depth. Likewise all curves that approach the normal or uniform depth line must approach it asymptotically, because uniform flow will only prevail at sections remote from disturbances, as pointed out in Sec. 8.1. Theoretically the curves that cross the critical-depth line must do so vertically, as the denominator of Eq. (8.41) becomes zero in this case. These end conditions govern the surface concavity. The critical-slope curves, for which $y_0 = y_c$, constitute exceptions, since it is not possible for a curve to be both tangent and perpendicular to the same critical-uniform depth line.

Profiles that get deeper downstream we call **backwater curves,** and those that decrease in depth downstream we call **drawdown curves.**

To the right of each water-surface profile is shown a representative example of how this particular curve can occur. Many of the examples show a rapid change from a depth below the critical to a depth above the critical. This is a local phenomenon, known as the hydraulic jump, which is discussed in detail in Sec. 8.18.

The qualitative analysis of water-surface profiles has been restricted to rectangular sections. The curve forms of Fig. 8.20 are, however, applicable to any channel of uniform cross section if y_0 is the depth for uniform flow and y_c is the depth that satisfies Eq. (8.33). The surface profiles can even be used qualitatively in the analysis of natural stream surfaces as well, provided that local variations in slope, shape, and roughness of cross section, etc., are taken into account. The step-by-step method presented in Sec. 8.15 for the solution of steady nonuniform-flow problems is not restricted to uniform channels, and is therefore suited to water-surface-profile computations for any stream whatever.

A very important precaution, when using the stepwise procedure to solve for water-surface profiles, is to *always first determine the critical and normal depths, y_c and y_0.* Then we can easily confirm that the profile sought does not cross either of these depths. For if the ends of a reach are on different sides of either of these depths, the results are always invalid, even though sometimes they may look numerically reasonable. Along similar lines, if part of a reach is at uniform flow, this can also lead to impossible results when solving for depth. We can avoid such a situation by first calculating the gradually varied profile distance to y_0.

8.17   Examples of Water-Surface Profiles

The M_1 Curve

The most common case of gradually varied flow is where the depth is already above the critical and is increased still further by a dam, as indicated in Fig. 8.21. Referring to the specific-energy diagram of Fig. 8.12, this case is found on the upper limb of the diagram, for here also, as the depth increases, the velocity diminishes without any abrupt

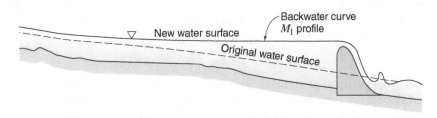

FIGURE 8.21   Backwater curve in a natural stream.

transitions, so that a smooth surface curve is obtained. Note that it is possible for the water surface *elevation* to rise downstream, in addition to the depth. In the case of flow in an artificial channel with a constant bed slope, the water-surface curve would be asymptotic at infinity to the surface for uniform flow, as we noted before. But the problems that are usually of more important interest are those concerned with the effect of a dam on a natural stream and the extent to which it raises the water surface at various points upstream. The resulting water-surface profile in such a case is commonly known as a ***backwater curve.***

For an artificial channel where the conditions are uniform, save for the variation in water depth, we can solve the problem using Eqs. (8.38) and (8.39). Usually, the solution commences at the dam, where conditions are assumed to be known, and the lengths of successive reaches upstream, corresponding to assumed increments of depth, are computed. A tabular type of solution (Sample Problem 8.11) is the most helpful, with column headings corresponding to the various elements of Eqs. (8.38) and (8.39), the last column being $\Sigma \Delta x$, which sums up the length from the dam to the point in question. It is important, if accuracy is desired, to keep the depth increment small within any reach; a depth change of 10% or less is fairly satisfactory. The smaller the depth increment used in this step-by-step procedure, the greater the accuracy of the final result. We can conveniently solve this type of problem that requires successive calculations with a computer.

For a natural stream, such as that shown in Fig. 9.11, the solution is not so direct, because the form and dimensions of a cross section cannot be assumed and then the distance to its location computed. As there are various slopes and cross sections at different distances upstream, we must assume the value of Δx in Eq. (8.37), and then we can compute the depth of stream at this section by trials, substituting the expression for S from Eq. (8.38) into Eq. (8.37). The solution is then pursued in similar fashion on a reach-by-reach basis. The accuracy of the results depends on the selection of a proper value for Manning's n, which is difficult when dealing with natural streams. For this reason and because of irregularities in the flow cross sections, the refinements of Eq. (8.37) are not always justified, and it is often satisfactory to assume uniform flow by applying Eq. (8.8) to each successive reach.

The M_2 Curve

This curve, representing accelerated subcritical flow on a slope that is flatter than critical, exists, like the M_1 curve, because of a control condition *downstream*. In this case, however, the control is not an obstruction but the removal of the hydrostatic resistance of the water downstream, as in the case of the free overfall shown in Fig. 8.17a. As in the M_1 curve, the surface will approach the depth for uniform flow at an infinite distance upstream. Practically, because of slight wave action and other irregularities, the distinction between the M_2, or ***drawdown curve,*** and the curve for uniform flow disappears within a finite distance.

The M_3 Curve

This occurs because of an upstream control, like the sluice gate shown in Fig. 8.20. The bed slope is not sufficient to sustain lower-stage flow, and at a certain point determined by energy and momentum relations, the surface will pass through a hydraulic jump unless this is made unnecessary by the existence of a free overfall before the M_3 curve reaches critical depth.

The S Curves

We can analyze these in much the same fashion as the M curves, having due regard for downstream control in the case of subcritical flow and for upstream control for supercritical flow. Thus a dam or an obstruction on a steep slope produces an S_1 curve upstream (Fig. 8.20), which approaches the horizontal asymptotically but cannot so approach the uniform depth line, which lies below the critical depth. Therefore the S_1 curve must be preceded by a hydraulic jump. The S_2 curve shows accelerated lower-stage flow, smoothly approaching uniform depth. Such a curve will occur whenever a steep channel receives flow at critical depth, as from an obstruction (as shown) or reservoir. The sluice gate on a steep channel will produce the S_3 curve, which also approaches smoothly the uniform depth line.

The C Curves

These curves, with the anomalous condition ($dy/dx = \infty$) at $y_0 = y_c$, do not occur frequently. As noted in Sec. 8.9, we should avoid creating conditions that produce them because of the inherent instability of such flows.

The *H* and the *A* Curves

These curves have in common the fact that there is no condition of uniform flow possible. The H_2 and A_2 drawdown curves are similar to the M_2 curve, but even more noticeable. The value of $y_b = 0.72y_c$ given in Sec. 8.12 applies strictly only to the H_2 curve, but it is approximately true for the M_2 curve also. The sluice gate on the horizontal and adverse slopes produces H_3 and A_3 curves that are like the M_3 curve, but they do not exist for as great a distance as the M_3 curve before a hydraulic jump occurs. Of course, it is not possible to have a channel of any appreciable length carry water on a horizontal grade, and even less so on an adverse grade.

Other Examples

Some other interesting water-surface profiles occur when the slope of a channel of uniform section changes abruptly from a mild to a milder slope or to a less mild slope. In this case the flow is everywhere subcritical. Similar water-surface profiles occur when a channel on a constant slope that is mild throughout its entire length has an abrupt change in width to an either narrower or wider channel. These possibilities are depicted in Fig. 8.22.

Other water-surface profiles include those that occur when the slope of a channel changes abruptly from steep to either steeper or less steep. In this case the flow is supercritical. Similar profiles occur when a channel on a constant slope that is steep throughout its entire length has an abrupt change in width to an either wider or narrower channel. As an exercise it is suggested that the reader sketch profiles similar to Fig. 8.22 for the steep-slope situations. In these cases it will be found that, with steep slopes (supercritical flow), uniform flow occurs upstream of the change in either slope or width, while with mild slopes (subcritical flow), uniform flow prevails downstream of the change.

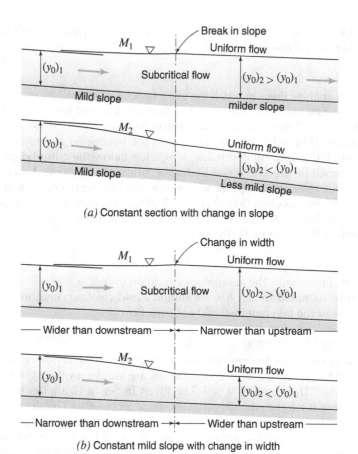

M_1

Break in slope

Uniform flow

$(y_0)_1$ → Subcritical flow $(y_0)_2 > (y_0)_1$ →

Mild slope

milder slope

M_2

$(y_0)_1$ → Uniform flow $(y_0)_2 < (y_0)_1$

Mild slope

Less mild slope

(a) Constant section with change in slope

Change in width

M_1

Uniform flow

$(y_0)_1$ → Subcritical flow $(y_0)_2 > (y_0)_1$

— Wider than downstream —|— Narrower than upstream —

M_2

Uniform flow

$(y_0)_1$ → $(y_0)_2 < (y_0)_1$

— Narrower than downstream —|— Wider than upstream —

(b) Constant mild slope with change in width

FIGURE 8.22   Subcritical flow water-surface profiles.

8.18   The Hydraulic Jump

By far the most important of the local nonuniform-flow phenomena is that which occurs when supercritical flow has its velocity reduced to subcritical. We have seen in the surface profiles of Fig. 8.20 that there is no ordinary means of changing from supercritical to subcritical flow with a smooth transition, because theory calls for a vertical slope of the water surface. The result, then, is a marked discontinuity in the surface, characterized by a steep upward slope of the profile, broken throughout with violent turbulence, and known universally as the *hydraulic jump.*

The specific reason for the occurrence of the hydraulic jump can perhaps best be explained by referring to the M_3 curve of Fig. 8.20. Downstream of the sluice gate the flow decelerates because the slope is not great enough to maintain supercritical flow. The specific energy decreases as the depth increases (proceeding to the left along the lower limb of the specific-energy diagram, Fig. 8.12). Were this condition to progress until the flow reached critical depth, the specific energy would need to increase as the depth increased from the critical to the uniform flow depth downstream. But this is a physical impossibility. Therefore the jump forms before the necessary energy is lost.

The hydraulic jump can also occur from an upstream condition of uniform supercritical flow to a nonuniform S_1 curve downstream when there is an obstruction on a steep slope, as illustrated in Fig. 8.20, or again from a nonuniform upstream condition to a nonuniform downstream condition, as illustrated by the H_3–H_2 or the A_3–A_2 combinations. An M_3–M_2 combination is also possible. In addition to the foregoing cases, where the channel bed continues at a uniform slope, a jump will form when the slope changes from steep to mild, as on the apron at the base of the spillway, illustrated in Fig. 8.23. This is an excellent example of the jump serving a useful purpose, for it dissipates much of the destructive energy of the high-velocity water, thereby reducing downstream erosion. The turbulence within hydraulic jumps has also been found to be very useful and effective for mixing fluids, and jumps have been used for this purpose in water treatment plants and sewage treatment plants.

Depth Relations—General

We will derive the equation relating the depths before and after a hydraulic jump for the case of a horizontal channel bottom (the H_3–H_2 combination of Fig. 8.20). For channels on a gradual slope (i.e., less than about 3° or $S_0 = 0.05$) the gravity component of the weight is relatively small and can be neglected without introducing significant error. The friction forces acting are negligible because of the short length of channel involved and therefore the only significant forces are hydrostatic forces. Applying momentum principle Eq. (6.7a) to the volume of fluid between sections 1 and 2 of Fig. 8.23, and using Eq. (3.16) for the end forces, we obtain for any shape of channel cross section

$$\sum F_x = \gamma h_{c_1} A_1 - \gamma h_{c_2} A_2 = \frac{\gamma}{g} Q(V_2 - V_1)$$

where h_c is the depth to the centroid of the end area. We can rearrange this equation to give

$$\frac{\gamma}{g} Q V_1 + \gamma h_{c_1} A_1 = \frac{\gamma}{g} Q V_2 + \gamma h_{c_2} A_2 \tag{8.42}$$

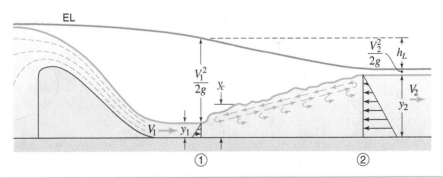

FIGURE 8.23 Hydraulic jump on horizontal bed following a spillway; horizontal scale foreshortened between sections 1 and 2 approximately 2½:1.

This states that the momentum plus the pressure force on the cross-sectional area is constant, or dividing by γ and observing that $V = Q/A$,

Any shape
cross-section:
$$F_m = \frac{Q^2}{Ag} + Ah_c = \text{constant} \qquad (8.43)$$

Substituting into Eq. (8.43) the geometric relationships for a trapezoidal channel (Sec. 8.3) and equations for hydrostatic force (Sec. 3.7),

Trapezoidal
cross-section:
$$F_m = \frac{by^2}{2} + \frac{my^3}{3} + \frac{Q^2}{gy(b + my)} = \text{constant}$$

The specific case of a rectangular cross section (a trapezoid with $m=0$) allows the development of relationships describing hydraulic jumps in rectangular channels.

Depth Relations—Rectangular Channel

In the case of a rectangular channel, we have $h_c = y/2$, $A = by$, and we can again use the flow q per unit width, so that $Q = bq$. Substituting for these into Eq. (8.43), it reduces for a unit width to

$$f_m = \frac{q^2}{gy_1} + \frac{y_1^2}{2} = \frac{q^2}{gy_2} + \frac{y_2^2}{2} = \text{constant} \qquad (8.44)$$

A curve of values of f_m for different values of y is plotted to the right of the specific-energy diagram shown in Fig. 8.24. Both curves are plotted for the condition of 2 cfs/ft of width. As the loss of energy in the jump does not affect the "force" quantity f_m, the latter is the same after the jump as before, and therefore any vertical line intersecting the f_m curve serves to locate two **conjugate depths**, y_1 and y_2. These depths represent possible combinations of depth that could occur before and after the jump.

Thus, in Fig. 8.24, the line for the initial water level y_1 intersects the f_m curve at a as shown, giving the value of f_m, which must be the same after the jump. The vertical line ab then fixes the value of y_2. This depth is then transposed to the specific-energy diagram to determine the value cd of $V_2^2/2g$. The value of $V_2^2/2g$ is the vertical distance ef.

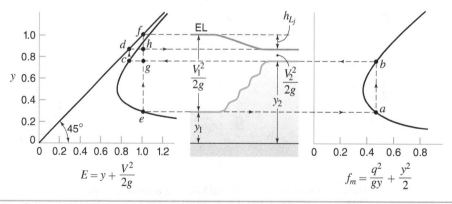

FIGURE 8.24 Energy and momentum relations in a hydraulic jump.

We can show that for a given flow rate in a rectangular channel the minimum value of f_m [Eq. (8.44)] occurs at the same depth as the minimum value of E. Differentiating f_m with respect to y and equating to zero gives

$$f_m = \frac{q^2}{gy} + \frac{y^2}{2}$$

$$\frac{df_m}{dy} = -\frac{q^2}{gy^2} + \frac{2y}{2} = 0$$

and

$$y = \left(\frac{q^2}{g}\right)^{1/3}$$

This expression is identical to Eq. (8.23). Thus we have shown that for a given q the minimum value of f_m occurs at the same depth as does the minimum value of E. We can see this in Fig. 8.24.

When the rate of flow and the depth before or after the jump are given, we see that Eq. (8.44) becomes a cubic equation when solving for the other depth. We can easily reduce this to a quadratic, however, by observing that $y_2^2 - y_1^2 = (y_2 + y_1)(y_2 - y_1)$, so that

$$\frac{q^2}{g} = y_1 y_2 \frac{y_1 + y_2}{2} \tag{8.45}$$

From Eq. (8.23) we also know that $q^2/g = y_c^3$, so if $y_1 < y_c$ then we must have $y_2 > y_c$. This proves that the hydraulic jump must always cross the critical depth.

Solving Eq. (8.45) by the quadratic formula gives

$$y_2 = \frac{y_1}{2}\left(-1 + \sqrt{1 + \frac{8q^2}{gy_1^3}}\right) = \frac{y_1}{2}(-1 + \sqrt{1 + 8F_1^2}) \tag{8.46a}$$

or

$$y_1 = \frac{y_2}{2}\left(-1 + \sqrt{1 + \frac{8q^2}{gy_2^3}}\right) = \frac{y_2}{2}(-1 + \sqrt{1 + 8F_2^2}) \tag{8.46b}$$

These equations relate the depths before and after a hydraulic jump (i.e., the conjugate depths) in a rectangular channel. They give good results if the channel slope is less than about 0.05 or 3°. For steeper channel slopes the effect of the gravity component of the weight of liquid between sections 1 and 2 of Fig. 8.23 must be considered.

Energy Loss

The head loss h_{L_j} caused by the jump is the drop in energy from 1 to 2. Or

Any channel: $\qquad h_{L_j} = E_1 - E_2 = \left(y_1 + \frac{V_1^2}{2g}\right) - \left(y_2 + \frac{V_2^2}{2g}\right) \tag{8.47}$

On Fig. 8.24 points e and c on the specific-energy diagram represent the conjugate depths, and the horizontal distance between them ($= cg = fh$) is the head loss. Replacing V by q/y and using Eq. (8.45), we can confirm that

Rectangular channel:
$$h_{L_j} = \frac{(y_2 - y_1)^3}{4y_1 y_2}$$
(8.48)

Sample Problem 8.12 includes examples of depth and energy loss calculations.

Jump Length

Although the length of a jump is difficult to predict, a good approximation for jump length is about $5y_2$. We can see this relation is approximately true by examining Fig. 8.25, a photograph of a hydraulic jump in a horizontal channel, caused by a sluice gate upstream. In most cases $4 < L_j/y_2 < 6$.

Types of Jump

Since the flow must be supercritical in order for a jump to occur, F_1, the Froude number of the flow just upstream of the jump, must be greater than 1.0. As F_1 increases, the jump becomes more turbulent and more energy is dissipated. Ranges of different jump behaviors have been identified, some more desirable than others. These are summarized in Table 8.4. In particular, the oscillating jump is to be avoided; the main stream of water flowing through the jump oscillates between the bed and the surface, generating waves that can be damaging for great distances downstream.

In some areas of the world with high tides, when they enter estuaries and the like, a small hydraulic jump forms, which travels upstream and can cause minor damage. This phenomenon is known as a ***surge*** or a ***tidal bore.*** It may be analyzed by similar methods to those used for gravity waves (Sec. 8.20).

FIGURE 8.25   Hydraulic jump. (Courtesy of the Archives, California Institute of Technology)

Name	F_1	Energy Dissipation	Characteristics
Undular jump	1.0–1.7	<5%	Standing waves
Weak jump	1.7–2.5	5–15%	Smooth rise
Oscillating jump	2.5–4.5	15–45%	Unstable; avoid
Steady jump	4.5–9.0	45–70%	Best design range
Strong jump	>9.0	70–85%	Choppy, intermittent

[a] U.S. Bureau of Reclamation, Research Studies on Stilling Basins, Energy Dissipators, and Associated Appurtenances, *Hydraul. Lab. Rept.* Hyd-399, 1955.

TABLE 8.4   Types of hydraulic jump[a]

Stilling Basins

When water from elevated water bodies like reservoirs flows over spillways, through outlet works, or down chutes, it can have tremendous energy. This energy can cause great damage to the structure itself, to the ground supporting the structure (and therefore ultimately to the structure), and to the downstream channels. The hydraulic jump is one of the most effective methods of dissipating such flow energy, as we noted earlier. But hydraulic jumps themselves can also be very destructive. The objective of a *stilling basin* is to initiate a jump for energy dissipation and to contain it within a structure that will minimize damage.

The basin floor may be recessed or have an adverse slope to provide sufficient tailwater depth; a *sill* across the downstream end is also commonly used for this purpose. To increase the turbulence and to encourage the jump to form immediately as the water enters the basin, rows of *chute blocks* and/or staggered *baffle blocks* are commonly arranged across the basin. Sometimes those blocks with an end sill can reduce the length of the jump, and so the basin. Design details for stilling basins, baffle blocks, and similar hydraulic features have been developed based on extensive experimentation.[14]

Because high entry velocities can cause cavitation damage (Sec. 5.10) and erosion by sediment, the basin must be finished with quality concrete brought to a smooth surface.

8.19 Location of Hydraulic Jump

The problem of determining the location of a hydraulic jump involves a combined application of the principles discussed in Secs. 8.17 and 8.18. Examples of the location of a hydraulic jump are shown in Figs. 8.20 and 8.26. In all these cases shown on mild and steep slopes one of the conjugate depths is a normal depth. In case (1) of Fig. 8.26 the jump occurs downstream of the break. The reasons for this are illustrated by the following example. For convenience here we shall designate y' to be the conjugate of y.

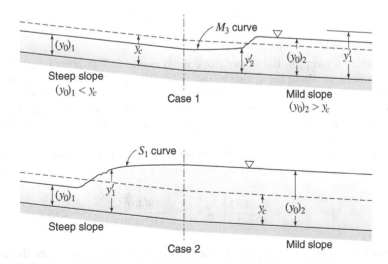

FIGURE 8.26 Examples of the location of a hydraulic jump where one conjugate depth is normal. [*Note:* y'_1 is the conjugate depth of $(y_0)_1$ and y'_2 is the conjugate depth of $(y_0)_2$.]

---

[14] U.S. Department of the Interior, Bureau of Reclamation, *Design of Small Dams*, Washington, D.C., 1987.

Sample Problem 8.12

Analyze the water-surface profile in a long rectangular channel lined with concrete ($n = 0.013$). The channel is 10 ft wide, the flow rate is 400 cfs, and the channel slope changes abruptly from 0.0150 to 0.0016. Find also the horsepower loss in the resulting jump.

Solution

Eq. (8.8a):
$$400 = \frac{1.486}{0.013}(10y_{0_1})\left(\frac{10y_{0_1}}{10 + 2y_{0_1}}\right)^{2/3}(0.015)^{1/2}$$

We can solve this equation by any of the methods described in Sample Problem 8.1. We obtain

$$y_{0_1} = 2.17 \text{ ft} \qquad \text{(normal depth on the upper slope)}$$

Using a similar procedure, we find the normal depth y_{0_2} on the lower slope to be 4.81 ft.

Eq. (8.23):
$$y_c = \left(\frac{q^2}{g}\right)^{1/3} = \left[\frac{(400/10)^2}{32.2}\right]^{1/3} = 3.68 \text{ ft}$$

Thus flow is supercritical ($y_{0_1} < y_c$) before the break in slope and subcritical ($y_{0_2} > y_c$) after the break, so a hydraulic jump must occur. One of the two profiles (case 1 or case 2) in Fig. 8.26 must occur.

Let us first explore case 2. Applying Eq. (8.46a) to determine the depth conjugate to the 2.17-ft normal depth on the upper slope, we get

$$y_1' = \frac{2.17}{2}\left\{-1 + \left[1 + \frac{8(40)^2}{32.2(2.17)^3}\right]^{1/2}\right\} = 5.77 \text{ ft}$$

Therefore a jump on the upper slope must rise to 5.77 ft, after which the surface must rise still more along an S_1 curve (Fig. 8.20). When the flow enters the lower slope, the depth would therefore be greater than $y_{0_2} = 4.81$ ft and so in zone 1 (the profiles in Fig. 8.20 with subscript 1). Because an M_1 curve cannot bring the water surface down from above 5.77 ft to 4.81 ft (normal depth), such a profile and jump cannot occur.

Let us now explore case 1. Applying Eq. (8.46b) to determine the depth conjugate to the 4.81-ft normal depth on the lower slope, we get

$$y_2' = \frac{4.81}{2}\left\{-1 + \left[1 + \frac{8(40)^2}{32.2(4.81)^3}\right]^{1/2}\right\} = 2.74 \text{ ft}$$

This lower conjugate depth of 2.74 ft will occur downstream of the break in slope. The water surface on the lower slope *can* rise from 2.17 ft to $y_2' = 2.74$ ft, via an M_3 curve (Fig. 8.20). So this case will occur.

We can find the location of the jump (i.e., its distance below the break in slope) by applying Eq. (8.39) to the M_3 curve between depths 2.17 ft at the break and 2.74 ft at the start of the jump:

$$\Delta x = \frac{E_1 - E_2}{S - S_0}$$

$$E_1 = 2.17 + \frac{(40/2.17)^2}{2(32.2)} = 7.45 \text{ ft}$$

$$E_2 = 2.74 + \frac{(40/2.74)^2}{2(32.2)} = 6.05 \text{ ft}$$

$$\bar{V} = \frac{1}{2}\left(\frac{40}{2.17} + \frac{40}{2.74}\right) = 16.53 \text{ fps}$$

$$\bar{R}_h = \frac{1}{2}\left(\frac{21.7}{14.34} + \frac{27.4}{15.47}\right) = 1.641 \text{ ft}$$

From Eq. (8.38a):

$$S = \left[\frac{(0.013)(16.53)}{1.486(1.641)^{2/3}}\right]^2 = 0.01081$$

Eq. (8.39):

$$\Delta x = \frac{7.452 - 6.054}{0.01081 - 0.00160} = 151.8 \text{ ft}$$

Note that we could compute this distance more accurately by dividing it into more reaches, i.e., by using more, smaller depth increments.

Thus the depth on the upper slope is 2.17 ft; downstream of the break the depth increases gradually (M_3 curve) to 2.74 ft over a distance of approximately 152 ft; then a hydraulic jump occurs from a depth of 2.74 ft to 4.81 ft; downstream of the jump the depth remains constant (i.e., normal) at 4.81 ft.   *ANS*

Eq. (8.48):

$$h_{L_j} = \frac{(4.81 - 2.74)^3}{4(4.81)2.74} = 0.1695 \text{ ft}$$

So from Eq. (5.40): $P \text{ loss} = \dfrac{\gamma Q h_{L_j}}{550} = \dfrac{62.4(400)0.1695}{550} = 7.69 \text{ hp}$ *ANS*

In the two examples shown in Fig. 8.26, and in some other examples in Fig. 8.20, the flow is at normal depth either before or after the hydraulic jump. These situations are straightforward and are easy to handle as shown in the preceding illustrative example. There are instances, however, where the flow is not normal either before or after the jump. Such a situation will occur, for example, when water flows out from under a sluice gate and there is a sufficiently high weir downstream of the gate. If the channel is on a steep slope, a hydraulic jump will occur between an S_3 and an S_1 water-surface profile, while if the channel is on a mild slope, the jump will occur between an M_3 and an M_1 profile (Fig. 8.20). Another instance would be between M_3 and M_2 profiles, as shown in Fig. 8.27. This combination is similar to the H_3–H_2 and A_3–A_2 combinations shown on the right-hand side of Fig. 8.20. To find the location of such a jump, varied

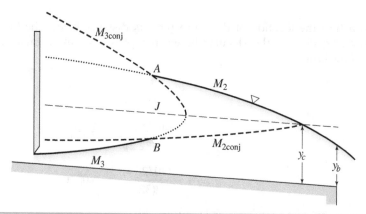

FIGURE 8.27 Example of the location of a hydraulic jump J when neither conjugate depth is normal.

flow calculations must be made in the downstream direction from the sluice gate and in the upstream direction from the free overfall (or from the weir in the other cases), at the same time computing the conjugate depths for one of these curves using Eq. (8.46). Where the conjugate of one profile crosses the other profile is the only place where the energies are correct for a hydraulic jump to occur. This is at point A or B (the same location) in Fig. 8.27. Graphically plotting these various curves greatly aids the solution process. Also, it will be helpful to reduce the size of the varied flow increments in the neighborhood of the intersection point, which will probably have to be found by interpolation.

8.20 Velocity of Gravity Waves

Consider a small wave of height Δy traveling at a velocity (or **celerity**) c to the left across the surface of stationary water whose depth is y (Fig. 8.28a). Let us now replace this with an equivalent steady flow to the right having a velocity $V_1 = -c$ (Fig. 8.28b). Because the wave *velocity or celerity is relative to the water*, the wave is now standing still with respect to the observer. Applying the principle of continuity to Fig. 8.28b we have $V_1 y = V_2 (y + \Delta y)$, where V_2 is the average velocity of flow past section 2 in Fig. 8.28b.

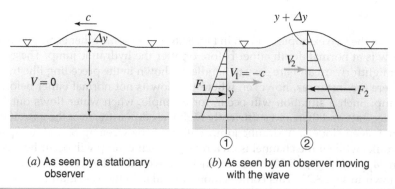

(a) As seen by a stationary observer

(b) As seen by an observer moving with the wave

FIGURE 8.28 Gravity wave of small amplitude.

Applying the momentum principle, $F = \rho Q \Delta V$, to the control volume between sections 1 and 2 for a unit width of channel gives

$$\gamma \frac{y^2}{2} - \gamma \frac{(y + \Delta y)^2}{2} = \frac{\gamma}{g} y V_1 (V_2 - V_1)$$

Substituting the expression for V_2 from continuity and letting $c = -V_1$ in the above gives

$$c = \sqrt{g(y + \Delta y)\left(\frac{y + \frac{1}{2}\Delta y}{y}\right)} \approx \sqrt{g(y + \Delta y)} \approx \sqrt{gy} \qquad (8.49)$$

The latter expressions apply if surface disturbance is relatively small, that is, $\Delta y << y$ We see that the velocity (celerity) of a wave will increase as the depth of the water increases. But this is the velocity relative to the water. If the water is flowing, the absolute speed of travel of the wave will be the resultant of the two velocities. Thus

$$c_{abs} = c \pm V \qquad (8.50)$$

Small Waves. Let us now consider only "small" waves, for which $c = \sqrt{gy}$. We are all familiar with the circular waves that spread on a still water surface when disturbed, as when we throw a stone into a pond. When the water is also moving, these spreading patterns are carried with the water, in accordance with Eq. (8.50). In Fig. 8.29 we see how the wave pattern produced by a repeated, regular disturbance varies with water velocity.

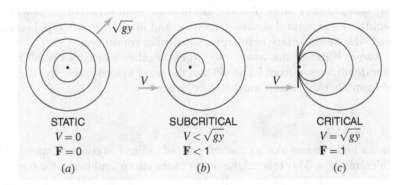

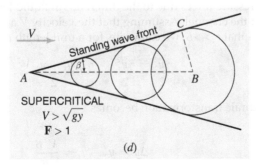

FIGURE 8.29 Variation of gravity wave patterns with flow conditions.

The familiar pattern of Fig. 8.29a is produced when the water is static. In Fig. 8.29b the water moving with a low, subcritical velocity carries the circular waves to the right. For the somewhat higher velocity of $V = \sqrt{gy}$, conditions are critical (Sec. 8.10), and $V = c$, so that waves trying to move upstream (to the left in Fig. 8.29c) are brought to a standstill as shown; they "hold fast." Thus we see that small disturbance waves can only travel upstream if $V < c$, i.e., if the flow is subcritical. For still higher, supercritical velocities, the circles are carried downstream as shown in Fig. 8.29d; they *cannot travel upstream.* As the center of the disturbance is carried in time t from A to B for example, the radius grows from B to C. So we see that

Small waves:
$$\sin \beta = \frac{BC}{AB} = \frac{ct}{Vt} = \frac{\sqrt{gy}}{V} = \frac{1}{\mathbf{F}} \tag{8.51}$$

If the disturbance is not intermittent but is continuous, as is usual, we shall not see the individual circles, but we shall see where they reinforce one another. This occurs perpendicular to the flow at the disturbance site for critical flow, see Fig. 8.29c, and along line AC in Fig. 8.29d for supercritical flow. The perpendicular wave front with critical flow is a special case of Fig. 8.29d, because when $V = c$ we have $\mathbf{F} = 1$ and $\beta = 90°$. AC is known as a **standing wave front** or a **disturbance line,** and is comparable to the oblique shock wave. The angle β is called the **wave angle,** and this obviously decreases as V increases. When the channel shape must change in any way, such as at bridge piers, bends, and transitions, these standing waves form with supercritical flow, and because they reflect off side walls, complex patterns of **cross waves** can result. Every small irregularity will similarly cause oblique standing waves in supercritical flow. There are many irregularities in natural streams of course, and in order to run rapids canoeists learn to "read" the water surface to interpret subsurface conditions.

Larger Waves. If the waves are larger, in other words not small compared to the water depth, we see from Eq. (8.49) that the wave velocity (celerity) is greater than $\sqrt{gy}$, and from Eq. (8.51) that β must also be larger.

8.21   Flow Around Channel Bends

When a body moves along a curved path of radius r at constant speed, it has a normal acceleration $a_n = V^2/r$ toward the center of the curve, and hence the body must be acted on by a force in that direction. In Fig. 8.30 we can see that this force comes from the unbalanced pressure forces due to the difference in liquid levels between the outer and inner banks of the channel. Assuming that the velocity V across the rectangular section is uniform and that $r >> B$, we can write, for a unit length of channel, $\Sigma F_n = ma_n$,

$$\frac{\gamma y_2^2}{2} - \frac{\gamma y_1^2}{2} = \frac{\gamma}{g} B \frac{y_1 + y_2}{2} \frac{V^2}{r}$$

which by algebraic transformation becomes

$$\Delta y = y_2 - y_1 = \frac{V^2 B}{gr} \tag{8.52}$$

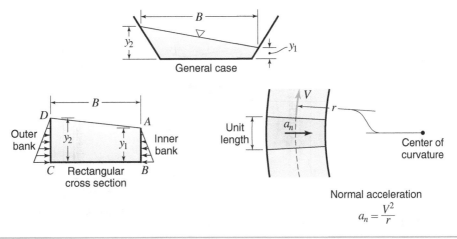

FIGURE 8.30   Flow in an open-channel bend.

where B is the top width of the water surface as shown in Fig. 8.30. We can show that Eq. (8.52) applies to any shape of cross section. If the effects of velocity distribution and variations in curvature across the stream are considered, the difference in water depths between the outer and inner banks may be as much as 20% more than that given by Eq. (8.52). If the actual velocity distribution across the stream is known, we can divide the width into sections and compute the difference in elevation for each section. The total difference in surface elevation across the stream is the sum of the differences for the individual sections.

With supercritical flow the complicating factor is the effect of disturbance waves, generated by the very start of the curve. These waves, one from the outside wall and one from the inside, traverse the channel, making an angle β with the original direction of flow, as discussed in Sec. 8.20. The result is a crisscross wave pattern. The water surface along the outside wall will rise from the beginning of the curve, reaching a maximum at the point where the wave from the inside wall reaches the outside wall. The wave is then reflected back to the inside wall and so on around the bend. Hence in supercritical flow around a bend the increase in water depth on the outer wall is equal to that created by centrifugal effects plus or minus the height of the waves. Field observations indicate that these waves change the subcritical depths by up to $\pm\Delta y/2$, where Δy is the value given by Eq. (8.52). Thus the water-depth increase on the outer wall varies between zero and Δy, i.e., double the subcritical rise, and the depth decrease on the inner wall varies similarly.

Several schemes to lessen the surface rise from wave effects have been investigated. The bed of the channel can be banked so that all elements of the flow are acted upon simultaneously, which is not possible when the turning force comes from the wall only. As in a banked-railway curve, this requires a transition section with a gradually increasing superelevation preceding the main curve. Another method is to introduce a counterdisturbance to offset the disturbance wave caused by the curve. Such a counterdisturbance can be provided by a section of curved channel of twice the radius of the main section, by a spiral transition curve, or by diagonal sills on the channel bed, all preceding the main curve.

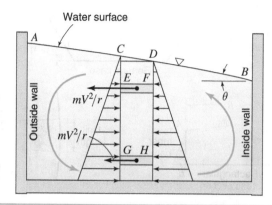

FIGURE 8.31 Schematic sketch of flow around a bend in a rectangular channel curving to the right, looking downstream, with spiral flow counterclockwise.

Flow around a bend in an open channel is complicated by the development of a secondary flow as the liquid travels around the curve. We discussed the development of secondary flow in the bends of pressure conduits in Sec. 7.27.

The flow around a channel bend is called *spiral flow* because superposition of the secondary flow on the forward motion of the liquid causes the liquid to follow a path like a corkscrew (or spiral). We can explain the occurrence of spiral flow by referring to Fig. 8.31. The water surface is superelevated at the outside wall. The element EF is subjected to a centrifugal force mV^2/r, which is balanced by an increased hydrostatic force on the left side, due to the superelevation of the water surface at C above that at D. The element GH is subjected to the same net hydrostatic force inward, but the centrifugal force outward is much less because the velocity is decreased by friction near the bottom. This results in a cross flow inward along the bottom of the channel, which is balanced by an outward flow near the water surface; hence the spiral. This spiral flow is largely responsible for the commonly observed erosion of the outside bank of a river bend, with consequent deposition and building up of a sandbar near the inside bank.

Sample Problem 8.13

A rectangular channel 2 m wide carries 4.2 m³/s in uniform flow at a depth of 0.5 m. What will be the maximum and minimum water depths at the inside and outside walls of a bend in the channel of radius 10 m to the centerline?

Solution

$$V = Q/A = 4.2/(2 \times 0.5) = 4.2 \text{ m/s}$$

Eq. (8.51):
$$\mathbf{F} = \frac{V}{\sqrt{gy}} = \frac{4.2}{\sqrt{9.81 \times 0.5}} = 1.896$$

$\mathbf{F} > 1$, so the flow is supercritical.

Eq. (8.52):
$$\Delta y = \frac{(4.2)^2 2}{9.81(10)} = 0.360 \text{ m}$$

Because of wave action with supercritical flow, Δy varies from zero to double the value computed. Thus the extreme water depths are as follows:

	Inside Wall	Outside Wall	
y_{max}	0.500 m	0.860 m	
y_{min}	0.140 m	0.500 m	*ANS*

8.22 Transitions

Special transition sections are often used to join channels of different size and shape in order to avoid undesirable flow conditions and to minimize head loss. If the flow is subcritical, a straightline transition (Fig. 8.32) with an angle θ of about 12.5° is fairly satisfactory. We also call this a *wedge transition.* Without the transition, i.e., with an abrupt change in section with square corners, the corresponding head losses are about two to four times larger. Intermediate between these two types is to replace the abrupt corners with vertical quarter-cylinder segments; this is known as a *cylinder-quadrant transition*. At Froude numbers between about 0.5 and 1.0, complex *warped transitions* that avoid any sharp angles are advisable.

Head loss in transitions is proportional to the change in the velocity head, and is traditionally expressed as

$$h_{L_t} = k_t \left| \frac{V_2^2}{2g} - \frac{V_1^2}{2g} \right| \tag{8.53}$$

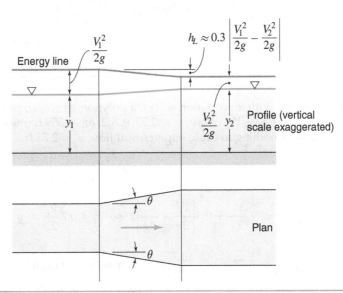

FIGURE 8.32 Simple open-channel wedge transition for decelerating flow.

	Loss Coefficient k_t	
Transition Type	Contracting	Expanding
Abrupt	0.4–0.5	0.75–1.00
Cylinder-quadrant	0.2	0.5
Wedge	0.1–0.2	0.3–0.5
Warped	0.1	0.3

TABLE 8.5   Loss coefficients for channel transitions with subcritical flow

Approximate values of the loss coefficient k_t for the transitions discussed are given in Table 8.5. Notice that the more efficient transitions are the ones that are more expensive to build, and that losses are greater with expanding flows, as usual. Experimenters have found that transition lengths need to be at least 2.25 times the change in the water surface width.

With supercritical flow ($\mathbf{F} > 1$), surface waves form as described in Sec. 8.20, and require special procedures for transition design.

At a channel entrance from a reservoir or from a larger channel the head loss for a square-edged entrance is about 0.5 times the velocity head. By rounding the entrance, we can reduce the head loss to slightly less than 0.2 times the velocity head.

Sample Problem 8.14

Refer to Fig. 8.32. A rectangular channel changes in width from 4 ft to 6 ft. Measurements indicate that $y_1 = 2.50$ ft and $Q = 50$ cfs. Determine the depth y_2 by (a) neglecting head loss, and (b) considering the head loss to be given as shown on the figure.

Solution

Eq. (8.23): $y_{c_1} = \left[\dfrac{(50/4)^2}{32.2}\right]^{1/3} = 1.693$ ft; $y_{c_2} = \left[\dfrac{(50/6)^2}{32.2}\right]^{1/3} = 1.292$ ft

$V_1 = 50/(2.5 \times 4) = 5$ fps; $V_2 = 50/6y_2 = 8.33/y_2$ fps

(a) $h_L = 0$, so $E_1 = E_2$, i.e., $2.5 + \dfrac{5^2}{2(32.2)} = y_2 + \dfrac{(8.33/y_2)^2}{2(32.2)}$

We could solve this with an equation solver, a polynomial solver, or by trial and error (see Sample Problem 8.9). We obtain $y_2 = 2.75, 0.702$, or -0.559 ft (meaningless).
As we know of nothing to cause supercritical flow, $y_2 = 2.75$ ft   *ANS*

(b) $E_1 = E_2 + h_L = E_2 + 0.3\left(\dfrac{V_1^2}{2g} - \dfrac{V_2^2}{2g}\right)$

$y_1 + \dfrac{V_1^2}{2g} = y_2 + \dfrac{V_2^2}{2g} + 0.3\left(\dfrac{V_1^2}{2g} - \dfrac{V_2^2}{2g}\right)$,   so $y_1 + 0.7\dfrac{V_1^2}{2g} = y_2 + 0.7\dfrac{V_2^2}{2g}$

$2.5 + 0.7\dfrac{5^2}{2(32.2)} = 2.77 = y_2 + \dfrac{0.755}{y_2^2}$;   by trial, $y_2 = 2.67, 0.584$, or -0.479 ft

As we know of nothing to cause supercritical flow, $y_2 = 2.67$ ft.   *ANS*

8.23   Hydraulics of Culverts

A culvert is a conduit passing under a road or highway. In section, culverts may be circular, rectangular, or oval. Culverts may operate with either a submerged entrance (Fig. 8.33) or a free (unsubmerged) entrance.

Submerged Entrance

In the case of a submerged entrance there are three possible regimes of flow as indicated in Fig. 8.33. Under conditions (a) and (b) of the figure the culvert is said to be flowing under **outlet control**, while condition (c) represents **entrance control**. In (a) the outlet is submerged, possibly because of inadequate channel capacity downstream or due to backwater from a connecting stream. In (b) the normal depth y_0 of the flow is greater than the culvert height D, causing the culvert to flow full.[15] Writing energy Eq. (5.28)

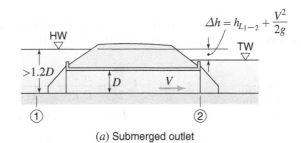

(a) Submerged outlet

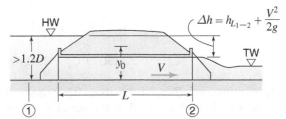

(b) Normal depth > barrel height

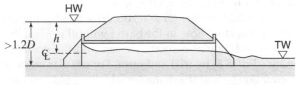

(c) Orifice flow, normal depth < barrel height

FIGURE 8.33   Flow conditions in culverts with submerged entrance.

---

[15] If there is a contraction of flow at the culvert entrance, reexpansion will require about six diameters. Hence a very short culvert may not flow full even though $y_0 > D$.

from the upstream water surface (1) to the water surface at the outlet (2), taking the latter as datum, for both cases (a) and (b) we obtain

$$0 + \Delta h + 0 - h_{L_{1-2}} = 0 + 0 + \frac{V^2}{2g}$$

or

$$\Delta h = h_{L_{1-2}} + \frac{V^2}{2g}$$

where Δh is defined[16] in Figs. 8.33a and 8.33b, and $V^2/2g$ is the velocity head loss at submerged discharge in case (a) or the residual velocity head at discharge in case (b). We can find Δh from $\Delta h = y_1 + LS_0 - y_2$, where y is the water depth above invert, and L and S_0 are the barrel length and slope, respectively. The head loss h_L between the two points is equal to h_e, the entrance loss, plus h_f, the friction loss in the barrel. Therefore

$$\Delta h = h_e + h_f + \frac{V^2}{2g} \tag{8.54}$$

Entrance loss is a function of the velocity head in the culvert, while friction loss may be computed using Manning's equation [Eq. (8.8)]. Thus,

Outlet control, case (a) or (b):
$$\Delta h = k_e \frac{V^2}{2g} + \frac{n^2 V^2 L}{C_3 R_h^{4/3}} + \frac{V^2}{2g} \tag{8.55}$$
$$C_3 = 1 \text{ (SI) or } C_3 = 2.21 \text{ (BG)}$$

Such an expression can be reduced to

Outlet control, case (a) or (b):
$$\Delta h = \left(k_e + \frac{C_4 n^2 L}{R_h^{4/3}} + 1 \right) \frac{V^2}{2g} \tag{8.56}$$
$$C_4 = 19.62 \text{ (SI) or } C_4 = 29.2 \text{ (BG)}$$

The entrance coefficient k_e (Fig. 7.13) is about 0.5 for a square-edged entrance and about 0.05 if the entrance is well rounded. If the outlet is submerged, the head loss may be reduced somewhat by flaring the culvert outlet so that the outlet velocity is reduced and some of the velocity head recovered. Tests show that the flare angle should not exceed about 6° for maximum effectiveness.

To determine which of case (b) or (c) occurs when the outlet is free (not submerged), we need to find if normal flow in the barrel will fill it. Usually the discharge is known or assumed. For rectangular culverts we can find the normal depth in the usual way from Manning's Eq. (8.8) by trial and error. For circular cross sections it is easier to use Manning's equation to find the diameter which would just flow full ($R_h = D/4$), and to compare that with the actual or proposed diameter. If we are considering alternative slopes with a given barrel diameter, we can rearrange the algebra to solve for the slope that just causes the barrel to flow full.

If normal depth in the culvert is less than the barrel height, with the inlet submerged and the outlet free, the condition (c) illustrated in Fig. 8.33c will normally result.

---

[16] The energy grade line at the entrance and exit is above the actual water surface by the velocity head of the approaching or leaving water. With water ponded at the entrance or exit, this velocity head is usually negligible but should be included in computations if it is of significant magnitude.

This culvert is said to be flowing under *entrance control,* i.e., the entrance will not admit water fast enough to fill the barrel, and the discharge is determined by the entrance conditions. The inlet then functions like an orifice for which

Entrance control, case (*c*): $Q = C_d A \sqrt{2gh}$ (8.57)

where h is the head on the center of the orifice[17] and C_d is the orifice coefficient of discharge. The head required for a given flow Q is therefore

Entrance control, case (*c*): $h = \dfrac{1}{C_d^2} \dfrac{Q^2}{2gA^2}$ (8.58)

It is impractical to cite appropriate values of C_d, because of the wide variety of entrance conditions which may occur; for a specific design we must determine C_d from model tests or tests of similar entrances. For a sharp-edged entrance without suppression of the contraction $C_d = 0.62$, while for a well-rounded entrance C_d approaches 1.0. If the culvert is set with its invert at stream-bed level, the contraction is suppressed at the bottom. Flared wingwalls at the approach may also cause partial suppression of the side contractions.

Case (*c*) can, in unusual conditions, have both inlet and outlets submerged. In this case, a hydraulic jump will form in the culvert barrel, which can create undesirable flow instabilities.

Free Entrance

Some box culverts may be designed so that the top of the box forms the roadway. In this case the headwater should not submerge the inlet, and one of the flow conditions of Fig. 8.34 (free entrance) will exist. In cases (*a*) and (*b*) critical depth in the barrel controls the headwater elevation, while in case (*c*) the tailwater elevation is the control. In all cases we can compute the headwater elevation using the principles of open-channel flow discussed in this chapter with an allowance for entrance and exit losses.

When the culvert is on a steep slope [case (*b*)], critical depth will occur at about $1.4y_c$ downstream from the entrance. The water surface will impinge on the headwall (Fig. 8.33) when the headwater depth is about $1.2D$ if y_c is $0.8D$ or more. Since it would be inefficient to design a culvert with y_c much less than $0.8D$, a headwater depth of $1.2D$ is approximately the boundary between free-entrance conditions (Fig. 8.34) and submerged-entrance conditions (Fig. 8.33).

Similar to Fig. 8.33(*c*), case (*b*) in Fig. 8.34 can experience a submerged outlet in some conditions, in which case a hydraulic jump would occur in the barrel since flow just downstream of the inlet would be supercritical. Determining the water-surface profile and location of the jump requires more advanced methods, like those in Secs. 8.18 and 8.19, but accounting for the specific culvert shape.

In the United States, federal design guidance for culverts is provided by the FHWA, which adopts culvert classification developed by the U.S. Geological Survey (USGS).[18] The flow conditions in Figs. 8.33 and 8.34 are summarized with the federal classification in Table 8.6.

---

[17] Equation (8.57) applies for an orifice on which the head h to the center of the orifice is large compared with the orifice height D. When the headwater depth is $1.20D$ (i.e., $h = 0.7D$), an error of 2% results. In the light of other uncertainties in the design this error can be ignored.
[18] U.S. Department of Transportation, Federal Highway Administration, *Hydraulic Design of Highway Culverts*, Hydraulic Design Series Number 5, FHWA-HIF-12-026, 2012.

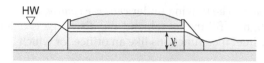

(a) Mild slope, low tailwater

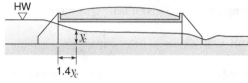

(b) Steep slope, low tailwater

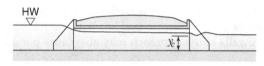

(c) Mild slope, tailwater submerges y_c

FIGURE 8.34   Flow conditions in culverts with free entrance.

Figure	FHWA/USGS Flow Type	Flow Control
8.33(a)	4	Outlet
8.33(b)	6 & 7[a]	Outlet
8.33(c)	5	Entrance
8.34(a)	2	Outlet
8.34(b)	1	Entrance
8.34(c)	3	Outlet

[a] Type 6 applies for a full barrel; Type 7 indicates a partially full barrel for at least part of the culvert length.

TABLE 8.6   Culvert flow classification

Sample Problem 8.15

A culvert under a road must be 30 m long, have a slope of 0.003, and carry 4.3 m³/s. If the maximum permissible headwater level is 3.6 m above the culvert invert, what size of corrugated-pipe culvert ($n = 0.025$) would you select? Neglect velocity of approach. Assume a square-edged inlet with $k_e = 0.5$ and $C_d = 0.65$. The outlet will discharge freely.

Solution
Sec. 8.23: Assume the headwater depth $> 1.2D$ so that the entrance is submerged per Fig. 8.33. Therefore, assume that $D <$ HW depth/1.2 = 3.6 m/1.2 = 3.0 m. Given the discharge is free, so Fig. 8.33a cannot apply. Conditions are those of either Fig. 8.33b or 8.33c.

Assume case (b), Fig. 8.33, i.e., that the barrel flows full.

$$V = \frac{Q}{A} = \frac{4.3}{\pi D^2/4} = \frac{5.47}{D^2}; \quad R = D/4$$

Fig. 8.33b: $\quad \Delta h = h_{L_{1-2}} + \frac{V^2}{2g} = (y_1 - y_2) + (z_1 - z_2) = y_1 - y_2 + S_0 L$

$$= 3.6 - D + 0.003(30) = 3.69 - D$$

Eq. (8.56): $\quad \Delta h = \left[0.5 + \frac{19.62(0.25)^2 \, 30}{(D/4)^{4/3}} + 1\right] \frac{5.47^2}{2(9.81)D^4}$

$$= \left(1.5 + \frac{2.34}{D^{4/3}}\right) \frac{1.528}{D^4}$$

Equating the two Δh expressions and simplifying,

$$3.69 = D + \left(1.5 + \frac{2.34}{D^{4/3}}\right) \frac{1.528}{D^4}$$

By trial and error (see Sample Problems 3.5 and 5.8) or by equation solver, $D = 1.196$ m. Thus the first assumption (submerged entrance with $D < 3$ m) is correct.

Now determine if we have case (b) or (c); to do this, find the maximum diameter d_0 that will just flow full with normal (uniform) flow:

Eq. (8.8b): $\quad 4.3 = \frac{1}{0.025} \frac{\pi d_0^2}{4} \left(\frac{d_0}{4}\right)^{2/3} (0.003)^{1/2}; \quad d_0 = 1.994$ m

As $D < d_0$, the culvert runs full, we do have case (b), as assumed. These assumptions and analysis are valid. $D = 1.196$ m. Refer to Table A.9 for available diameters.
Use standard $D = 1.2$ m   *ANS*

8.24 Further Topics in Open-Channel Flow

The basics of open-channel flow discussed in this chapter are just an introduction to many more advanced subjects.

With steady gradually varied flow, a variety of specialized techniques have been devised to solve for water-surface profiles in irregular, natural channels. Ways have been developed to treat the dividing of flow around islands, and the joining of flows at *confluences*. When a canal connects two reservoirs having varying levels, the discharge in the canal, which depends on these levels, is called its *delivery*, and we can develop *delivery curves* to relate the discharge to the depth at each end. When flow from a reservoir or lake enters a channel with a mild slope, the discharge may depend on the length of the channel if there is a downstream control such as a contraction or a free overfall; we call this *the discharge problem*. Steady flow along a channel may vary with distance, such as in gutters and with side-channel spillways. We refer to this as *spatially varied flow*, for which the water-surface profiles are usually solved by trial-and-error procedures.

Near channel *appurtenances* like weirs (Secs. 9.11 and 9.12), spillways (Sec. 9.13), bridge piers, flumes, transition structures (Sec. 8.22), culverts (Sec. 8.23), and stilling basins (Sec. 8.18), an analysis of the resulting steady rapidly varied flow conditions is often quite involved. These are specialized subjects, which depend strongly on details of the appurtenances. *Standing waves* and *cross waves*, introduced in Sec. 8.20, are rapidly varying flow phenomena that often occur with supercritical flow. They can be very complex, and are strongly influenced by the channel shape.

With *unsteady flow* in open channels, the depth and/or the velocity varies with time at a point. This is an advanced subject not addressed in this text. We can vary the flow gradually or rapidly. Gradually varied unsteady flow includes flood waves, tidal flows, and waves resulting from the gradual change of channel control structures such as gates. It is governed by differential equations that usually require numerical methods and computers to solve them. Rapidly varied unsteady flow includes such flows as *pulsating flow* (*roll waves*), which occurs in supercritical flow on very steep slopes, and *surges* or *surge waves* (moving hydraulic jumps) caused by the rapid operation of a control structure or the sudden failure of a dam. In some locations surges are caused by tidal action, when the surge is known as a *tidal bore.* Although we can solve some very simplified problems of these types analytically, in most cases more realistic problems must be solved by numerical methods using computers.

Phenomena governed by density include *salt wedges* at estuaries where fresh river water overlies saline ocean water, and similar *cooling water wedges* in rivers caused by waste heat from power plants. Density effects in lakes and reservoirs in many cases result in an annual overturning of the contents due to seasonal temperature changes. *Density currents* can also be caused by sediment loads, which in turn relate to the subjects of *scour* and *sediment transport.*

Other more advanced phenomena of interest arise when we are analyzing the fate of pollutants, and include *advection* and mixing processes known as *turbulent diffusion* and *dispersion.* Again, solution of these processes usually requires numerical methods using computers.

Excellent books specializing in open-channel flow include references 20, 30, and 38 in Appendix C.

CHAPTER 9

Fluid Measurements

Both the engineer and the scientific investigator are often faced with the problem of measuring various fluid properties such as density, viscosity, and surface tension. Also, measurements are often required of various fluid phenomena, as pressure, velocity, and flow rate. In this chapter only the principles and theory of such measurements will be discussed. Detailed information on the various measuring devices can be found elsewhere in the literature.[1] This chapter draws on many concepts presented in previous chapters; thus, it also serves as a good review.

9.1 Measurement of Fluid Properties

Most commonly we measure *liquid density* by weighing a known volume of the liquid to find its specific weight (Sec. 2.3), and then dividing by the acceleration due to gravity. Another technique makes use of hydrostatic weighing, where a nonporous solid object of known volume is weighed (*a*) in air and then (*b*) in the liquid whose density is to be determined. The *hydrometer* is a variation of this technique. We calculate the densities from fluid statics. Another, though not very accurate, way to determine liquid density is to place two immiscible liquids in a U tube, one of known density, the other of unknown density. From fluid statics we can find the unknown density. These various techniques of measuring liquid density are illustrated in the exercises that accompany this section.

The measurement of *viscosity* is generally made with a device known as a *viscometer.* Various types of viscometers are available. They all depend on the creation of laminar-flow conditions. We shall confine this discussion to the measurement of the viscosity of liquids. Since viscosity varies considerably with temperature, it is essential that the fluid be at a constant temperature when a measurement is being made. This is generally accomplished by immersing the device in a constant-temperature bath.

Several types of *rotational viscometers* are available. These generally consist of two concentric cylinders that are rotated with respect to one another. We fill the narrow space between them with liquid whose viscosity is to be measured. The rate of rotation under the influence of a given torque is indicative of the viscosity of the liquid. One difficulty with this type of viscometer is that we must account for mechanical friction, and this is difficult to deal with accurately.

The *tube-type viscometer* is perhaps the most reliable. Figure 9.1 shows the Saybolt viscometer. In this device the liquid is originally at *M*, with the bottom of the tube plugged.

---

[1] See, for example, *Fluid Meters: Their Theory and Application,* 6th ed., American Society of Mechanical Engineers, New York, 1971.

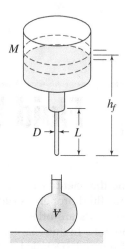

FIGURE 9.1 Falling-head tube-type viscometer.

The plug is removed, and the time required for a certain volume of liquid to pass through the tube is a measure of the kinematic viscosity of the liquid. In this device the flow is unsteady and the tube is of such small diameter that we can assume that the flow is laminar. Substituting the expression for mean velocity V for laminar flow in tubes [Eq. (7.28)] into the continuity expression [Eq. (4.7)] gives

$$Q = \frac{\pi D^4 \gamma h_f}{128 \mu L} \tag{9.1}$$

As an approximation, let h_f be the average imposed head during the flow period and let $Q \approx V\!\!\!\!/\,/t$ where $V\!\!\!\!/\,$ is the volume of liquid that flows out of the tube in time t. Substituting $Q = V\!\!\!\!/\,/t$, into Eq. (9.1) and making use of Eq. (2.11), we get

$$\nu = \frac{\pi D^4 h_f}{128 V\!\!\!\!/\, L} gt \tag{9.2}$$

Since $D, L, V\!\!\!\!/\,$, and h_f are constants of the device, $\nu = Kgt$, and the kinematic viscosity is seen to be proportional to the measured time. Equation (9.2) gives good results if the tube is relatively long. However, for a short tube, as with the Saybolt viscometer, we need to apply a correction factor if the tube is too short for the establishment of laminar flow (Sec. 7.9).

There are several other types of tube viscometers, but they are all based on the same principle. Some come with a set of tubes of various diameters so that measurements can be made on liquids with a wide range of viscosities in a convenient time period. Because the dimensions of such fine tubes cannot be perfectly duplicated, each tube is individually calibrated by measuring the time for a liquid of known viscosity at a given temperature to discharge the standard volume.

The *falling-sphere viscometer* is a third type. In such a device we place the liquid in a tall transparent cylinder and drop in a sphere of known weight and diameter. If the sphere is small enough, Stokes' law will prevail and the fall velocity of the sphere will be approximately inversely proportional to the absolute viscosity of the liquid. We can

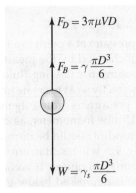

$$F_D = 3\pi\mu VD$$

$$F_B = \gamma \frac{\pi D^3}{6}$$

$$W = \gamma_s \frac{\pi D^3}{6}$$

FIGURE 9.2 Free-body diagram of sphere falling at terminal velocity.

see that this is so by examining the free-body diagram of such a falling sphere (Fig. 9.2). The forces acting include gravity, buoyancy, and drag. Stokes' law states that if $DV/\nu < 1$, the drag force on a sphere is given by $F_D = 3\pi\mu VD$, where V is the velocity of the sphere and D is its diameter. When the sphere is dropped in a liquid, it will quickly accelerate to terminal velocity, at which $\Sigma F_z = 0$. Then

$$W - F_B - F_D = \gamma_s \frac{\pi D^3}{6} - \gamma \frac{\pi D^3}{6} - 3\pi\mu VD = 0$$

where γ_s and γ represent the specific weight of the sphere and liquid, respectively. Solving this equation, we get

$$\mu = \frac{D^2(\gamma_s - \gamma)}{18V} \tag{9.3}$$

In the preceding development we assumed that the sphere was dropped into a liquid of infinite extent. In actuality, the liquid will be contained in a tube and a *wall effect* will influence the drag force and hence the fall velocity. Experimenters have found that the wall effect[2] can be expressed approximately as

$$\frac{V}{V_t} \approx 1 + \frac{9D}{4D_t} + \left(\frac{9D}{4D_t}\right)^2 \tag{9.4}$$

where D_t is the tube diameter, and V_t represents the fall velocity in the tube. Equation (9.4) is reliable only if $D/D_t < 1/3$.

Other fluid properties such as surface tension, elasticity, vapor pressure, specific heats at constant pressure and constant temperature, and gas constant are commonly determined by physicists, and the techniques for their measurement will not be discussed here.

---

[2] J. S. McNown, H. M. Lee, M. B. McPherson, and S. M. Engez, Influence of Boundary Proximity on the Drag of Spheres, *Proc. 7th Intern. Congr. Appl. Mech.*, 1948.

9.2 Measurement of Static Pressure

We first defined the static pressure at a point in a flowing fluid in Secs. 5.4 and 5.11; it is the pressure in the fluid unchanged by the measuring instrument. To get an accurate measurement of static pressure in a flowing fluid, it is important that the measuring device fit the streamlines perfectly so as to create no disturbance to the flow. In a straight reach of conduit, the static pressure is ordinarily measured by attaching to the piezometer a pressure gage or a U-tube manometer, as described in Sec. 3.5. The piezometer opening in the side of the conduit should be normal to and flush with the surface. Any projection, such as (c) in Fig. 9.3, will result in error. It has been found, for example, that a projection of 0.10 in (2.5 mm) will cause a 16% change in the local velocity head. In this case the recorded pressure is depressed below the pressure in the undisturbed fluid because the disturbance of the streamline pattern increases the velocity, hence decreasing the pressure according to the Bernoulli equation.

When measuring the static pressure in a pipe, it is desirable to have two or more openings around the periphery of the section to account for possible imperfections of the wall. For this purpose a *piezometer ring* (Fig. 9.4) is used.

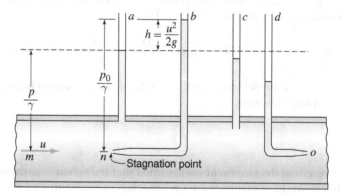

FIGURE 9.3   Arrangements of pitot tubes and piezometers for measuring flow in a pipe.

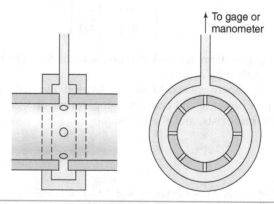

FIGURE 9.4   Piezometer ring.

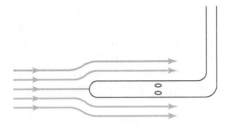

FIGURE 9.5   Static tube.

To measure the static pressure in a flow field, we use a *static tube* (Fig. 9.5). In this device the pressure is transmitted to a gage or manometer through piezometric holes that are evenly spaced around the circumference of the tube. This device will give good results if it is perfectly aligned with the flow. Actually, the mean velocity past the piezometer holes will be slightly larger than that of the undisturbed flow field; hence the pressure at the holes will generally be somewhat below the pressure of the undisturbed fluid. We can minimize this error by making the diameter of the tube as small as possible. If the direction of the flow is unknown for two-dimensional flows, we can use a *direction-finding tube* (Fig. 9.6). This device is a cylindrical tube having two piezometer holes located as shown. Each piezometer is connected to its own measuring device. We can rotate the tube until each tube shows the same reading. Then, from symmetry, we can determine the direction of flow. If the piezometer openings are located as shown, the recorded pressures will correspond very closely to those of the undisturbed flow.

To obtain a reading of the fluid pressure, we can connect a piezometer tube to a Bourdon gage (Sec. 3.5) or to a pressure transducer (Fig. 3.9). The latter is sometimes connected to a strip-chart recorder or the pressure reading may be displayed on a panel in digital form.

9.3   Measurement of Velocity with Pitot Tubes

One means of measuring the local velocity u in a flowing fluid is the pitot tube, named after its inventor Henri Pitot (1695–1771), a French physicist who used a bent tube in 1732 to measure velocities in the River Seine. In Sec. 5.4 we saw that the pressure at the forward stagnation point of a stationary body in a flowing fluid is $p_0 = p + \rho u^2/2$, where

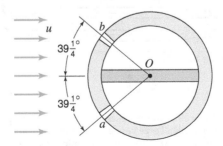

FIGURE 9.6   Direction-finding tube.

p and u are the pressure and velocity, respectively, in the undisturbed flow upstream from the body. If $p_0 - p$ can be measured, the velocity at a point is determined by this relation. The stagnation pressure can be measured by a tube facing upstream, such as (*b*) in Fig. 9.3. For a liquid jet or open stream with parallel streamlines, only this single tube is necessary, since the height h to which the liquid rises in the tube above the surrounding free surface is equal to the velocity head in the stream approaching the tip of the tube.

For a closed conduit under pressure we must measure the static pressure also, as shown by tube (*a*) in Fig. 9.3, and subtract this from the total pitot reading to secure the differential head h. We can measure the differential pressure with any suitable manometer arrangement. We can derive the formula for the pitot tube for incompressible flow by writing the (ideal) energy equation between points m and n of Fig. 9.3,

$$\frac{p}{\gamma} + \frac{u^2}{2g} = \frac{p_0}{\gamma} \tag{9.5}$$

from which

$$u_i = \sqrt{2g\left(\frac{p_0}{\gamma} - \frac{p}{\gamma}\right)}$$

This equation gives the ideal velocity of flow[3] u_i at the point in the stream where the pitot tube is located. In actuality, directional velocity fluctuations due to turbulence increase pitot-tube readings, so that we must multiply the right side of Eq. (9.6) by a factor C varying from 0.98 to 0.995 to give the true velocity,

$$u = Cu_i = C\sqrt{2g\left(\frac{p_0}{\gamma} - \frac{p}{\gamma}\right)} \tag{9.6}$$

Where conditions are such that it is impractical to measure static pressure at the wall, we can use a combined *pitot-static tube,* as in Fig. 9.7. We measure the static pressure through two or more holes drilled through an outer tube into an annular space.

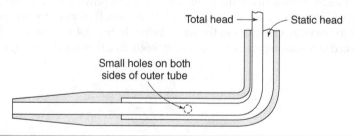

Total head → Static head

Small holes on both sides of outer tube

FIGURE 9.7   Pitot-static tube.

[3] Equations (9.5)–(9.7) as well as those presented in Secs. 9.6–9.9 apply strictly to incompressible fluids. However, these equations will all give very good results when applied to compressible fluids if $\mathbf{M} < 0.1$, where $\mathbf{M}$ is the Mach number; $\mathbf{M} = V/c$, and c = sonic velocity (Sec. 5.4). At high values of $\mathbf{M}$ we must consider the effects of compressibility as we will in Sec. 9.10.

Rarely are the piezometer holes located in precisely the correct position to indicate the true value of p/γ. So we modify Eq. (9.6) into:

$$u = C_I \sqrt{2g\left(\frac{p_0}{\gamma} - \frac{p}{\gamma}\right)} \tag{9.7}$$

where we introduce C_I, a coefficient of instrument, to account for this discrepancy. We can use either BG or SI units with this equation, since C_I is dimensionless. However, when a coefficient possesses dimensions [see Eq. (9.27), for example], we must modify an equation developed for BG units for application to SI units, and vice versa. A particular type of pitot-static tube with a blunt nose, the ***Prandtl tube,*** is designed so that $C_I = 1$. For other pitot-static tubes, we must determine the coefficient C_I by calibration in the laboratory.

Another instrument, the ***pitometer,*** consists of two tubes, one pointing upstream and the other downstream, such as tubes (b) and (d) of Fig. 9.3. The reading for tube (d) will be considerably below the level of the static head. The equation applicable to a pitometer is identical to Eq. (9.7), except that p/γ is replaced by the pressure head sensed by the downstream tube.

Most of these devices will give reasonably accurate results even if the tube is as much as $\pm 15°$ out of alignment with the direction of flow.

Still greater insensitivity to angularity is possible if we guide the flow past the pitot tube by means of a shroud, as shown in Fig. 9.8. Such an arrangement, called a ***Kiel probe,*** is used extensively in aeronautics. The stagnation-pressure measurement with this device is accurate to within 1% of the dynamic pressure for yaw angles up to $\pm 54°$. A disadvantage is that the static pressure must be measured independently.

We can use the direction-finding tube (Fig. 9.6) to determine velocity. The procedure is to orient it properly so that both piezometers give the same reading. This reading is the static head. Then turn the tube through $39\frac{1}{4}°$ to obtain the stagnation pressure head. The difference in the two readings is the velocity head. This device has been used extensively in wind tunnels and in investigations of hydraulic machinery.

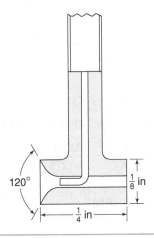

FIGURE 9.8   Kiel probe.

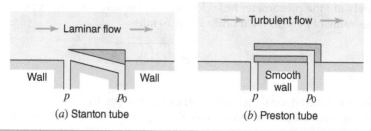

(a) Stanton tube   (b) Preston tube

FIGURE 9.9   Wall pitot tubes.

Specialized, very small pitot tubes are used in experimental work to measure veloc-ities very close to a boundary. The **Stanton tube** (Fig. 9.9a) uses the wall to form one side of the tube, and we can use it only within the viscous sublayer of turbulent flows (Fig. 7.8) or with laminar flow. The **Preston tube** (Fig. 9.9b) is designed for use in the transition zone (see Fig. 7.8), not submerged in the viscous sublayer, and we can use it for turbulent flow over smooth surfaces. Engineers use both to collect data for the deter-mination of the boundary shear stress τ_0, which has been found to depend on $p_0 - p$. We calculate the boundary shear stress by substituting the measured velocities into the theory of velocity profiles in circular pipes. We can assume that this theory holds for flat boundaries as well as curved boundaries in the region very close to the boundary.

Sample Problem 9.1

Air at 20°C is flowing through the pipe shown in Fig. S9.1, resulting in pressure gage readings of 70.2 kPa at A and 71.1 kPa at B. Atmospheric pressure is 684 mmHg. (a) Find the air velocity u. (b) What is the largest pressure difference (kPa) between the two gages for which compressibility effects can be safely neglected?

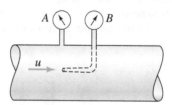

FIGURE S9.1

Solution

Table A.5 for air: $R = 287 \ \text{m}^2/(\text{s}^2 \cdot \text{K})$

(a) $p_{\text{abs}} = p_{\text{at}} + p_{\text{gage}} = 101.32(684/760) + 70.2 = 161.4 \ \text{kPa}$

Eq. (2.5): $\gamma = \dfrac{gp}{RT} = \dfrac{9.81(161.4)}{287(273 + 20)} = 0.01883 \ \text{kN/m}^3$

Eq. (9.6): $u = \sqrt{2(9.81)\left[\dfrac{71.1 - 70.2}{0.01883}\right]} = 30.6 \ \text{m/s}$ **ANS**

(b)   Sec. 9.3, footnote 3:   We need $\mathbf{M} < 0.1$

Sec. 5.4: $\mathbf{M} = V/c$; at 20°C, $c = 345$ m/s.

So we need $V < 0.1c = u_{max} = 34.5$ m/s

From Eq. (9.5): $$\Delta p = p_0 - p = \gamma \frac{u^2}{2g}$$

So $$\Delta p_{max} = \frac{\gamma (u_{max})^2}{2g} = \frac{0.01883(34.5)^2}{2(9.81)} = 1.142 \text{ kPa} \qquad \textit{ANS}$$

Note: This Δp_{max} compares with the observed Δp of $71.1 - 70.2 = 0.9$ kPa. Therefore, it was safe to neglect compressibility effects in part (*a*).

9.4   Measurement of Velocity by Other Methods

Other methods for measuring local velocity will be discussed in this section.

Current Meter and Rotating Anemometer

These two instruments, which are the same in principle, determine the velocity as a function of the speed at which a series of cups or vanes rotate about an axis either parallel or normal to the flow. The instrument used in water is called a *current meter,* and when designed for use in air, we call it an *anemometer.* As the force exerted depends on the density of the fluid as well as on its velocity, the anemometer must be so made as to operate with less friction than the current meter.

If the meter is made with cups that move in a circular path about an axis perpendicular to the flow, it always rotates in the same direction and at the same rate regardless of the direction of the velocity, whether positive or negative, and it even rotates when the velocity is at right angles to its plane of rotation. Thus this type is not suitable where there are eddies or other irregularities in the flow. If the meter is constructed of vanes rotating about an axis parallel to the flow, like a propeller, it will register the component of velocity along its axis, especially if it is surrounded by a shielding cylinder. It will rotate in an opposite direction for negative flow, and is therefore a more dependable type of meter.

Hot-Wire and Hot-Film Anemometer

The *hot-wire anemometer* measures the instantaneous velocity at a point. It consists of a small sensing element that is placed in the flow field at the point where the velocity is to be measured. The sensing element is a short thin wire, which is generally of platinum or tungsten, connected to a suitable electronic circuit. The operation depends on the fact that the electrical resistance of a wire is a function of its temperature; that the temperature, in turn, depends on the heat transfer to the surrounding fluid; and that the rate of heat transfer increases with increasing velocity of flow past the wire.

In one type of hot-wire anemometer the wire is maintained at a constant temperature by a variable voltage, which changes the current through the wire. Thus, when an increase in velocity tends to cool the wire, a balancing device creates an increase in voltage to increase the current through the wire. This tends to heat up the wire to counteract the cooling and thus maintain it at constant temperature. The voltage provides a measure of the velocity of the fluid. The hot-wire anemometer is a very sensitive instrument particularly adapted to the measurement of turbulent velocity fluctuations as in Fig. 4.6. A *hot-film anemometer,* though similar to the hot-wire, is more rugged in that its sensing element consists of a metal film laid over a glass rod and provided with a protecting coating.

Float Measurements

A crude technique for estimating the average velocity of flow in a river or stream is to observe the velocity at which a float will travel down a stream. To get good results the reach of stream should be straight and uniform with a minimum of surface disturbances. The average velocity of flow V will generally be about (0.85 ± 0.05) times the float velocity.

Photographic and Optical Methods

The camera is one of the most valuable tools in a fluid-mechanics research laboratory. In studying the motion of water, for example, researchers can introduce a series of small spheres consisting of a mixture of benzene and carbon tetrachloride adjusted to the same specific gravity as the water into the flow through suitable nozzles. When illuminated from the direction of the camera, these spheres will stand out in a picture. If successive exposures are taken on the same film, the velocities and accelerations of the particles can be determined. A similar technique involves the use of hydrogen bubbles generated through use of a fine wire which serves as the negative electrode of a dc electric circuit. By pulsing the voltage across the wire, the water is electrolyzed, thus releasing hydrogen bubbles. Short uninsulated sections of wire will permit the bubbles to be emitted at fixed points along the wire. This, when combined with intermittent pulsing, will aid in flow visualization.

More recently developed instruments employ lasers. *Laser Doppler velocimetry* (LDV), also called *laser Doppler anemometry* (LDA), directs laser light into a fixed, very small zone, practically a point, within moving fluid. When small particles (about micron sized) or bubbles moving with the flow pass through the measuring zone, the LDV measures the Doppler shift of light they scatter. This enables accurate determination of the flow velocity, or its three components, without disturbing the flow. If there are sufficient particles, a practically continuous time history is achievable.

Somewhat similar, *particle image velocimetry* (PIV) uses laser light emitted in rapid pulses to illuminate a plane or volume through which flow seeded with neutral-density, micron-sized particles or bubbles passes. Photographs or images of the successive positions of the particles can provide good visualization of the flow field, and enable the calculation of the particle (fluid) velocity vectors. PIV has been used for single-phase and two-phase flows. It is accurate and does not disturb the flow; but because three-dimensional applications are complex, they are still in early stages of development. The main advantage of PIV over LDV is that a much larger region can be measured.

In the study of compressible fluids many techniques have been devised to measure optically the variations in density, as given by the *interferometer,* or the rate at which density changes in space, as determined in the *shadowgraph* and *schlieren* methods.[4]

---

[4] For an excellent discussion of optical methods used in the study of fluid flow, see Sec. 11.6 of Ref. 75 in Appendix C.

From such measurements of density and density gradient, it is possible to locate shock waves. Although of great importance, these photographic methods are too complex to warrant further description here.

Other Methods

Other devices for measuring velocity of flow include magnetic flowmeters and acoustic flowmeters. *Magnetic flowmeters* are used to measure velocity of flow in liquids. The liquid serves as a conductor and develops a voltage as it travels through a magnetic field. With proper calibration, this device can be used to measure the average velocity in pipes. Small magnetic flowmeters can be used to measure local velocities in a flowing liquid. However, their accuracy drops off in the vicinity of boundaries.

Acoustic flowmeters depend on the effect of the moving fluid on sound waves. These devices are expensive, and are used primarily for research. One of their advantages is that they can be used without disturbing the flow.

9.5 Measurement of Discharge

There are various ways of measuring discharge. In a pipe, for example, we can determine the velocity at various radii using a pitot-static tube or a pitot tube in combination with a wall piezometer. We can then consider the cross section of a pipe as a series of concentric rings, each with a known velocity, as in Fig. 9.10. The flow, $A_i V_i$, through each of these rings is summed up to determine the total flow rate.

To determine the flow in a river or stream, we use a similar technique. We divide the stream into a number of convenient sections, as in Fig. 9.11, and measure the average velocity in each section. A pitot tube could be used for such measurements, but a current meter is more commonly used. It has been found that the average velocity occurs at about $0.6 \times$ depth (Sec. 8.4), so the velocity is generally measured at that level. As a result, in Fig. 9.11

$$Q_{2-3} = \left(\frac{y_2 + y_3}{2}\right) b_{2-3} \frac{V_2 + V_3}{2}$$

and $Q_{total} = \Sigma Q$. A second widely used method is to replace the velocity at $0.6 \times$ depth with the average of the velocities at $0.2 \times$ depth and $0.8 \times$ depth. A third, crude estimate

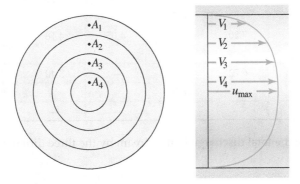

FIGURE 9.10 Determination of pipe discharge.

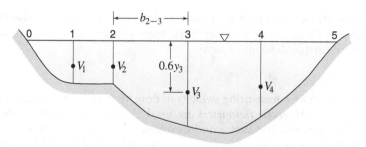

Figure 9.11 Determination of discharge in a stream.

of the flow in a river or stream can be made by multiplying (0.85 × float velocity) times the area of the average cross section in the reach of stream over which the float measurement was made.

Devices for the direct measurement of discharge can be divided into two categories, those that measure by weight or positive displacement a certain quantity of fluid and those that employ some aspect of fluid mechanics. An example of the first type of device is the household water meter in which a nutating disk oscillates in a chamber. On each oscillation a known quantity of water passes through the meter. We will discuss the second type of flow-measuring device, which depends on basic principles of fluid mechanics combined with empirical data, in Secs. 9.6–9.14.

Sample Problem 9.2

Measurements made on the cross section of Fig. 9.11 are:

Position	y, ft	b, ft	Velocity V, fps, at			
			Surface	$0.2y$	$0.6y$	$0.8y$
0	—		—	—	—	—
		4.09				
1	3.60		—	2.88	2.80	1.50
		3.53				
2	3.44		—	3.15	3.02	2.50
		6.55				
3	7.20		4.05	4.18	3.48	3.10
		6.45				
4	6.72		—	3.40	3.15	2.85
		6.58				
5	—		—	—	—	—

Calculate the total discharge in the stream by the three different methods described in the text.

Solution

Using $Q = AV$:

Method:		(1)	(1)	(2) $\dfrac{V_{0.2} + V_{0.8}}{2}$	(2)	(2)
Position	$A = b\bar{y}$ ft²	$V_{0.6}$ fps	$Q_{0.6}$ cfs	fps	$\bar{V}_{0.2,0.8}$ fps	$Q_{0.2,0.8}$ cfs
0				—		
	7.36	1.40	10.31		1.10	8.06
1				2.19		
	12.43	2.91	36.16		2.51	31.16
2				2.83		
	34.85	3.25	113.25		3.23	112.64
3				3.64		
	43.28	3.32	143.47		3.38	146.39
4				3.13		
	20.46	1.58	32.23		1.56	31.97
5				—		
Total	118.38		335.4 *ANS*			330.2 *ANS*

(3) Assume float velocity = surface velocity at position 3:

$$Q = 0.85AV_s = 0.85(118.38)4.05 = 407.5 \text{ cfs} \qquad ANS$$

Methods (1) and (2) agree well and are more reliable than method (3), which overestimates the discharge in this example.

9.6   Orifices, Nozzles, and Tubes

Among the devices used for the measurement of discharge are orifices and nozzles. Tubes are rarely used in this way but we include them here because their theory is the same and experiments on tubes provide information about entrance losses from reservoirs into pipelines. An *orifice* is an opening (usually circular) in the wall of a tank or in a plate normal to the axis of a pipe, the plate being either at the end of the pipe or in some intermediate location. An orifice is characterized by the fact that the thickness of the wall or plate is very small relative to the size of the opening. A *standard orifice* is one with a sharp edge as in Fig. 9.12*a* or an absolutely square shoulder as in Fig. 9.12*b*, so that there is only a line contact with the fluid. Those shown in Figs. 9.12*c* and *d* are not standard, because the flow through them is affected by the thickness of the plate, the roughness of the surface, and, for *d*, the radius of curvature. Hence such orifices should be calibrated if high accuracy is desired.

A *nozzle* is a tube of changing diameter, usually converging as in Fig. 9.13 if it is used for liquids; but for a gas or a vapor a nozzle may first converge and then diverge to produce supersonic flow. In addition to possible use as a flow measuring device a nozzle has other important uses, such as providing a high-velocity stream for firefighting

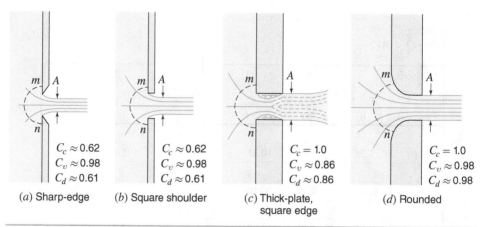

$C_c \approx 0.62$
$C_v \approx 0.98$
$C_d \approx 0.61$

(a) Sharp-edge

$C_c \approx 0.62$
$C_v \approx 0.98$
$C_d \approx 0.61$

(b) Square shoulder

$C_c = 1.0$
$C_v \approx 0.86$
$C_d \approx 0.86$

(c) Thick-plate, square edge

$C_c = 1.0$
$C_v \approx 0.98$
$C_d \approx 0.98$

(d) Rounded

FIGURE 9.12   Orifices.

or for power in a steam turbine or an impulse turbine. The **base** of a nozzle is the point (or strictly the plane, perpendicular to its centerline) at which the diameter starts to change from that of the supporting pipe. It is the nozzle entry point, and usually the point of attachment to the supporting pipe.

A **tube** is a short pipe whose length is not more than two or three diameters. There is no sharp distinction between a tube and the thick-walled orifices of Figs. 9.12c and d. A tube may be of uniform diameter, or it may diverge.

A **jet** is a stream issuing from an orifice, nozzle, or tube. It is not enclosed by solid boundary walls but is surrounded by a fluid whose velocity is less than its own. The two fluids may be different or they may be of the same kind. A **free jet** is a stream of *liquid* surrounded by a *gas* and is therefore directly under the influence of gravity. A **submerged jet** is a stream of any fluid surrounded by a fluid of the same type, that is, a gas jet discharging into a gas or a liquid jet discharging into a liquid. A submerged jet is buoyed up by the surrounding fluid and is not directly under the action of gravity.

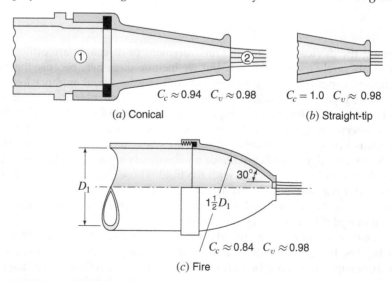

$C_c \approx 0.94$ $C_v \approx 0.98$

(a) Conical

$C_c = 1.0$ $C_v \approx 0.98$

(b) Straight-tip

$30°$

D_1

$1\frac{1}{2}D_1$

$C_c \approx 0.84$ $C_v \approx 0.98$

(c) Fire

FIGURE 9.13   Nozzles.

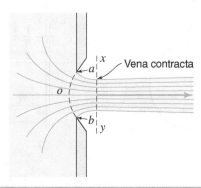

FIGURE 9.14   Jet contraction.

Jet Contraction

Where the streamlines converge in approaching an orifice, as shown in Fig. 9.14, they continue to converge beyond the upstream section of the orifice until they reach the section xy, where they become parallel. Commonly this section is about $0.5D_o$ from the upstream edge of the opening, where D_o is the diameter of the orifice. The section xy is then a section of minimum area, and is called the *vena contracta*. Beyond the vena contracta the streamlines commonly diverge because of frictional effects.[5] In Fig. 9.12c the minimum section is referred to as a *submerged vena contracta*, since it is surrounded by its own fluid. In Fig. 9.12d there is no vena contracta as the rounded entry to the opening permits the streamlines to gradually converge to the cross-sectional area of the orifice.

Jet Velocity and Pressure

Jet velocity is defined as the average velocity at the vena contracta in Figs. 9.12a and b, and at the downstream edge of the orifices in Figs. 9.12c and d. The velocity at these sections is practically constant across the section except for a small annular region around the outside (Fig. 9.15b). In all four of the jets of Fig. 9.12 the pressure is practically constant across the diameter of the jet wherever the streamlines are parallel, and this pressure must be equal to that in the medium surrounding the jet at that section. At sections *mn* in Fig. 9.12 where the streamlines are curved, the effective cross-sectional area of the flow (at right angles to streamlines) is greater than at the minimum section, and hence the average velocities at sections *mn* are considerably less than the jet velocities. The same is true of section *aob* of Fig. 9.14. In Fig. 9.15a the velocity and pressure distributions at section *aob* of Fig. 9.14 are shown. These variations are the result of the curvature of the streamlines and centrifugal effects (Sec. 5.16).

Coefficient of Contraction C_c

The ratio of the area A of a jet (Fig. 9.12), to the area A_o of the orifice or other opening, is called the *coefficient of contraction.* Thus $A = C_c A_o$.

---

[5] If a free jet is discharged vertically downward, the acceleration due to gravity will cause its velocity to increase and the area to decrease continuously, so that there may be no apparent section of minimum area. In such special cases, we should take the vena contracta as the place where marked contraction ceases and before the place where gravity has increased the velocity to any appreciable extent above the true jet velocity.

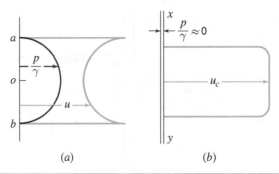

Figure 9.15 **Figure 9.15** Pressure and velocity variations in a jet. (a) At section *aob* of Fig. 9.14. (b) At vena contracta (section *xy*) in Fig. 9.14.

Coefficient of Velocity C_v

The velocity that would be attained in the jet if friction did not exist may be termed the ideal velocity V_i.[6] It is practically the value of u_c in Fig. 9.15. Because of friction, the actual average velocity V is less than the ideal velocity, and the ratio V/V_i is called the *coefficient of velocity.* Thus $V = C_v V_i$.

Coefficient of Discharge C_d

The ratio of the actual rate of discharge Q to the ideal rate of discharge Q_i (the flow that would occur if there were no friction and no contraction) is defined as the coefficient of discharge. Thus $Q = C_d Q_i$. By observing that $Q = AV$ and $Q_i = A_o V_i$, it is seen that $C_d = C_c C_v$.

Determining the Coefficients

The coefficient of contraction can be determined by using outside calipers to measure the jet diameter at the vena contracta and then comparing the jet area with the orifice area. The contraction coefficient is very sensitive to small variations in the edge of the orifice or in the upstream face of the plate. Thus slightly rounding the edge of the orifice in Fig. 9.12*b* or roughening the orifice plate will increase the contraction coefficient materially.

The average velocity V of a free jet may be determined by a velocity traverse of the jet with a fine pitot tube or it may be obtained by measuring the flow rate and dividing by the cross-sectional area of the jet. We can also compute the velocity approximately from the coordinates of the trajectory of the jet, as discussed in Sec. 5.15. The ideal velocity V_i is computed by the Bernoulli theorem. So we can compute C_v for an orifice, nozzle, or tube by dividing V by V_i.

The coefficient of discharge is the one that can most readily be obtained and with a high degree of accuracy. It is also the one that is of the most practical value. For a liquid, we can determine the actual Q by some standard method such as a volume or a weight measurement over a known time. For a gas we can note the change in pressure and temperature in a container of known volume from which the gas may flow. Obviously, if we measure any two of the coefficients, we can compute the third from them. Thus, in equation form,

---

[6] We frequently call this the *theoretical velocity,* but this may be a misuse of the word "theoretical." Any correct theory should allow for the fact that friction exists and affects the result. Otherwise, it is not correct theory but merely an incorrect hypothesis.

Ideal flow rate: $$Q_i = A_i V_i = A_o \sqrt{2g(\Delta H)}$$ (9.8)

Actual flow rate: $$Q = AV = C_c A_o (C_v \sqrt{2g(\Delta H)})$$ (9.9)

or $$Q = C_d Q_i = C_d A_o \sqrt{2g(\Delta H)}$$ (9.10)

and $$C_d = \frac{Q}{Q_i} = C_c C_v$$ (9.11)

where ΔH is the total difference in energy head between the upstream section and the minimum section of the jet (section A of Fig. 9.12). Recall that the total energy head $H = z + p/\gamma + V^2/2g$. If the flow is from a tank, the velocity of approach is negligible and can be neglected. If the discharge is to the atmosphere (free jet), the downstream pressure head is zero, whereas if the jet is submerged, the downstream pressure head is equal to the depth of submergence[7] (Fig. 9.18) in the case of a liquid or to the pressure head surrounding the jet in the case of a gas.

Typical values of the coefficients for orifices, nozzles, and tubes are as indicated in Figs. 9.12,[8] 9.13, and 9.16, respectively. It is apparent from Fig. 9.16 that rounding the entrance to a tube increases the coefficient of velocity. Any device that provides a uniform diameter for a long enough distance before exit, such as the tubes of Fig. 9.16 or the nozzle tip of Fig. 9.13b, will usually create a $C_c = 1.0$. Although this increases the size of the jet from the given area, it also tends to produce more friction.

If the geometry of the orifice, nozzle, or tube is standard such as those of Figs. 9.12, 9.13, and 9.16, the coefficients should be very close to the values indicated on the figures. However, the best way to determine the coefficients of a device, particularly one of unusual shape, is by experiment in the laboratory. Also, we can make a fair estimate of the contraction by sketching the flow net. If we wish to estimate the coefficient of discharge of an orifice, nozzle, or tube it is usually best to estimate velocity and contraction coefficients separately and calculate the discharge coefficient from them.

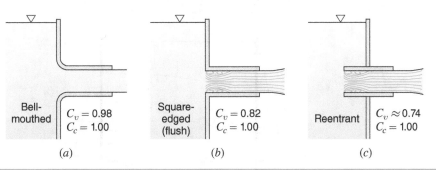

| Bell-mouthed | $C_v = 0.98$ $C_c = 1.00$ | Square-edged (flush) | $C_v = 0.82$ $C_c = 1.00$ | Reentrant | $C_v \approx 0.74$ $C_c = 1.00$ |

(a) (b) (c)

FIGURE 9.16 Coefficients for tubes.

---

[7] If the jet discharges into a different liquid, we must convert the depth of submergence to an equivalent depth of the *flowing liquid*.

[8] Surface tension can become important when orifices operate under low heads. The coefficient of contraction of small, sharp-edged, and square-edged orifices such as those of Figs. 9.12a and b have values of C_c as high as 0.72 rather than the usual 0.62 when operating under heads less than about 0.5 ft.

Borda Tube

Tubes (b) and (c) in Fig. 9.16 are shown as flowing full, and because of the turbulence, the jets issuing from them will have a "broomy" appearance. Because of the contraction of the jet at entrance to these tubes the local velocity in the central portion of the stream will be higher than that at exit from the tubes, and hence the pressure will be lower. If we lower the pressure to that of the vapor pressure of the liquid, the streamlines will then no longer follow the walls, and the jet springs clear. In such a case tube (b) of Fig. 9.16 becomes equivalent to orifice (b) in Fig. 9.12, while tube (c) of Fig. 9.16 behaves as shown in Fig. 9.17. If its length is less than its diameter, the reentrant tube is called a **Borda mouthpiece.** Because of the greater curvature of the streamlines for a reentrant tube, if the tube flows full the velocity coefficient is lower (Fig. 9.16c) than for any other type of entry. But if the jet springs clear as in Fig. 9.17, the velocity coefficient is as high as for a sharp-edged orifice.

The Borda mouthpiece is of interest because it is one device for which the contraction coefficient can be very simply calculated. For all other orifices and tubes there is a reduction of the pressure on the walls adjacent to the opening, but the exact pressure values are unknown. But for the reentrant tube the velocity along the wall of the tank is almost zero at all points, and hence the pressure is essentially hydrostatic. In the case of a Borda tube with the jet springing clear, the only unbalanced pressure is that on an equal area A_o opposite to the tube (Fig. 9.17), and the force due to this pressure is $\gamma h A_o$. The time rate of change of momentum due to the flow out of the tube is $\rho QV = \gamma AV^2/g$, where A is the area of the jet. Equating force to time rate of change of momentum, $\gamma h A_o = \gamma AV^2/g$, and thus $V^2 = ghA_o/A$. Ideally, $V^2 = 2gh$, and thus, ideally, $C_c = A/A_o = 0.5$. The actual values of the coefficients for a Borda tube with the jet springing clear are $C_c = 0.52$, $C_v = 0.98$, and $C_d = 0.51$.

Head Loss

We can find the relationship between the head loss and the coefficient of velocity of an orifice, nozzle, or tube by comparing the ideal energy equation with the actual (or real) energy equation between points 1 and 2 in Fig. 9.13a. The ideal energy equation given by Eq. (5.29) is

$$\frac{p_1}{\gamma} + z_1 + \frac{V_1^2}{2g} = \frac{p_2}{\gamma} + z_2 + \frac{V_2^2}{2g}$$

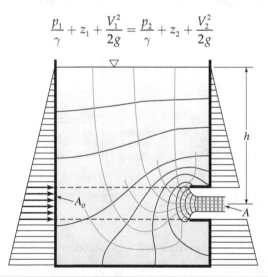

FIGURE 9.17   Borda tube.

In the case of a free jet $p_2 = 0$, while for the most general case of a submerged jet $p_2 \neq 0$. From continuity, $A_1V_1 = A_2V_2$; hence we can write

$$\frac{p_1}{\gamma} + z_1 + \left(\frac{A_2}{A_1}\right)^2 \frac{V_2^2}{2g} = \frac{p_2}{\gamma} + z_2 + \frac{V_2^2}{2g}$$

which leads to

$$(V_2)_{ideal} = \frac{1}{\sqrt{1 - (A_2/A_1)^2}} \sqrt{2g\left[\left(\frac{p_1}{\gamma} + z_1\right) - \left(\frac{p_2}{\gamma} + z_2\right)\right]} \tag{9.12}$$

The real energy equation given by Eq. (5.28) accounts for head loss and is expressed as

$$\frac{p_1}{\gamma} + z_1 + \frac{V_1^2}{2g} - h_{L_{1-2}} = \frac{p_2}{\gamma} + z_2 + \frac{V_2^2}{2g}$$

which leads to the actual velocity

$$V_2 = \frac{1}{\sqrt{1 - (A_2/A_1)^2}} \sqrt{2g\left[\left(\frac{p_1}{\gamma} + z_1\right) - \left(\frac{p_2}{\gamma} + z_2\right) - h_{L_{1-2}}\right]} \tag{9.13}$$

Recalling that the actual $V = C_v V_{ideal}$, and combining this with the above expressions for V_{ideal} and actual V gives

$$h_{L_{1-2}} = \left(\frac{1}{C_v^2} - 1\right)\left[1 - \left(\frac{A_2}{A_1}\right)^2\right]\frac{V_2^2}{2g} \tag{9.14}$$

where V_2 is an actual velocity. This equation is perfectly general; it expresses the head loss between a section upstream of an orifice and the jet (section A in Fig. 9.12) or between sections 1 and 2 in Fig. 9.13a, etc. If the orifice or nozzle takes off directly from a tank where $A_1 \gg A_2$, the velocity of approach is negligible and Eq. (9.14) reduces to

$$h_{L_{1-2}} = \left(\frac{1}{C_v^2} - 1\right)\frac{V_2^2}{2g} \tag{9.15}$$

Note that for the tubes of Fig. 9.16 with $C_v = 0.98, 0.82,$ and 0.74, Eq. (9.15) yields $h_L = 0.04V_2^2/2g, 0.5V_2^2/2g,$ and $0.8V_2^2/2g$, respectively. These correspond to the values for minor loss at entrance shown in Fig. 7.13.

Submerged Jet

For the case of a submerged jet, as in Fig. 9.18, the ideal energy equation is written between 1 and 2, realizing that the pressure head on the jet at 2 is equal to h_3. Thus

$$h_1 = h_3 + \frac{V_i^2}{2g}$$

or
$$V_i = \sqrt{2g(h_1 - h_3)} = \sqrt{2g(\Delta H)}$$

where V_i is the ideal velocity at the vena contracta of the submerged jet and ΔH is the net head differential expressed in terms of the flowing liquid. Hence $Q = C_c C_v A_o \sqrt{2g(\Delta H)}$ as in Eq. (9.9).

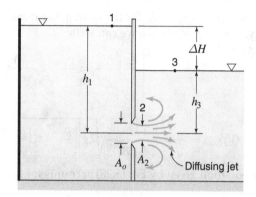

FIGURE 9.18   Submerged jet.

For a submerged orifice, nozzle, or tube the coefficients are practically the same as for a free jet, except that, for heads less than 10 ft (3 m) and for very small openings, the discharge coefficient may be slightly less. It is of interest to observe that, if the energy equation is written between 1 and 3, the result is $h_{L_{1-3}} = h_1 - h_3 = \Delta H$. Actually, the head loss in this case is that of Eq. (9.15) plus that of a submerged discharge, as described in Sec. 7.23. Hence, for submerged orifices, nozzles, and tubes,

$$h_{L_{1-3}} = \left(\frac{1}{C_v^2} - 1\right)\frac{V_2^2}{2g} + \frac{V_2^2}{2g} = \frac{1}{C_v^2}\frac{V_2^2}{2g} = \frac{V_i^2}{2g} = \Delta H$$

where the actual $V_2 = C_v V_i$, the velocity at the vena contracta.

Sample Problem 9.3

A 2-in circular orifice (not standard) at the end of the 3-in-diameter pipe shown in Fig. S9.3 discharges into the atmosphere a measured flow of 0.60 cfs of water when the pressure in the pipe is 10.0 psi. The jet velocity is determined by a pitot tube to be 39.2 fps. Find the values of the coefficients C_v, C_c, and C_d. Find also the head loss from inlet to vena contracta.

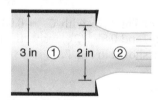

FIGURE S9.3

Solution

Define the inlet as section 1 and the throat as section 2.

$$\frac{p_1}{\gamma} = 10\left(\frac{144}{62.4}\right) = 23.1 \text{ ft}$$

$$V_1 = \frac{Q}{A_1} = \frac{0.60}{\pi(1.5/12)^2} = 12.22 \text{ fps}, \qquad \frac{V_1^2}{2g} = 2.32 \text{ ft}$$

Express the ideal energy equation from 1 to 2 to determine the ideal velocity V_{2i} at 2

$$\frac{p_1}{\gamma} + \frac{V_1^2}{2g} = \frac{V_{2i}^2}{2g}$$

$$\frac{V_{2i}^2}{2g} = 23.1 + 2.32 = 25.4; \quad V_{2i} = 40.4 \text{ fps}$$

$$C_v = \frac{V_2}{V_{2i}} = \frac{39.2}{40.4} = 0.969 \qquad ANS$$

Area of jet
$$A_2 = \frac{Q}{V_2} = \frac{0.60}{39.2} = 0.01531 \text{ ft}^2$$

$$C_c = \frac{A_2}{A_o} = \frac{0.01531}{\pi(1/12)^2} = 0.702 \qquad ANS$$

Hence
$$C_d = C_c C_v = 0.680 \qquad ANS$$

From Eq. (9.14):

$$h_{L_{1-2}} = \left(\frac{1}{(0.969)^2} - 1\right)\left[1 - \left(\frac{2}{3}\right)^4\right]\frac{V_2^2}{2g} = 0.0517\frac{V_2^2}{2g}$$

$$= 0.0517\frac{(39.2)^2}{2(32.2)} = 1.233 \text{ ft} \qquad ANS$$

Check: Calculate the actual velocity V_2 at 2 by expressing the real energy equation from 1 to 2:

$$\frac{p_1}{\gamma} + \frac{V_1^2}{2g} - h_{L_{1-2}} = \frac{V_2^2}{2g}$$

$$23.1 + 2.32 - 1.233 = \frac{V_2^2}{2g}; \quad \text{actual } V_2 = 39.4 \text{ fps}$$

This checks well with the measured velocity of 39.2 fps.

9.7 Venturi Meter

The converging tube is an efficient device for converting pressure head to velocity head, while the diverging tube converts velocity head to pressure head. The two may be combined to form a *venturi tube,* named after Giovanni B. Venturi (1746–1822), an Italian physicist who investigated its principle about 1791. It was applied to the measurement of water by an American engineer, Clemens Herschel, in 1886. As shown in Fig. 9.19, it consists of a tube with a constricted *throat,* which produces an increased velocity accompanied by a reduction in pressure, followed by a gradually diverging portion in which the velocity is transformed back into pressure with slight friction loss. As there is a definite relation between the pressure differential and the rate of flow, the tube may be made to serve as a metering device known as a *venturi meter.* The venturi meter is used

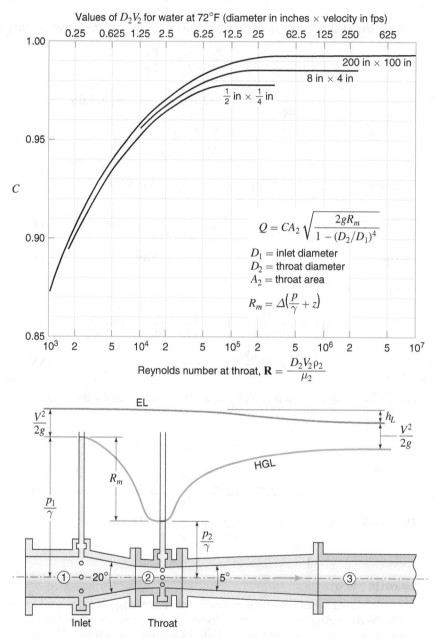

Values of D_2V_2 for water at 72°F (diameter in inches × velocity in fps)

$$Q = CA_2 \sqrt{\frac{2gR_m}{1 - (D_2/D_1)^4}}$$

D_1 = inlet diameter
D_2 = throat diameter
A_2 = throat area

$$R_m = \Delta\left(\frac{p}{\gamma} + z\right)$$

Reynolds number at throat, $\mathbf{R} = \dfrac{D_2V_2\rho_2}{\mu_2}$

FIGURE 9.19   Venturi meter with conical entrance and flow coefficients for $D_2/D_1 = 0.5$.

for measuring the rate of flow of both compressible and incompressible fluids.[9] In this section we shall consider the application of the venturi meter to incompressible fluids; its application to compressible fluids will be discussed in Sec. 9.10.

---

[9] As mentioned in Sec. 9.3, if $\mathbf{M} < 0.1$, a compressible fluid can be treated as if it were incompressible without introducing much error.

Writing the Bernoulli equation between sections 1 (inlet) and 2 (throat) of Fig. 9.19, we have, for the ideal case,

$$\frac{p_1}{\gamma} + z_1 + \frac{V_1^2}{2g} = \frac{p_2}{\gamma} + z_2 + \frac{V_2^2}{2g}$$

Substituting the continuity equation, $V_1 = (A_2/A_1)V_2$, we get for the ideal throat velocity

$$V_{2i} = \sqrt{\frac{1}{1-(A_2/A_1)^2}} \sqrt{2g\left[\left(\frac{p_1}{\gamma}+z_1\right)-\left(\frac{p_2}{\gamma}+z_2\right)\right]}$$

Recalling, from Eq. (3.12*b*), that we often find it convenient to write

$$\left(\frac{p_1}{\gamma}+z_1\right)-\left(\frac{p_2}{\gamma}+z_2\right) = \Delta\left(\frac{p}{\gamma}+z\right)$$

then here we have

$$V_{2i} = \sqrt{\frac{1}{1-(A_2/A_1)^2}} \sqrt{2g\Delta\left(\frac{p}{\gamma}+z\right)}$$

As there is some friction loss between sections 1 and 2, the true velocity V_2 is slightly less than the ideal value given by this expression. Therefore we introduce a discharge coefficient C, so that the flow is given by CQ_i or

$$Q = A_2 V_2 = CA_2 V_{2i} = \frac{CA_2}{\sqrt{1-(D_2/D_1)^4}} \sqrt{2g\Delta\left(\frac{p}{\gamma}+z\right)} \tag{9.16}$$

In this situation the discharge coefficient C is identical to the velocity coefficient C_v since the coefficient of contraction $C_c = 1.0$. If we use a differential manometer with piezometric connections at sections 1 and 2, then from Fig. 3.14 and Eq. (3.12*b*) we note that

$$\Delta\left(\frac{p}{\gamma}+z\right) = R_m\left(\frac{s_M}{s_F}-1\right) \tag{9.17}$$

where R_m is the manometer reading, and s_M and s_F are the specific gravities of the manometer and flowing fluids, respectively.

The venturi tube provides an accurate means for measuring flow in pipelines. With a suitable recording device, we can integrate the flow rate so as to give the total quantity of flow. Aside from the installation cost, the only disadvantage of the venturi meter is that it introduces a permanent frictional resistance in the pipeline. Practically all this loss occurs in the diverging part (as depicted in Fig. 5.12) between sections 2 and 3, and is ordinarily from $0.1h$ to $0.2h$, where h is the static-head differential between the upstream section and the throat, as indicated in Fig. 9.19.

Values of D_2/D_1 may vary from $\frac{1}{4}$ to $\frac{3}{4}$, but a common ratio is $\frac{1}{2}$. A small ratio gives increased accuracy of the gage reading, but is accompanied by a higher friction loss and may produce an undesirably low pressure at the throat, sufficient in some cases to cause liberation of dissolved air or even vaporization of the liquid at this point. This phenomenon, called *cavitation,* has been described in Sec. 5.10. The angles of convergence and divergence indicated in Fig. 9.19 are those considered optimum, though somewhat larger angles are sometimes used to reduce the length and cost of the tube.

For accuracy in use, the venturi meter should be preceded by a straight pipe whose length is at least 5 to 10 pipe diameters. The approach section becomes more important as the diameter ratio increases, and the required length of straight pipe depends on the conditions preceding it. For example, the vortex formed from two short-radius elbows in planes at right angles is not eliminated within 30 pipe diameters (see Sec. 7.27). We can alleviate such a condition by installing straightening vanes preceding the meter. The pressure differential should be obtained from piezometer rings (Fig. 9.4) surrounding the pipe, with a number of suitable openings in the two sections. In fact, these openings are sometimes replaced by very narrow slots extending most of the way around the circumference.

Unless specific information is available for a given venturi tube, we can assume the value of C is about 0.99 for large tubes and about 0.97 or 0.98 for small ones, provided the flow is such as to give Reynolds numbers greater than about 10^5 (Fig. 9.19). A roughening of the surface of the converging section from age or scale deposit will reduce the coefficient slightly. Venturi tubes in service for many years have shown a decrease in C of the order of 1–2%. Dimensional analysis of a venturi tube indicates that the coefficient C should be a function of Reynolds number and of the geometric parameters D_1 and D_2. Values of venturi-tube coefficients are shown in Fig. 9.19. This diagram is for a diameter ratio of $D_2/D_1 = 0.5$, but is reasonably valid for smaller ratios also. For best results we should calibrate a venturi meter by conducting a series of tests in which the flow rate is measured over a wide range of Reynolds numbers.

Occasionally, the precise calibration of a venturi tube has given a value of C greater than 1. Such an abnormal result is sometimes due to improper piezometer openings. But another possible explanation is that the α's at sections 1 and 2 are such that this is really so.

Sample Problem 9.4

Find the discharge rate of 20°C water through the venturi tube shown in Fig. S9.4 if $D_1 = 800$ mm, $D_2 = 400$ mm, $\Delta z = 2.00$ m, and $R_m = 150$ mmHg. Assume Fig. 9.19 is applicable.

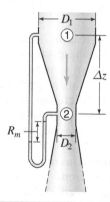

FIGURE S9.4

Solution

Table A.1 for water at 20°C: $\nu = 1.003 \times 10^{-6}$ m²/s

Venturi size is 800 mm × 400 mm = 31.5 in × 15.75 in.

This is about midway between the 8 in × 4 in and 200 in × 100 in curves on Fig. 9.19. So, from Fig. 9.19: Maximum $C \approx 0.988$; assume this value. Equations (9.16) and (9.17):

$$Q = \frac{0.988\pi(0.40/2)^2}{\sqrt{1-(400/800)^4}}\sqrt{2(9.81)0.15\left(\frac{13.55}{1}-1\right)} = 0.779 \text{ m}^3/\text{s}$$

$$V_2 = \frac{Q}{A_2} = \frac{0.779}{\pi(0.40/2)^2} = 6.20 \text{ m/s}$$

Eq. (7.1): $\mathbf{R} = \dfrac{D_2 V_2}{\nu} = \dfrac{0.40(6.20)}{1.003 \times 10^{-6}} = 2.47 \times 10^6$

Check Fig. 9.19 for this $\mathbf{R}$: $C = 0.988$. Thus $Q = 0.779$ m³/s   *ANS*

9.8 Flow Nozzle

If the diverging discharge cone of a venturi tube is omitted, the result is a *flow nozzle* of the type shown in Fig. 9.20. This is simpler than the venturi tube and can be installed between the flanges of a pipeline. It will serve the same purpose, though at the expense of an increased frictional loss in the pipe. Although we could use venturi-meter Eq. (9.16) for the flow nozzle, it is more convenient and customary to include the correction for velocity of approach with the coefficient of discharge, so that

$$Q = KA_2\sqrt{2g\Delta\left(\frac{p}{\gamma}+z\right)} \tag{9.18}$$

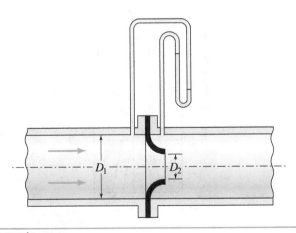

FIGURE 9.20 Flow nozzle.

where we call K the *flow coefficient* and A_2 is the area of the nozzle throat. Comparison with Eq. (9.16) establishes the relation

$$K = \frac{C}{\sqrt{1-(D_2/D_1)^4}} \qquad (9.19)$$

Although there are many designs of flow nozzles, the ISA (International Standards Association) nozzle (Fig. 9.21) has become an accepted standard form in many countries. The quoted "nozzle diameter" is the throat diameter D_2. Values of K for various diameter ratios of the ISA nozzle are given in Fig. 9.22 as a function of Reynolds number. Note that in this case the Reynolds number is computed for the approach pipe rather than for the nozzle throat, which is a convenience since **R** in the pipe is frequently needed for other computations also.

As shown in Fig. 9.22, many of the values of K are greater than unity, which results from including the correction for approach velocity with the conventional coefficient of discharge. Many attempts have been made to design a nozzle for which the velocity-of-approach correction would just compensate for the discharge coefficient, leaving a value of the flow coefficient equal to unity, principally using so-called *long-radius nozzles.* Usually such a coefficient of unity is approached over only a limited range.

As in the case of the venturi meter, the flow nozzle should be preceded by at least 10 diameters of straight pipe for accurate measurement. Two alternative arrangements for the pressure taps are shown in Fig. 9.21.

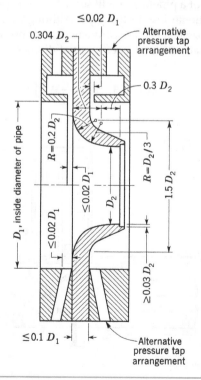

FIGURE 9.21   ISA flow nozzle.

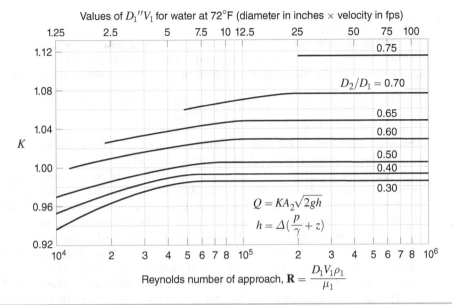

Values of $D_1''V_1$ for water at 72°F (diameter in inches × velocity in fps)

$$Q = KA_2\sqrt{2gh}$$

$$h = \Delta\left(\frac{p}{\gamma} + z\right)$$

Reynolds number of approach, $\mathbf{R} = \dfrac{D_1 V_1 \rho_1}{\mu_1}$

FIGURE 9.22 Flow coefficients for ISA nozzle. (Adapted from *ASME Flow Measurement*, 1959.)

Sample Problem 9.5

A 2-in ISA flow nozzle is installed in a 3-in pipe carrying water at 72°F. If a water-air manometer shows a differential of 2 in, find the flow.

Solution

This can be solved using a trial-and-error type of solution. First, assume a reasonable value of K. From Fig. 9.22, for $D_2/D_1 = 0.67$, and for the level part of the curve, $K = 1.06$.

$$A_1 = \frac{\pi}{4}\left(\frac{3}{12}\right)^2 = 0.0491 \text{ ft}^2; \quad A_2 = \frac{\pi}{4}\left(\frac{2}{12}\right)^2 = 0.0218 \text{ ft}^2$$

This air-water manometer is like Fig. 3.14b with $z_B - z_A = 0$. For air and water, $s_M/s_F \approx 0.001$, so it can be neglected in Eq. (3.13), and

$$\Delta\left(\frac{p}{\gamma} + z\right) = R_m = \frac{2}{12} = 0.1667 \text{ ft}$$

Eq. (9.18): $\qquad Q = 1.06 \times 0.0218\sqrt{2(32.2) \times 0.1667} = 0.0746 \text{ cfs}$

With this first determination of Q,

$$V_1 = \frac{Q}{A} = \frac{0.0746}{0.0491} = 1.519 \text{ fps}$$

Then $\qquad\qquad\qquad D_1''V_1 = 3 \times 1.519 = 4.56$

From Fig. 9.22, $K = 1.04$ and

$$Q = \frac{1.04}{1.06} \times 0.0746 = 0.0732 \text{ cfs} \qquad \textbf{\textit{ANS}}$$

No further correction is necessary.

9.9   Orifice Meter

We can use an orifice in a pipeline, as in Fig. 9.23, as a meter in the same manner as the venturi tube or the flow nozzle. We can also place it on the end of the pipe so as to discharge a free jet. The flow rate through an orifice meter is commonly expressed as

$$Q = KA_o\sqrt{2g\Delta\left(\frac{p}{\gamma} + z\right)}$$
(9.20)

This is the same form as Eq. (9.18), except that A_2 is replaced by A_o, the cross-sectional area of the orifice opening. Typical values of K for a standard orifice meter are given in Fig. 9.24. The variation of K with Reynolds number is quite different from the trend of the flow coefficients for venturi tubes and flow nozzles. At high Reynolds numbers K is essentially constant, but as the Reynolds number is lowered, the value of K for the orifice increases, with the maximum value of K occurring at Reynolds numbers between 200 and 600, depending on the D_o/D_1 ratio of the orifice. The lowering of the Reynolds number increases viscous action, which causes a decrease in C_v and an increase in C_c. The latter apparently predominates over the former until C_c reaches a maximum value of about 1.0. With a further decrease in Reynolds number, K then becomes smaller because C_v continues to decrease.

The difference between an orifice meter and a venturi tube or flow nozzle is that for both of the latter there is no contraction, so that A_2 is also the area of the throat and is fixed, while for the orifice A_2 is the area of the jet and is a variable and is less than A_o, the area of the orifice. For the venturi tube or flow nozzle the discharge coefficient is practically a velocity coefficient, while for the orifice it is much more affected by variations in C_c than it is by variations in C_v.

The pressure differential may be measured between a point about one pipe diameter upstream of the orifice and the vena contracta, approximately one-half the pipe diameter downstream. The distance to the vena contracta is not constant, but decreases as D_o/D_1 increases. The differential can also be measured between the two corners on each side of the orifice plate. These *flange taps* have the advantage that the orifice meter is self-contained; the plate may be slipped into a pipeline without the necessity of making piezometer connections in the pipe.

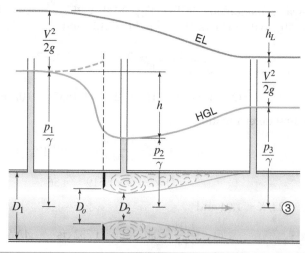

FIGURE 9.23   Thin-plate orifice in a pipe. (Scale distorted: the region of eddying turbulence will usually extend $4D_1$ to $8D_1$ downstream, depending on the Reynolds number.)

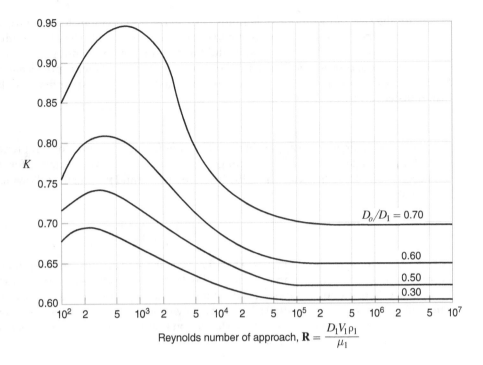

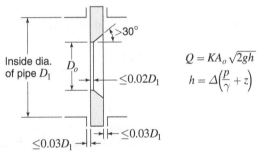

$$Q = KA_o \sqrt{2gh}$$

$$h = \Delta\left(\frac{p}{\gamma} + z\right)$$

FIGURE 9.24   VDI orifice meter and flow coefficients for flange taps. (Adapted from *NACA Tech. Mem.* 952.)

The particular advantage of an orifice as a measuring device is that it may be installed in a pipeline with a minimum of trouble and expense. Its principal disadvantage is the greater frictional loss it causes as compared with the venturi meter or the flow nozzle.

9.10   Flow Measurement of Compressible Fluids

Strictly speaking, most of the equations that we have presented earlier in this chapter apply only to incompressible fluids, but, practically, they can be used for all liquids and even for gases and vapors where the pressure differential is small relative to the total pressure. As this is the condition usually encountered in the metering of all fluids, even compressible ones, the preceding treatment has extensive application. However, there are conditions in metering fluids where we must consider compressibility.

As for the case of incompressible fluids, we can derive equations for ideal friction-less flow and then introduce a coefficient to obtain a correct result. The ideal condition that we will impose on the compressible fluid is that the flow be isentropic, i.e., a frictionless adiabatic process (no transfer of heat, see Sec. 2.7). The latter is practically true for metering devices, as the time for the fluid to pass through is so short that very little heat transfer can take place.

Pitot Tubes

We can derive an expression applicable to pitot tubes for subsonic flow of compressible fluids for perfect gases (Sec. 2.7). Without going into the details, we can derive for stagnant conditions at the upstream tip of the tube (i.e., $V_2 = 0$ and $p_2 = p_0$) the following for pitot tubes:

$$\frac{V_1^2}{2} = c_p T_1 \left[\left(\frac{p_2}{p_1} \right)^{(k-1)/k} - 1 \right] = c_p T_2 \left[1 - \left(\frac{p_1}{p_2} \right)^{(k-1)/k} \right] \tag{9.21}$$

We can obtain the static pressure p_1 (see Sec. 9.2) from the side openings of the pitot tube or from a regular piezometer, and the pitot tube itself indicates the stagnation pressure $p_0 (= p_2$, Sec. 9.3). We must apply a coefficient if the side openings do not measure the true static pressure. Equation (9.21) does not apply to supersonic conditions, because a shock wave would form upstream of the stagnation point. In such a case a special analysis considering the effect of the shock wave is required.

Venturi Meters

To develop an expression applicable to compressible flow through venturi tubes, we can employ an approach similar to that for pitot tubes, combining it with continuity $(g\dot{m} = G = \gamma_1 A_1 V_1 = \gamma_2 A_2 V_2)$ to get

$$g\dot{m}_{ideal} = G_{ideal} = A_2 \sqrt{2g \frac{k}{k-1} p_1 \gamma_1 \left(\frac{p_2}{p_1} \right)^{2/k} \frac{1 - (p_2/p_1)^{(k-1)/k}}{1 - (A_2/A_1)^2 (p_2/p_1)^{2/k}}} \tag{9.22}$$

We can transform this equation into an equation for the actual weight rate of flow through venturi tubes by introducing the discharge coefficient C (Fig. 9.19) and an expansion factor Y. The resulting equation is

$$g\dot{m} = G = CYA_2 \sqrt{2g\gamma_1 \frac{p_1 - p_2}{1 - (D_2/D_1)^4}} \tag{9.23}$$

where C has the same value as for an incompressible fluid at the same Reynolds number and we can replace γ_1 by p_1/RT_1 if desired. Values of Y for $k = 1.4$ are plotted in Fig. 9.25. For a venturi or nozzle throat where $C_c = 1$,

$$Y = \sqrt{\frac{[k/(k-1)](p_2/p_1)^{2/k}[1 - (p_2/p_1)^{(k-1)/k}]}{1 - (p_2/p_1)}} \sqrt{\frac{1 - (D_2/D_1)^4}{1 - (D_2/D_1)^4 (p_2/p_1)^{2/k}}}$$

In Fig. 9.25 we see that for the venturi meter no values for Y are shown for p_2/p_1 ratios less than 0.528. This is so because, for air and other gases having adiabatic constant $k = 1.4$, the p_2/p_1 ratio will always be greater than 0.528 if the flow is subsonic.

Equation (9.23) is directly applicable to the flow of compressible fluids through venturi tubes where $C_c = 1.0$, provided the flow is subsonic.

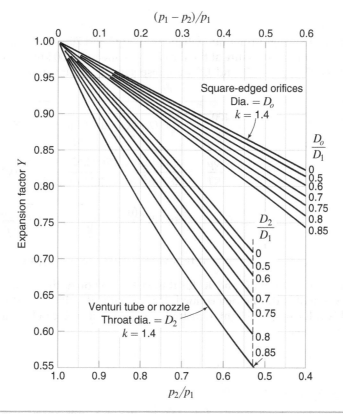

FIGURE 9.25 Expansion factors.

Flow Nozzles and Orifice Meters

We can also use Eq. (9.23) for flow nozzles and orifice meters, though for flow nozzles C should be replaced by $K\sqrt{1 - (D_2/D_1)^4}$ [from Eq. (9.19)], so that we can use Fig. 9.22 directly. For orifice meters we should replace the C of Eq. (9.23) by $K\sqrt{1 - (D_o/D_1)^4}$, we should replace D_2 by D_o, and we should replace A_2 by A_o where D_o is the diameter of the orifice opening and A_o is its area.

For compressible fluids the C_c of an orifice meter depends on the ratio p_2/p_1; hence Y varies in a different manner than in the case of a venturi. Values of Y for orifice meters are shown in Fig. 9.25. In the case of an orifice meter the maximum jet velocity is the sonic velocity c, but this does not impose a limit on the rate of discharge because the jet area continues to increase with decreasing values of p_2/p_1. For this reason the values of Y for the orifice are extended in Fig. 9.25 to lower values of p_2/p_1.

Supersonic Conditions

The general case of flow measurement under supersonic conditions will not be discussed in this text.

Sample Problem 9.6

Determine the weight flow rate when air at 20°C and 700 kN/m² abs flows through a venturi meter if the pressure at the throat of the meter is 400 kN/m² abs. The diameters at inlet and throat are 250 and 125 mm, respectively. Assume that $C = 0.985$.

Solution

$$p_2/p_1 = \frac{400}{700} = 0.571; D_2/D_1 = 0.50. \text{ Fig. 9.25: } Y \approx 0.72$$

$$\text{Eq. (2.5): } \gamma_1 = \frac{gp}{RT} = \frac{(9.81 \text{ m/s}^2)(700 \text{ kN/m}^2)}{[287 \text{ m}^2/(\text{s}^2 \cdot \text{K})](273 + 20)\text{K}} = 0.0817 \text{ kN/m}^3$$

$$\text{Eq. (9.23): } G = 0.985(0.72)\frac{\pi(0.125)^2}{4}\sqrt{2(9.81)0.0817\frac{700 - 400}{1 - (0.5)^4}}$$

$$G = 0.1971 \text{ kN/s} = 197.1 \text{ N/s} \quad \textit{ANS}$$

If the relation between C and $\mathbf{R}_2$ for this meter is known, the value of $\mathbf{R}_2$ for the computed value of G can be determined. If the assumed value of C does not correspond with this value of $\mathbf{R}_2$, a slight adjustment in the value of C can be made to give a more accurate answer.

9.11   Thin-Plate Weirs

Weirs have long been standard devices for the measurement of water flow. One category, known as *thin-plate weirs*, includes those constructed from a thin plate, usually of metal and erected perpendicular to the flow. The upstream face of the weir plate should be smooth, and the plate must be strictly vertical. All thin-plate weirs are sharp-crested; the upstream edge is formed by a horizontal top surface usually less than $\frac{1}{16}$ in (1.6 mm) long in the flow direction, followed by a bevel on the downstream edge. Such a design causes the *nappe* (sheet of overflowing water) to spring clear as in Fig. 9.26a, with only a line contact at the crest, for all but the very lowest heads. If the nappe does not spring clear, we cannot consider the flow as true weir flow and the experimentally determined coefficients do not apply. The approach channel should be long enough so that a normal velocity distribution exists, and the water surface should be as free from waves as possible.

The weir crest can be straight and horizontal, or it may have a variety of shapes; a number of these are discussed below.

Suppressed Rectangular Weir

In its simplest form, water flows over the top of a plate with a straight and horizontal crest as shown in Fig. 9.26. The weir is as wide as the channel, so the width of the nappe is the same as the length of the crest. As there are no contractions of the stream at the sides, we say that end contractions are *suppressed*. It is essential that the sides of the channel upstream be smooth and regular. It is common to extend the sides of the channel downstream beyond the crest so that the nappe is confined laterally. The flowing

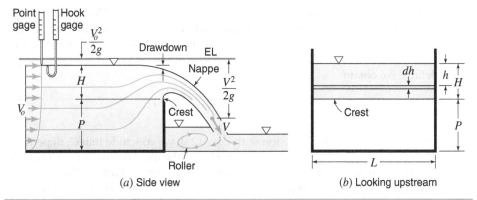

FIGURE 9.26 Flow over sharp-crested weir.

water tends to entrain air from this enclosed space under the nappe, and unless this space is adequately ventilated, there will be a partial vacuum and perhaps all the air may eventually be swept out. The water will then cling to the downstream face of the plate, and the discharge will be greater for a given head than when the space is vented. Therefore venting of a suppressed weir is necessary if the standard formulas are to be applied.

We determine the rate of flow by measuring the water height H (head), above the crest, at a distance upstream from the crest equal to at least four times the maximum head to be employed. The amount of surface drawdown at the crest is typically about $0.15H$. The velocity at any point in the nappe is related to the energy line as shown in Fig. 9.26a.

To derive the flow equation for a rectangular weir having a crest of length L, consider an elementary area $dA = L\,dh$ in the plane of the crest, as shown in Fig. 9.26b. This elementary area is in effect a horizontal slot of length L and height dh. Neglecting velocity of approach, the ideal velocity of flow through this area will be equal to $\sqrt{2gh}$. The apparent flow through this area is

$$dQ_i = L\,dh\sqrt{2gh} = L\sqrt{2g}h^{1/2}dh$$

Because each element is at a different h, the flow through it has a different velocity, so we must integrate over the whole area, from $h = 0$ to $h = H$, to obtain the total ideal flow Q_i, i.e.,

$$Q_i = \sqrt{2g}L\int_0^H h^{1/2}\,dh = \frac{2}{3}\sqrt{2g}LH^{3/2}$$

The actual flow over the weir will be less than the ideal flow, because the effective flow area is considerably smaller than LH due to drawdown from the top and contraction of the nappe from the crest below (see Fig. 9.26a). Introducing a coefficient of discharge C_d to account for this,

$$Q = C_d\frac{2}{3}\sqrt{2g}LH^{3/2} \tag{9.24}$$

Dimensional analysis of weir flow leads to some interesting conclusions that provide a basis for an understanding of the factors that influence the coefficient of discharge. The physical variables that influence the flow Q over the weir of Fig. 9.26 include L, H, P, g, μ, σ, and ρ. Using the Buckingham Π theorem (Sec. 7.2), and without going into the details, we obtain

$$Q = \phi\left(\mathbf{W}, \mathbf{R}, \frac{H}{P}\right)L\sqrt{g}H^{3/2}$$

Thus, comparing this expression with Eq. (9.24), we conclude that C_d depends on $\mathbf{W}$, $\mathbf{R}$, and H/P. Investigators have found that H/P is the most important of these. The Weber number, defined as $\mathbf{W} = V\sqrt{\rho L/\sigma}$, accounts for surface-tension effects, and is important only at low heads. In the flow of water over weirs the Reynolds number is generally quite high, so viscous effects are generally insignificant. If one were to calibrate a weir for the flow of oil, however, $\mathbf{R}$ would undoubtedly affect C_d substantially. Typical values of C_d for sharp-crested weirs with water flowing range from about 0.62 for $H/P = 0.10$ to about 0.75 for $H/P = 2.0$.

Small-scale but precise experiments covering a wide range of conditions led T. Rehbock of the Karlsruhe Hydraulic Laboratory in Germany to the following expressions for C_d in Eq. (9.24):

BG units, H and P in ft:
$$C_d = 0.605 + \frac{1}{305H} + 0.08\frac{H}{P} \qquad (9.25a)$$

SI units, H and P in m:
$$C_d = 0.605 + \frac{1}{1000H} + 0.08\frac{H}{P} \qquad (9.25b)$$

These equations were obtained by fitting a curve to the plotted values of C_d for a great many experiments and are purely empirical. Capillarity is accounted for by the second term, while velocity of approach (assumed to be uniform) is responsible for the last term. Rehbock's formula has been found to be accurate within 0.5% for values of P from 0.33 to 3.3 ft (0.10 to 1.0 m) and for values of H from 0.08 to 2.0 ft (0.025 to 0.60 m) with the ratio H/P not greater than 1.0. It is even valid for ratios greater than 1.0 if the bottom of the discharge channel is lower than that of the approach channel so that backwater does not affect the head.

It is convenient to express Eq. (9.24) as

$$Q = C_W LH^{3/2} \qquad (9.26)$$

where C_W, the **weir coefficient**,[10] replaces $C_d\frac{2}{3}\sqrt{2g}$.

Using a value of 0.62 for C_d in Eq. (9.24), we can write

$$Q \approx \begin{cases} 3.32LH^{3/2} & \text{in BG units} \\ 1.83LH^{3/2} & \text{in SI units} \end{cases} \qquad (9.27)$$

These equations give good results if $H/P < 0.4$, which is well within the usual operating range. If the velocity of approach V_0 is appreciable, a correction must be applied to the

---

[10] Since C_W is not dimensionless, its value in BG units is different from that in SI units, as indicated in Eq. (9.27).

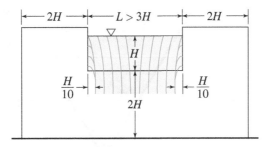

Limiting proportions of standard contracted weirs.

preceding equations either by changing the form of the equation or, more commonly, by changing the value of the coefficient.

Rectangular Weir with End Contractions

When the length L of the crest of a rectangular weir is less than the width of the channel, the nappe will have **end contractions** so that its width is less than L. Experiments have indicated that under the conditions depicted in Fig. 9.27 the effect of each side contraction is to reduce the effective width of the nappe by $0.1H$. Hence for such a situation the flow rate may be computed by employing any of Eqs. (9.26) and (9.27) and substituting $(L - 0.1nH)$ for L, where n is the number of end contractions, normally 2 but sometimes 1. We call the results **Francis formulas.**

Cipolletti Weir

To avoid correcting for end contractions we often use a Cipolletti weir. It has a trapezoidal shape with side slopes of four vertical on one horizontal. The additional area adds approximately enough to the effective width of the stream to offset the lateral contractions.

V-notch, or Triangular, Weir

For relatively small flows the rectangular weir must be very narrow and thus of limited maximum capacity, or else the value of H will be so small that the nappe will not spring clear but will cling to the plate. For such a case the V-notch or triangular weir has the advantage that it can function for a very small flow and also measure reasonably large flows as well. The vertex angle is usually between 10° and 90° but rarely larger.

In Fig. 9.28 is a V-notch weir with a vertex angle θ. The ideal rate of discharge through an elementary area dA is $dQ = \sqrt{2gh}\, dA$. Now $dA = 2x\, dh$, and $x/(H - h) = \tan(\theta/2)$. Substituting in the foregoing, and introducing a coefficient of discharge C_d, we get this result for the entire notch:

$$Q = C_d 2\sqrt{2g} \tan\frac{\theta}{2} \int_0^H (H - h)h^{1/2}\, dh$$

Integrating between limits and reducing, the fundamental equation for all V-notch weirs is:

$$Q = C_d \frac{8}{15}\sqrt{2g} \tan\frac{\theta}{2} H^{5/2} \tag{9.28}$$

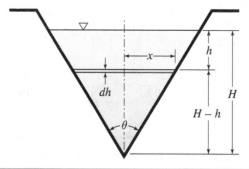

FIGURE 9.28 V-notch weir.

For a given angle θ and assuming C_d is constant, this may be reduced to

$$Q = KH^{5/2} \tag{9.29}$$

The value of the constant K in English units will be different from that in SI units.

In Fig. 9.29 we present experimental values for C_d for water flowing over V-notch weirs with central angles varying from 10° to 90°. The rise in C_d at heads less than 0.5 ft is due to incomplete contraction. At lower heads the frictional effects reduce the coefficient. At very low heads, when the nappe clings to the weir plate, the phenomenon can no longer be classed as weir flow and Eqs. (9.28) and (9.29) are inapplicable.

Proportional Weirs

While, for given simple geometric shapes of thin-plate weirs, head-discharge (H–Q) relations can easily be obtained using elementary calculus, the reverse problem of working out the shapes of weirs that produce desired H–Q relations is more challenging and

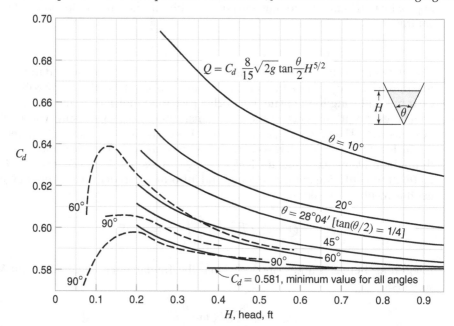

FIGURE 9.29 Coefficients for V-notch weirs obtained by three different experimenters.

generally requires the solution of integral equations. Research[11] has led to the development of new types of nonlinear weirs like the logarithmic weir ($Q \propto \log H$) and the quadratic weir ($Q \propto H^{1/2}$), called *proportional weirs.* Many of these proportional weirs measure discharge more accurately than traditional weirs, and we use them as efficient velocity controllers in sediment settling chambers and in industrial dosing facilities.

Sample Problem 9.7
Water is flowing in a rectangular channel at a velocity of 3 fps and depth of 1.0 ft. Neglecting the effect of velocity of approach and employing Eq. (9.27), determine the height of a sharp-crested suppressed weir that must be installed to raise the water depth upstream of the weir to 4 ft.

Solution

$$L = \text{length of weir crest} = \text{width of channel}$$

Eq. (9.27): $\qquad Q = AV = LyV = L(1)(3) = 3.33LH^{3/2}$

$$H^{3/2} = \frac{3.0}{3.33} = 0.901, \quad H = 0.933 \text{ ft}$$

Eq. (9.26): $\qquad P = \text{height of weir required} = 4.00 - 0.933 = 3.07 \text{ ft} \qquad ANS$

9.12 Streamlined Weirs and Free Overfall

In contrast to thin-plate weirs, streamlined weirs have large dimensions in the flow direction and they must have rounded crests and/or approach edges (they are not sharp-crested). Shapes of streamlined weir cross sections include rectangular with a rounded upstream edge, triangular with a rounded crest, and hydrofoil shapes (similar to airfoils).

Streamlined weirs are usually built of concrete, giving them the advantage of being rugged and able to stand up well under field conditions.

Broad-Crested Rectangular Weir

The broad-crested weir (Fig. 9.30), as noted in Sample Problem 8.9, is a critical-depth meter; that is, if the weir is high enough (Sec. 8.13) critical depth occurs on the crest of the weir. In Eq. (8.22) we saw that for a rectangular channel $V_c = \sqrt{gy_c}$, while Eq. (8.26) stated that when the flow is critical $y_c = \frac{2}{3}E$. Using these relations, we can write for the flow over a broad-crested weir,

$$Q = AV = (Ly_c)\sqrt{gy_c} = L\sqrt{g}y_c^{3/2} = L\sqrt{g}(2/3)^{3/2}E^{3/2} \qquad (9.30)$$

where L is the width of the weir (length of crest), and E is the total head over the weir, equal to $H + V_0^2/2g$, in which H is the measured head and V_0 is the approach velocity

[11] K. Keshava Murthy and D. P. Giridhar, Improved Inverted V-Notch or Chimney Weir, *J. Irrigation and Drainage Engineering, ASCE,* 116(3), 1990; K. Keshava Murthy and M. N. S. Prakash, Practical Constant Accuracy Linear Weir, *J. Irrigation and Drainage Engineering, ASCE,* 120(3), 1994.

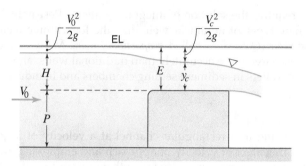

Figure 9.30 Broad-crested rectangular weir.

from upstream. Let us now substitute Eq. (9.30) into Eq. (9.24), which is applicable to broad-crested weirs as well as sharp-crested suppressed rectangular weirs, since both have rectangular flow cross sections. This yields

$$C_d = \frac{1}{\sqrt{3}}\left(\frac{E}{H}\right)^{3/2} \tag{9.31}$$

For very high weirs (that is, H/P small) the velocity of approach becomes small, so that $E \to H$ (Fig. 9.30) and thus $E/H \to 1$ and $C_d \to 1/\sqrt{3} = 0.577$. With a lower weir and the same flow rate the velocity of approach becomes larger and hence E/H increases in magnitude, resulting in larger values of C_d for lower broad-crested weirs. The actual value of C_d for broad-crested weirs depends on the length of the weir and whether or not the upstream corner (edge) of the weir is rounded. The foregoing discussion, of course, assumes critical flow (Sec. 8.9) on the weir. If the approaching flow is supercritical, the presence of surface waves causes broad-crested rectangular weirs to be rather impractical for use as metering devices.

Other Streamlined Weirs

Many different streamlined weirs have been tested[12] and found not only to have high discharge coefficients but also to function efficiently under conditions of high submergence. Weir *submergence* σ is defined as the downstream water depth divided by the upstream water depth, i.e., $\sigma = y_d/y_u$. For sufficiently low values of σ the discharge over the weir is unaffected by the downstream depth; for increasing σ, when the discharge is first reduced by a prescribed small amount like 2 or 5% the submergence is defined to be the *critical submergence* σ_c.

The coefficient of discharge C_d of a streamlined weir, according to the International Standards Organization (ISO), is defined by the equation

$$Q = C_d (2/3)^{3/2}\sqrt{g}LE^{3/2} \tag{9.32}$$

where the variables are as for Eq. (9.30) (see also Figs. 9.30 and 8.13). Note that this equation is the same as Eq. (9.30) with a discharge coefficient added.

---

[12] N. S. Lakshmana Rao, Theory of Weirs, in *Advances in Hydroscience*, V. T. Chow, ed., Vol. 10, Academic Press, New York, 1975.

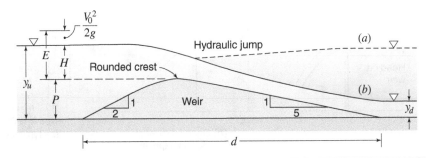

FIGURE 9.31 Streamlined triangular weir, $d/P = 7.5$, $C_d = 1.11$. Free surface at (a) critical submergence, $\sigma_c = 0.82$, and (b) $\sigma = 0.2$.

Over 25 streamlined weirs have been classified into nine categories, depending on (a) whether σ_c is low (<0.65), medium (0.65–0.80), or high (>0.80), and (b) whether the weir discharge coefficient C_d is low (<1.0), medium (1.0–1.3), or high (>1.3). For example, several triangular-shaped streamlined weirs, and what we call hydrofoil weirs, fall into the category of medium C_d and high σ_c types. The streamlined triangular weir of Fig. 9.31 has a medium C_d and a high σ_c of 0.82 when we specify a discharge reduction of 6%.

Free Overfall

Another method for determining the flow rate in a rectangular channel is to measure the depth of flow y_b at the brink of a free overfall (Fig. 8.17). Substituting $y_c = y_b/0.72$ in Eq. (8.23) permits an approximate determination of the rate of flow per foot of channel width. This method can be used only if the flow is subcritical.

9.13   Overflow Spillway

An overflow spillway is a section of dam designed to permit water to pass over its crest. Overflow spillways are widely used on gravity, arch, and buttress dams. The ideal spillway should take the form of the underside of the nappe of a sharp-crested weir (Fig. 9.32a) when the flow rate corresponds to the maximum design capacity of the spillway (Fig. 9.32b). Figure 9.32c defines the shape of an *ogee spillway,* which closely approximates the ideal[13]; ogee means having an S-shaped curve. The reverse curve on the lower downstream face of the spillway should be smooth and gradual; a radius of about one-fourth of the spillway height has proved satisfactory.

The discharge of an overflow spillway is given by the weir Eq. (9.26):

$$Q = C_W L H^{3/2} \tag{9.33}$$

where Q = discharge, cfs or m³/s
   C_w = weir coefficient
   L = length of the crest, ft or m
   H = head on the spillway (vertical distance from the crest of the spillway to the reservoir level), ft or m

---

[13] Hydraulic Models as an Aid to the Development of Design Criteria, U.S. Waterways Expt. Sta., Bull. 37, Vicksburg, Miss., 1951.

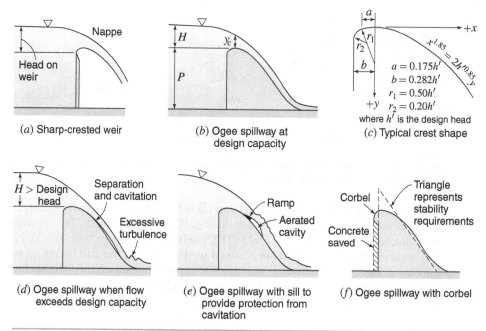

(a) Sharp-crested weir

(b) Ogee spillway at design capacity

(c) Typical crest shape

(d) Ogee spillway when flow exceeds design capacity

(e) Ogee spillway with sill to provide protection from cavitation

(f) Ogee spillway with corbel

FIGURE 9.32   Characteristics of an ogee spillway.

The coefficient C_w varies with the design and head. For the standard overflow crest of Fig. 9.32c the variation of C_w is given in Fig. 9.33. Experimental models are often used to determine spillway coefficients. End contractions on a spillway reduce the effective length below the actual length L. Square-cornered piers disturb the flow considerably and reduce the effective length by the width of the piers plus about $0.2H$ for each pier. Streamlining the piers or flaring the spillway entrance minimizes the flow disturbance. If the cross-sectional area of the reservoir just upstream from the spillway is less than five times the area of flow over the spillway, the approach velocity will increase the discharge a noticeable amount. The effect of approach velocity can be accounted for by the equation

$$Q = C_W L \left(H + \frac{V_0^2}{2g} \right)^{3/2}$$ (9.34)

where V_0 is the approach velocity.

On high spillways, if the overflowing water breaks contact with the spillway surface, a vacuum will form at the point of separation (Fig. 9.32d) and cavitation may occur. Cavitation and vibration from the alternate making and breaking of contact between the water and the face of the spillway may result in serious structural damage. A ramp of proper shape and size when properly located (Fig. 9.32e) will direct the water away from the spillway surface to form a cavity.[14] To be effective, air must be freely admitted to the cavity. The result is that air is entrained in the water, the water bulks up,

---

[14] K. Zagustin and N. Castillejo, Model-Prototype Correlation for Flow Aeration in Guru Dam Spillway, *Proceedings International Association for Hydraulic Research*, Vol. 3, 1983.

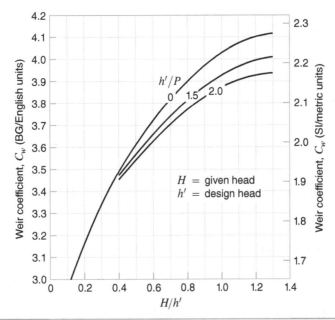

FIGURE 9.33   Variation of discharge coefficient with head for an ogee spillway crest such as shown in Fig. 9.32.

and when it returns to the spillway surface there is no problem with cavitation. On very high spillways these ramps may be used in tandem.

Other types of spillways include chute, side-channel, ski-jump, and shaft spillways.[15]

9.14   Sluice Gate

The sluice gate shown in Fig. 9.34 is a device used to control the passage of water in an open channel. When properly calibrated, it may also serve as a means of flow measurement. As the lower edge of the gate opening is flush with the floor of the channel, contraction of the bottom surface of the issuing stream is entirely suppressed. Side contractions will of course depend on the extent to which the opening spans the width of the channel. The complete contraction on the top side, however, because of the larger velocity components parallel to the face of the gate, will offset the suppressed bottom contraction, resulting in a coefficient of contraction nearly the same as for a slot with contractions at top and bottom.

Flow through a sluice gate differs fundamentally from flow through a slot in that the jet is not free but guided by a horizontal floor. Consequently, the final jet pressure is not atmospheric, but distributed hydrostatically in the vertical section. Writing the energy equation with respect to the stream bed as datum from point 1 to point 2 in the free-flow case (Fig. 9.34a) and neglecting head loss,

$$\frac{V_1^2}{2g} + y_1 = \frac{V_2^2}{2g} + y_2$$

---

[15] For a brief discussion of the different types of spillways see Ref. 58 in Appendix C.

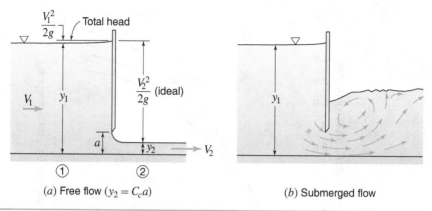

(a) Free flow ($y_2 = C_c a$)   (b) Submerged flow

FIGURE 9.34 Flow through sluice gate.

from which, introducing continuity, the ideal velocity downstream is

$$V_{2i} = \frac{1}{\sqrt{1 - (A_2/A_1)^2}} \sqrt{2g(y_1 - y_2)} \qquad (9.35)$$

The actual flow rate $Q = C_d Q_i = C_c C_v (A V_{2i})$, where $A = aB$ is the area of the gate opening and B is the width of the opening. For free flow, $y_2 = C_c a$.

Absorbing the effects of flow contraction, friction, velocity of approach, and the downstream depth y_2 into an experimental flow coefficient, we obtain a simple discharge equation for flow under a sluice gate:

$$Q = K_s A \sqrt{2g y_1} \qquad (9.36)$$

where K_s is defined as the ***sluice coefficient***.[16]

Values of K_s depend on a and y_1 (Fig. 9.34a) and are usually between 0.55 and 0.60 for free flow, but they are significantly reduced when the flow conditions downstream cause submerged flow, as shown in Fig. 9.34b.

9.15 Measurement of Liquid-Surface Elevation

We can determine the elevation of the surface of a liquid at rest through use of a piezometer column, manometer, or pressure gage (Sec. 3.5). These will also give accurate results when applied to a stationary liquid contained in a tank that is moving, provided the tank is not undergoing an acceleration (Sec. 3.10). Staff gages, such as those used at reservoirs, provide approximate liquid surface elevation data.

Various methods are used to measure the surface elevation of moving liquids. To determine the head H on a weir, we must measure the elevation difference between the crest of the weir and the liquid surface. In the field the elevation of the liquid surface is often determined through use of a ***stilling well*** connected by a pipe to the main liquid body. A ***float*** in the well is used to actuate a clock-driven liquid-level recorder so that a continuous record of the liquid-surface elevation is obtained. In the laboratory a ***hook gage***

---

[16] Values of discharge coefficients for sluice and other types of gates may be found in Hunter Rouse (ed.), *Engineering Hydraulics*, John Wiley & Sons, Inc., New York, 536–543, 1950.

or *point gage* (Fig. 9.26) is commonly employed for liquid-surface level determinations. The point gage is particularly suitable for fast-moving liquids where a hook gage would create a local disturbance in the liquid surface. In all liquid-surface level determinations care should be taken to make the measurements in regions where there is no curvature of streamlines; otherwise centrifugal effects will give a false reading of the piezometric head.

Other methods for determining the elevation of a liquid surface include the use of the radar level sensors, sonic devices, electric gages, and bubblers. A *radar level sensor* can be mounted to an overhead structure such as a bridge, and measures the time for a radar pulse to travel to the water surface and return. Similarly, a *sonic device* is mounted at some convenient location above the liquid surface and the time required for a sound pulse to travel vertically downward to the liquid surface and return is indicative of the relationship between the elevation of the liquid surface and the device. *Electrical gages* include those with capacitive sensors and those with resistive sensors. In the resistive type two parallel bare-wire conductors are partially immersed in the liquid as a component of an electrical system. The electrical resistance between them is a function of the liquid depth. With proper circuitry and calibration, the device can be used to provide data on liquid-surface elevation. The sonic devices and the electrical gages are equally applicable to liquids at rest or in motion. The *bubbler* is used primarily for liquids at rest. In the bubbler system, the minimum pressure required to drive a gas into a liquid (i.e., to form bubbles) at a depth is a measure of the depth of the liquid. If we know the elevation at which the bubbles are emitted we can determine the elevation of the liquid surface.

9.16 Other Methods of Measuring Discharge

In addition to the foregoing "standard" devices for measuring the flow of fluids, a number of supplementary devices are less amenable to exact theoretical analysis but are worthy of brief mention. One of the simplest for measuring flow in a pipeline is the *elbow meter,* which consists of nothing more than piezometer taps at the inner and outer walls of a 90° elbow in the line (see Sec. 7.27). The pressure difference, due to the centrifugal effects at the bend, will vary approximately as the velocity head in the pipe. Like other meters, the elbow should have sections of a straight pipe upstream and downstream and should be calibrated in place.

The *rotameter* (Fig. 9.35) consists of a vertical glass tube that is slightly tapered, in which the metering *float* is suspended by the upward motion of the fluid around it.

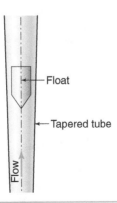

Float

Tapered tube

Flow

FIGURE 9.35   Rotameter type of flow meter.

Directional notches cut in the float keep it rotating and thus free of wall friction. The rate of flow determines the equilibrium height of the float, and the tube is graduated to read the flow directly. We also use the rotameter for gas flow, but the weight of the float and the graduation must be changed accordingly.

The *inferential meter* or *turbine meter* consists of a propeller or wheel with curved blades, shaped so that the flow of gas or liquid passing through it causes it to rotate. The rotational speed indicates the flow rate, after calibration. Such meters are often provided with guide vanes that we can adjust to change the calibration.

Other techniques for measuring flow rate include the *salt-velocity* method. In this method a charge of concentrated salt is injected into the flow at an upstream station. Its arrival at a downward station is detected from conductivity measurements. In flowing between the two stations, the salt disperses and its arrival at the downstream station is spread out over a considerable period of time. The time of travel between the two stations is taken as the time from the instant of injection at the upstream station to the time at which the centroid of the conductivity-time curve passes the downstream station. Knowing the travel time and the distance, the velocity can be calculated, and then by multiplying by the cross-sectional area, we get the flow rate. In special circumstances other substances such as dye or tritium may be used instead of salt.

CHAPTER 10

Unsteady-Flow Problems

10.1 Introduction

This text deals mostly with steady flow, since the majority of cases of engineering interest are of this nature.[1] However, a number of cases of unsteady flow are very important, and some of them are discussed in this chapter. Earlier we learned that turbulent flow is unsteady in the strictest sense of the word, but if the mean temporal values are constant over a period of time, it is called mean steady flow. Here we shall consider cases where the mean temporal values continuously vary.

We shall investigate two main types of unsteady flow here. The first is where the water level in a reservoir or pressure tank is steadily rising or falling, so that the flow rate varies continuously, but where change takes place slowly. The second is where the velocity in a pipeline is changed rapidly by the fast closing or opening of a valve.

In the first case, of slow change, the flow is subject to the same forces as have previously been considered. Fast changes, of the second type, require the consideration of elastic forces.

Unsteady flow also includes such topics as oscillations in connected reservoirs and in U tubes and such phenomena as tidal motion and flood waves in open channels. Likewise, the field of machinery regulation by servomechanisms is intimately connected with unsteady motion. However, we shall not consider any of these topics here.

10.2 Discharge with Varying Head

When flow occurs under varying head, the rate of discharge will continuously vary. Let us consider the situation depicted in Fig. 10.1 in which V represents the volume of liquid contained in the tank at a particular instant of time. There is inflow at the rate Q_i and outflow at rate Q_o. The change in volume during a small time interval dt can be expressed as

$$dV = Q_i dt - Q_o dt$$

---

[1] Where the unsteadiness is not too rapid, unsteady flow can usually be approximated by assuming the flow is steady at different rates over successive time periods of short duration.

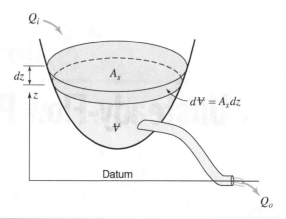

Q_i

dz

A_s

z

$d\forall = A_s dz$

$\forall$

Datum

Q_o

FIGURE 10.1   A draining tank with varying cross section and head.

If A_s = area of the surface of the volume while dz is the change in level of the surface then $d\forall = A_s dz$. Equating these two expressions for $d\forall$,

$$A_s dz = Q_i dt - Q_o dt \tag{10.1}$$

Either Q_i or Q_o or both may be variable. The outflow Q_o is usually a function of z. For example, if liquid is discharged through an orifice or a pipe of area A under a differential head z, $Q_o = C_d A\sqrt{2gz}$, where C_d is a numerical discharge coefficient and z is a variable. If the liquid flows out over a weir or a spillway of length L, $Q_o = CLh^{3/2}$, where C is the appropriate coefficient and h is the head on the weir or spillway (Secs. 9.11–9.13). In either case z or h is the variable height of the liquid surface above the appropriate datum. The inflow Q_i commonly varies with time; however, such problems will not be considered here. We shall consider only the cases where $Q_i = 0$ or where Q_i = constant.

Rewriting Eq. (10.1) and integrating gives an expression for t, the time for the water level to change from z_1 to z_2. Thus

$$t = \int_{z_1}^{z_2} \frac{A_s dz}{Q_i - Q_o} \tag{10.2}$$

The right side of this expression can be integrated if Q_i is zero or constant and if A_s and Q_o can be expressed as functions of z. In the case of natural reservoirs, we cannot express the surface area as a simple mathematical function of z, but can obtain values of it from a topographic map. In such a case, we can solve Eq. (10.2) graphically by plotting values of $A_s/(Q_i - Q_o)$ against simultaneous values of z. The area under such a curve to some scale is the numerical value of the integral.

We must note here that instantaneous values for Q_o have been expressed in the same manner as for steady flow. This is not strictly correct, since for unsteady flow the energy equation should also include an acceleration head [see Eq. (10.6)]. The introduction of such a term renders the solution much more difficult. In cases where the value of z does not vary rapidly, no appreciable error will result if this acceleration term is disregarded. Therefore, the equations can be written as for steady flow.

Sample Problem 10.1

The open wedge-shaped tank in Fig. S10.1 has a length of 15 ft perpendicular to the sketch. It is drained through a 3-in-diameter pipe 10 ft long whose discharge end is at elevation zero. The coefficient of loss at pipe entrance is 0.50, the total of the bend loss coefficients is 0.20, and f for the pipe is 0.018. Find the time required to lower the water surface in the tank from elevation 8 to 5 ft. Neglect the possible change of f with $\mathbf{R}$, and assume that the acceleration effects in the pipe are negligible.

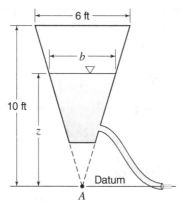

Solution

Energy Eq. (5.28) from water surface to jet at discharge:

$$0 + z + 0 - \left[0.5 + 0.2 + 0.018\left(\frac{10}{0.25}\right)\right]\frac{V^2}{2g} = 0 + 0 + \frac{V^2}{2g}$$

$$z - 1.420\frac{V^2}{2g} = \frac{V^2}{2g}; \quad V = 5.16z^{1/2}$$

$$Q_o = AV = \frac{\pi}{4}(0.25)^2 5.16z^{1/2} = 0.253z^{1/2}$$

Similar triangles:
$$\frac{b}{z} = \frac{6}{10}; \quad \text{so} \quad b = 0.6z$$

So the area of the water surface is

$$A_s = 15b = 15(0.6)z = 9z$$

Eq. (10.2):
$$t = \int_8^5 \frac{9z\,dz}{0 - 0.253z^{1/2}} = -\frac{9}{0.253}\int_8^5 z^{1/2}\,dz$$

$$= -35.5\left[\frac{2}{3}z^{3/2}\right]_8^5 = 271 \text{ sec} \quad ANS$$

Note: If the pipe had discharged at an elevation other than zero, the integral would have been different, because the head on the pipe would then have been $z + h$, where h is the vertical distance of the discharge end of the pipe below (h positive) or above (h negative) point A of the figure.

10.3   Unsteady Flow of Incompressible Fluids in Pipes

When the flow in a pipe is unsteady, we shall show that the energy equation has a term, the ***accelerative head*** $h_a = (L/g)(dV/dt)$, which accounts for the effect of the acceleration of the fluid. Let us consider an elemental length of the flow in a pipe, as in Fig. 10.2. We shall follow the same procedure we used in Sec. 7.5 by writing $\Sigma F = ma$; however, in this situation, with unsteady flow, at a particular point in the flow at a particular instant, we express the acceleration as $V(dV/ds) + dV/dt$. This comes from the general expression for acceleration in unsteady flow, Eq. (4.29). Applying $\Sigma F = ma$ to the cylindrical fluid element of Fig. 10.2, we get for unsteady flow

$$F_1 - F_2 - dW\cos\theta - \tau_0 A_s = ma$$

i.e., $$pA - (p + dp)A - \rho g A \, ds\left(\frac{dz}{ds}\right) - \tau_0(2\pi r)ds = \rho A \, ds\left(V\frac{dV}{ds} + \frac{dV}{dt}\right)$$

Dividing by $\gamma = \rho g$, dividing by $A = \pi r^2$, and simplifying, we get

$$-\frac{dp}{\gamma} - dz - \frac{2\tau_0 ds}{\gamma r} = \frac{VdV}{g} + \frac{ds}{g}\frac{dV}{dt} \tag{10.3}$$

Noting that $VdV = \frac{1}{2}d(V^2)$ this can be written.

$$-\frac{dp}{\gamma} - dz - d\frac{V^2}{2g} - \frac{2\tau_0 ds}{\gamma r} - \frac{ds}{g}\frac{dV}{dt} = 0 \tag{10.4}$$

This equation applies to unsteady flow of both compressible and incompressible real fluids. However, for compressible fluids, γ is a variable, and we must introduce a gas law equation relating γ to p and T before integration.

For incompressible fluids that we are considering here γ is a constant, so we can integrate directly. Integrating from some section 1 to another section 2, where the distance between them is L, we get

$$\frac{p_1 - p_2}{\gamma} + z_1 - z_2 + \frac{V_1^2 - V_2^2}{2g} - \frac{2\tau_0 L}{\gamma r} - \frac{L}{g}\frac{dV}{dt} = 0$$

or, noting from Sec. 7.4 that $R_h = D/4 = r/2$,

$$\left(\frac{p}{\gamma} + z + \frac{V^2}{2g}\right)_1 - \frac{\tau_0 L}{\gamma R_h} - \frac{L}{g}\frac{dV}{dt} = \left(\frac{p}{\gamma} + z + \frac{V^2}{2g}\right)_2 \tag{10.5}$$

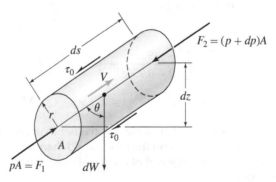

FIGURE 10.2   Element in pipe flow.

But when there is no acceleration ($a = 0$), i.e., the flow is steady and V is constant, from Eq. (10.5) we see that the dV/dt term and the V^2 terms drop out, leaving

$$\left(\frac{p}{\gamma} + z\right)_1 - \frac{\tau_0 L}{\gamma R_h} = \left(\frac{p}{\gamma} + z\right)_2$$

which, by comparison with Eqs. (7.7) and (7.8), indicates that the term $\tau_0 L/\gamma R_h = h_{f_{1-2}}$, the pipe friction head loss over length L. Thus Eq. (10.5) becomes

$$\left(\frac{p}{\gamma} + z + \frac{V^2}{2g}\right)_1 - h_f - \frac{L}{g}\frac{dV}{dt} = \left(\frac{p}{\gamma} + z + \frac{V^2}{2g}\right)_2 \qquad (10.6a)$$

or

$$H_1 - h_f - h_a = H_2 \qquad (10.6b)$$

This is the same as the steady flow Eq. (5.28), with the addition of the third term, the accelerative head $h_a = (L/g)(dV/dt)$. Equation (10.6) states that the energy loss between sections 1 and 2 is equal to the energy head required to overcome the friction plus that required to produce the acceleration. We are presuming that the head loss at any instant is equal to the steady-flow head loss for the flow rate at that instant. Experimental evidence indicates that this presumption is reasonably valid.

If the pipe consists of two or more pipes in series, an $(L/g)(dV/dt)$ term for each pipe should appear in the equation just as there would be a separate term for the head loss in each pipe. To clarify the discussion further, the simple case of unsteady flow of an incompressible fluid in a horizontal pipe is shown in Fig. 10.3. The left sketch shows the

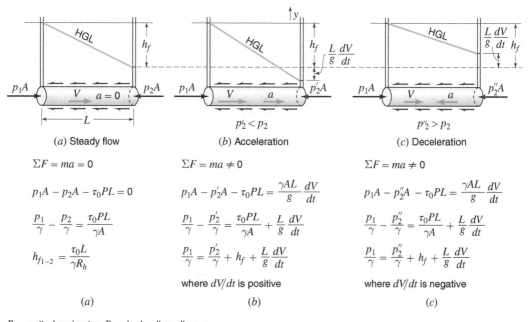

P = wetted perimeter; R_h = hydraulic radius.

FIGURE 10.3 Steady and unsteady flow of incompressible fluid in a horizontal pipe. (Flow is instantaneously equal in all three pipes.) (a) Steady flow ($dV/dt = 0$). (b) Unsteady flow (dV/dt is positive, p_2 is reduced). (c) Unsteady flow (dV/dt is negative, p_2 is increased).

steady-flow case, while unsteady flow is depicted in the two right sketches. The analysis below the sketches indicates that, with the same instantaneous flow rates, the pressure is depressed at section 2 if the acceleration is positive or increased if it is negative.

Sample Problem 10.2

Although the assumptions of instantaneous change in pump speed and head made in this example are unrealistic, it will serve to illustrate application of Eq. (10.6). When the centrifugal pump in Fig. S10.2 is rotating at 1650 rpm, the steady flow rate is 1600 gpm. Suppose the pump speed can be increased instantaneously to 2000 rpm. Determine the flow rate as a function of time. Assume that the head developed by the pump is proportional to the square of the rotative speed (see Sec. 11.4).

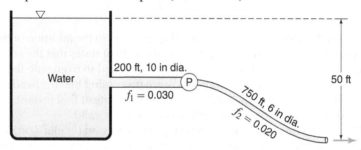

FIGURE S10.2

Solution

Writing the unsteady-flow energy equation from the water surface to the jet,

$$0 + 50 + 0 - 0.5\frac{V_1^2}{2g} - f_1\frac{L_1}{D_1}\frac{V_1^2}{2g} + h_p - f_2\frac{L_2}{D_2}\frac{V_2^2}{2g} - \frac{L_1}{g}\frac{dV_1}{dt} - \frac{L_2}{g}\frac{dV_2}{dt} = 0 + 0 + \frac{V_2^2}{2g}$$

where the subscripts 1 and 2 refer to the 10- and 6-in-diameter pipes, respectively. Note that the accelerative head for each pipe depends on the respective L and dV/dt values.

$$\text{From continuity:}\quad V_1 = \frac{A_2V_2}{A_1} = \left(\frac{6}{10}\right)^2 V_2 = 0.36V_2$$

So

$$\frac{dV_1}{dt} = \frac{A_2}{A_1}\frac{dV_2}{dt} = 0.36\frac{dV_2}{dt}$$

Thus

$$50 - 0.5\frac{(0.36V_2)^2}{2g} - 0.030\left(\frac{200}{10/12}\right)\frac{(0.36V_2)^2}{2g} + h_p - 0.020\left(\frac{750}{6/12}\right)\frac{V_2^2}{2g}$$

$$= \frac{V_2^2}{2g} + \frac{200}{g}(0.36)\frac{dV_2}{dt} + \frac{750}{g}\frac{dV_2}{dt}$$

Evaluating and combining terms: $50 + h_p = 32.0\frac{V_2^2}{2g} + \frac{822}{g}\frac{dV_2}{dt}$ (1)

With the original steady-flow conditions ($dV/dt = 0$):

$$V_2 = \frac{Q}{A_2} = \frac{1600/449}{(\pi/4)(0.5)^2} = 18.16 \text{ fps}$$

and
$$h_p = 32\frac{V_2^2}{2g} - 50 = 113.8 \text{ ft}$$

Given that $h_p \propto (\text{rpm})^2$, so after the speed is increased to 2000 rpm

$$h_p = 113.8\left(\frac{2000}{1650}\right)^2 = 167.2 \text{ ft}$$

Substituting into (1):
$$50 + 167 = 32\frac{V_2^2}{2g} + \frac{822}{g}\frac{dV_2}{dt}$$

Expressing this in terms of Q, using $V_2 = Q/A_2 = 5.09Q$:

$$217 = 12.89Q^2 + 130.0\frac{dQ}{dt} \qquad (2)$$

Solving (2) for dt and integrating, noting that at $t = 0$, $Q = 3.57$ cfs (1600 gpm):

$$\int_0^t dt = 130\int_{3.57}^Q \frac{dQ}{217 - 12.89Q^2}$$

$$t = 1.229 \ln\frac{4.10 + Q}{4.10 - Q} - 3.27$$

$$e^{0.814t + 2.66} = \frac{4.10 + Q}{4.10 - Q}$$

Finally
$$Q = 4.10\frac{e^{0.814t + 2.66} - 1}{e^{0.814t + 2.66} + 1} \qquad ANS$$

Notes:
1. As t gets larger, Q approaches 4.10 cfs (1840 gpm), the steady-state flow rate for the condition where $h_p = 169$ ft.
2. We cannot instantaneously change the speed of a pump from one value to another, as we assumed in this example. To solve this problem correctly, we would have to know the operating characteristics of the pump and motor and the moment of inertia of the rotating system.

10.4 Approach to Steady Flow

Determining the time for the flow to become steady in a pipeline when a valve is suddenly opened at the end of the pipe can be accomplished through application of Eq. (10.6). Immediately after the valve is opened (Fig. 10.4), the head h is available to accelerate the flow. Thus, flow commences, but as the velocity increases the accelerating head is reduced by fluid friction and minor losses. Let us assume the total head loss h_L can be expressed as $k_L V^2/2g$, where k_L is constant, although it may vary somewhat with velocity unless the pipe is very rough. Then

$$k_L = f\frac{L}{D} + \Sigma k$$

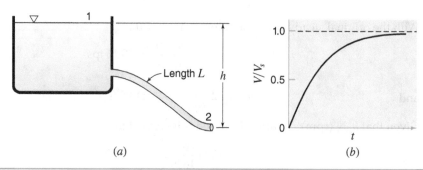

FIGURE 10.4   Approach to steady flow (V_s = velocity at steady flow).

where Σk is the sum of the coefficients of all minor losses in the pipeline (Secs. 7.21–7.27). Writing Eq. (10.6) between sections 1 and 2 in Fig. 10.4a, while noting that $p_1 = p_2 = V_1 = 0$ and $z_1 - z_2 = h$, gives

$$h - \frac{V_2^2}{2g} = h_L + \frac{L}{g}\frac{dV}{dt}$$

or

$$h = \left(\frac{fL}{D} + \Sigma k\right)\frac{V^2}{2g} + \frac{V^2}{2g} + \frac{L}{g}\frac{dV}{dt}$$

i.e.,

$$h = K\frac{V^2}{2g} + \frac{L}{g}\frac{dV}{dt}, \quad \text{where } K = \frac{fL}{D} + 1 + \Sigma k$$

For the free discharge at point 2 in Fig. 10.4, note that in the expression for $K = k_L + 1$ the 1 represents the residual velocity head and Σk does *not* include a submerged discharge loss coefficient $k_d = 1$ (Sec. 7.23). If, however, there is submerged discharge into still water in another reservoir, there is no residual velocity head so that $K = k_L$, but Σk now *does* include $k_d = 1$. As a result, the value of K is the same for both situations if minor losses are included. It is not quite the same if minor losses are *neglected* (see Sec. 7.28), because the residual velocity head is not considered a minor loss.

Let us represent the steady-flow velocity by V_s. Noting that for steady flow $(dV/dt) = 0$, we get

$$V_s = \sqrt{\frac{2gh}{K}} \tag{10.7}$$

Substituting the value of h from this expression into the above energy equation gives

$$dt = \frac{2L}{K}\frac{dV}{V_s^2 - V^2}$$

Treating f and therefore K as constants, we can integrate. Noting that the constant of integration = zero, since $V = 0$ at $t = 0$ and $\ln(V_s/V_s) = 0$, we get

$$t = \frac{L}{KV_s}\ln\frac{V_s + V}{V_s - V} \tag{10.8}$$

This equation indicates that V approaches V_s asymptotically and that equilibrium will be attained only after an infinite time (Fig. 10.4b), but it must be remembered that this is an idealized case. In reality there will be elastic waves and damping, so that true equilibrium will be reached in a finite time. Also, as noted earlier, we can apply Eqs. (10.7) and (10.8) to a submerged discharge as well as to a free discharge.

Sample Problem 10.3

Two large water reservoirs are connected to one another with a 100-mm-diameter pipe ($f = 0.02$) of length 15 m. The water-surface elevation difference, h, between the reservoirs is 2.0 m. A valve in the pipe, initially closed, is suddenly opened. (a) Determine the times required for the flow to reach $\frac{1}{4}, \frac{1}{2}$, and $\frac{3}{4}$ of the steady-state flow rate. Assume the water-surface elevations remain constant. (b) Repeat for pipe lengths of 150 m and 1500 m with all other data remaining the same. Assume a square-edged entrance.

Solution
(a) For $L = 15$ m, square-edged entrance, and submerged discharge: $L/D = 15/0.10 = 150$, so minor losses are significant (Sec. 7.28). $\Sigma k = 0.5 + 1.0 = 1.5$

$$K = k_L = \frac{fL}{D} + \Sigma k = 0.02\frac{15}{0.10} + 1.5 = 4.5$$

For steady flow:

Eq. (10.7): $$V_s = \sqrt{\frac{2gh}{K}} = \sqrt{\frac{2(9.81)2}{4.5}} = 2.95 \text{ m/s}$$

For unsteady flow use Eq. (10.8):

$$t = \frac{L}{KV_s}\ln\frac{V_s + V}{V_s - V} = \frac{15}{4.5(2.95)}\ln\frac{2.95 + V}{2.95 - V} = 1.129\ln\frac{2.95 + V}{2.95 - V}$$

For $Q = \frac{1}{4}Q_s$ substitute $V = \frac{1}{4}V_s$, etc.:

Q	V, m/s	$\frac{2.95 + V}{2.95 - V}$	ln	For $L = 15$ m t, s	
$0.25Q_s$	0.74	1.667	0.511	0.577	
$0.50Q_s$	1.48	3.00	1.099	1.240	
$0.75Q_s$	2.21	7.00	1.946	2.197	*ANS*

(b) For the other two lengths the minor losses are insignificant (Sec. 7.28), and the results are as follows:

Q	For $L = 150$ m	For $L = 1500$ m	
$0.25Q_s$	2.18 s	7.04 s	
$0.50Q_s$	4.69 s	15.15 s	
$0.75Q_s$	8.30 s	26.84 s	*ANS*

10.5   Velocity of Pressure Wave in Pipes

Unsteady phenomena, with rapid changes taking place, frequently involve the transmission of pressure in waves or surges. Considerations of velocity, pressure, density, and elasticity reveal that the velocity or *celerity* of a pressure (sonic) wave is

$$c = \sqrt{\frac{E_v}{\rho}} = \sqrt{\frac{g}{\gamma} E_v}$$ (10.9)

where E_v is the volume (or bulk) modulus of elasticity of the liquid medium. For water, a typical value of E_v is 300,000 psi (2.07×10^6 kN/m²), and thus the velocity (or celerity) c of a pressure wave in free water is about 4720 fps (1440 m/s), depending on the temperature.

For water in an elastic pipe, this velocity is reduced due to the stretching of the pipe walls. E_v is replaced by the joint (or combined) modulus E_j, such that

$$E_j = \frac{E_v}{1 + \frac{D}{t}\frac{E_v}{E}} = \frac{1}{\frac{1}{E_v} + \frac{D}{tE}}$$

where D and t are the inside diameter and wall thickness of the pipe (see Table A.9), respectively, and E is the modulus of elasticity of the pipe material; E_j is less than E_v. As the ratios D/t and E_v/E are dimensionless, any consistent units may be used in each.

The velocity of a pressure wave in an elastic fluid inside an elastic pipe is then

$$c_j = \sqrt{\frac{E_j}{\rho}} = \sqrt{\frac{g}{\gamma} E_j} = \frac{c}{\sqrt{1 + \frac{D}{t}\frac{E_v}{E}}} = \sqrt{\frac{1}{\rho\left(\frac{1}{E_v} + \frac{D}{tE}\right)}}$$ (10.10)

The subscript j, used for water inside a pipe, corresponds to that used in E_j. Values of the modulus of elasticity E for steel, cast iron, and concrete are about 30,000,000, 15,000,000, and 3,000,000 psi, respectively.[2] Values of the volume (bulk) modulus E_v for various liquids are given in Appendix A, Table A.4, and for water in Sec. 2.5 and Table A.1.

For normal pipe dimensions the speed c_j of a pressure wave in a water pipe usually ranges between 2000 and 4000 fps (600 and 1200 m/s), but it will always be less than $c \approx 4720$ fps (1440 m/s), the velocity of a pressure wave in free water. Note, e.g., from E_v values in Table A.1, that both these velocities vary with temperature, by up to 7.5%.

10.6   Water Hammer

In the preceding unsteady-flow cases in this chapter, the changes of velocity were presumed to take place slowly. But if the velocity of a liquid in a pipeline is abruptly decreased by a valve movement, the phenomenon encountered is called *water hammer.* This is a very important problem in the case of hydroelectric plants, where the flow of water must be rapidly varied in proportion to the load changes on the turbine. Water hammer has burst large penstocks, causing great damage to hydraulic and power generating facilities, in

---

[2] Corresponding values of E for steel, cast iron, and concrete in SI units are 207×10^6, 103×10^6, and 20.7×10^6 kPa (kN/m²), respectively.

addition to loss of life and power generation. Water hammer occurs in liquid-flow pressure systems whenever a valve is closed or opened, fully or partially. The terminology "water hammer" is perhaps misleading, since this phenomenon can occur with any liquid.

Instantaneous Closure

Although it is physically impossible to close a valve instantaneously, such a concept is useful as an introduction to the study of real cases. For convenience let us start by considering steady flow in a horizontal pipe (Fig. 10.5a) with a partly open valve. Then let us assume that the valve at N is closed completely and instantaneously. The lamina of liquid next to the valve is compressed by the rest of the column of liquid flowing against it. At the same time the walls of the pipe surrounding this lamina are stretched by the

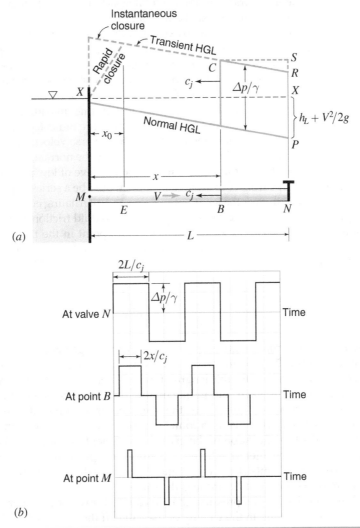

(a)

(b)

FIGURE 10.5   Water hammer with pipe friction and damping neglected. (a) Valve N at the end of the pipeline is abruptly closed and the pressure wave has traveled part way up the pipe. (b) Idealized water-hammer pressure heads at N, B, and M as a function of time for instantaneous valve closure.

excess pressure produced; in this way the kinetic energy of the moving water transforms into these two forms of potential (elastic) energy. The next upstream lamina will then be brought to rest, and so on. The liquid in the pipe does not behave as a rigid incompressible body, but the phenomenon is affected by the elasticity of both the liquid and the pipe. The cessation of flow and the resulting pressure increase move upstream together along the pipe as a wave with the velocity c_j as given by Eq. (10.10).

After a short interval of time the liquid column BN will have been brought to rest at an increased pressure, while the liquid in the length MB will still be flowing with its initial velocity and initial pressure. When the pressure wave finally reaches the inlet at M, the entire mass in the length L will be at rest but will be under an excess pressure throughout. During travel of the pressure wave from N to M, there will be a transient hydraulic grade line parallel to the original steady-flow grade line XP but at a height $\Delta p/\gamma$ above it, where Δp represents the water hammer pressure.

It is impossible for a pressure to exist at M that is greater than that due to depth MX, and so when the pressure wave arrives at M, the pressure at M drops instantly to the value it would have for zero flow. But the entire pipe is now under an excess pressure; the liquid in it is compressed, and the pipe walls are stretched. So some liquid starts to flow back into the reservoir and a wave of pressure unloading travels along the pipe from M to N. Assuming there is no damping, at the instant this unloading wave reaches N, the entire mass of liquid will be under the normal pressure indicated by the line XP, but the liquid is still flowing back into the reservoir. This reverse velocity will provide a suction or pressure drop at N that ideally will be as far below the normal, steady-flow pressure as the pressure an instant before was above it. Then a wave of low pressure or suction travels back up the pipe from N to M. Ideally, there would be a series of pressure waves traveling back and forth over the length of the pipe and alternating equally between high and low pressures. Actually, because of damping due to fluid friction and imperfect elasticity of liquid and pipe, the pressure extremes at any point in the pipe will gradually tend toward the pressure for the no-flow condition indicated by XX in Fig. 10.5a.

The time for a round trip of the pressure wave from N to M and back again is

$$T_r = 2\frac{L}{c_j} \tag{10.11}$$

where L is the actual length of pipe, and not the length adjusted for minor losses as described in Sec. 7.21. So for an instantaneous valve closure the excess pressure at the valve remains constant for this length of time T_r before it is affected by the return of the unloading pressure wave; and in like manner the pressure defect during the period of low pressure remains constant for the same length of time. At a distance x from the inlet, such as at B, the time for a round trip of a pressure wave is only $2x/c_j$, and so at that point the time duration of the excess or deficient pressure will be $2x/c_j$, as shown in Fig. 10.5b. At the inlet M, where $x = 0$, the excess pressure occurs for only an instant. Thus, we find at points closer to the inlet M that the duration of the excess or deficient pressure is less, but its magnitude is the same.

We can calculate the pressure change caused by water hammer as follows. In Fig. 10.6 we see a close-up in the vicinity of the valve. If the valve is closed abruptly, but not necessarily completely, a pressure wave travels up the pipe with a celerity c_j. In a short interval of time Δt an element of liquid of length $c_j \Delta t$ is decelerated. Therefore the mass m decelerated in the same time is $m = \rho V\!\!\!/ = \rho A(c_j \Delta t)$. Applying Newton's second law,

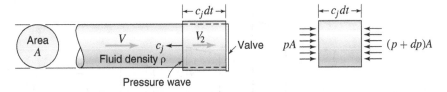

FIGURE 10.6 Definition sketch for analysis of water hammer in pipes.

$F\,\Delta t = m\,\Delta V$, recalling from Sec. 6.1 that $\Delta V = V_{\text{out}} - V_{\text{in}} = V_2 - V_1$, and writing the initial V_1 as V and neglecting friction, we get

$$[pA - (p + \Delta p)A]\Delta t = (\rho A c_j \Delta t)(V_2 - V)$$

from which

Instantaneous, partial or complete closure:
$$\Delta p = \rho c_j(V - V_2) \tag{10.12}$$

which indicates the change in pressure Δp that results from an instantaneous change in flow velocity $(V - V_2)$. Partially closing a valve will affect the size of $(V - V_2)$, and Eq. (10.12) indicates that the pressure change is proportional to the velocity change. We may also eliminate c_j between Eqs. (10.10) and (10.12) to obtain

Instantaneous, partial or complete closure:
$$\Delta p = (V - V_2)\sqrt{\rho E_j} = (V - V_2)\sqrt{\dfrac{\rho}{\dfrac{1}{E_v} + \dfrac{D}{tE}}} \tag{10.13}$$

We see that the pressure increase is independent of the length of the pipe and depends solely upon the celerity of the pressure wave in the pipe and the change in the velocity of the water. The total pressure at the valve immediately after closure is $\Delta p + p$, where p is the pressure in the pipe just upstream of the valve prior to closure.

In the case of instantaneous and *complete* closure of a valve the velocity is reduced from V to zero, that is, $V_2 = 0$. Substituting this into Eqs. (10.12) and (10.13), we get

Instantaneous, *complete* closure:
$$\Delta p = \rho c_j V = V\sqrt{\dfrac{\rho}{\dfrac{1}{E_v} + \dfrac{D}{tE}}} \tag{10.14}$$

Consider now conditions at the valve as affected by both pipe friction and damping. In Fig. 10.5a, when the pressure wave from N has reached B, the water in BN will be at rest and for zero flow the hydraulic grade line CR should be a horizontal line. There is thus a tendency for the grade line to flatten out for the portion BN. Hence, instead of the transient gradient having the slope imposed by friction, as shown in the figure, it will approach the horizontal line CS as the pressure at C cannot change. Thus the pressure head at N will be raised to a slightly higher value than NR shortly after the valve closure.

This slight increase in pressure head at the valve over the theoretical value $c_j(V - V_2)/g$ has been borne out by tests. As the wave front (B in Fig. 10.5a) moves toward M, C moves with it and so RS gets larger. As a result, on the diagram of the pressure history at the valve in Fig. 10.7, the line ab is shown as sloping upward, and for the same reason ef may slope

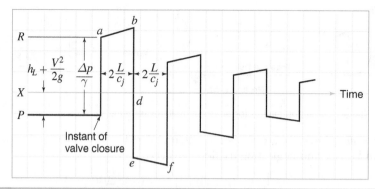

Figure 10.7 **Figure 10.7** Pressure history at the valve *N* of Fig. 10.5 caused by instantaneous complete closure, considering pipe friction and damping.

slightly downward, as all conditions are now reversed. Also, because of damping, the waves will be of decreasing amplitude until the final equilibrium pressure is reached.

All of the preceding analysis assumes that the wave of low pressure (suction) will not cause the minimum pressure at any point to drop down to or below the vapor pressure. If it should do so, the water would separate and produce a discontinuity. When the pressure recovers and the water rushes back into the vapor cavity, its momentum and the resulting implosion on the valve could be very damaging.

Sample Problem 10.4

Water flows at 10 fps in a 400-ft-long steel pipe of 8-in diameter with 0.25-in-thick walls. Calculate the pulse interval and maximum pressure rise theoretically caused by instantaneously closing the end valve (*a*) completely and (*b*), partially, reducing the velocity to 6 fps.

Solution

Eq. (10.10): $$c_j = \frac{c}{\sqrt{1 + \dfrac{D}{t}\dfrac{E_v}{E}}} = \frac{4720}{\sqrt{1 + \dfrac{8}{0.25}\left(\dfrac{3 \times 10^5}{3 \times 10^7}\right)}} = 4110 \text{ fps}$$

Eq. (10.11): $$T_r = \frac{2(400)}{4110} = 0.195 \text{ sec} \qquad \textbf{ANS}$$

(*a*) Eq. (10.14): $$\Delta p = \rho c_j V = \frac{62.4}{32.2}(4110)10 = 79,600 \text{ psf} \qquad \textbf{ANS}$$

$$= 553 \text{ psi}$$

for which $$\Delta p/\gamma = 79,600/62.4 = 1276 \text{ ft of water}$$

For this pipe, the water-hammer impact initially equivalent to the pressure of 1276 ft of water (553 psi) occurs about five times every second!

(*b*) With partial closure, c_j and T_r are unchanged.

Eq. (10.12): $$\Delta p = \frac{62.4}{32.2}(4110)(10 - 6) = 31,800 \text{ psf} \qquad \textbf{ANS}$$

Rapid Closure ($t_c < T_r$)

It is physically impossible for a valve to be closed instantaneously; so we shall now consider the real case where the valve is closed, completely or partially, in a finite time t_c that is more than zero but less than $T_r = 2L/c_j$. Figure 10.8 shows actual pressure recordings, made at the valve, for such a case.[3] The slope of the curve during the time t_c depends entirely on the operation of the valve and its varying effect on the velocity in the pipe. But the maximum pressure rise is still the same as for instantaneous closure. The only differences are that it endures for a shorter period of time and the vertical lines of Fig. 10.7 are changed to the sloping lines of Fig. 10.8. The duration of the maximum pressure is $T_r - t_c$ (Fig. 10.8). If the time of valve closure were exactly T_r, the maximum pressure rise at the valve would still be the same, but the curves in Fig. 10.8 would all end in sharp points for both maximum and minimum values, since the time duration of maximum pressure would be reduced to zero.

Earlier we noted that at distance x from the inlet, such as B in Fig. 10.5, the time for a round trip of a pressure wave is $2x/c_j$. Therefore, no matter how rapid the valve closure, since t_c cannot be zero there must be some distance x_0 such that $t_c = 2x_0/c_j$. Clearly

$$x_0 = \frac{c_j t_c}{2} \tag{10.15}$$

Thus, at the point E in Fig. 10.5, a distance x_0 from the inlet, the time for a round trip of a pressure wave is exactly t_c. At points closer to the inlet ($x < x_0$) this travel time is shorter. But, as we see in Fig. 10.8, it takes time t_c for the pressure at a point to rise from zero to its maximum value. Therefore, at points nearer to the inlet than E, the reflected relief (unloading) pressure wave front will arrive before the maximum pressure is achieved. This effect will reduce the maximum water-hammer pressure at points where $x < x_0$. Nearer the inlet, at smaller x, the reflected wave front arrives sooner and the reduction is greater. As a result, in any real case the maximum water-hammer pressure experienced will be Δp [Eqs. (10.12–10.14)] at all points where $x > x_0$, and it will decrease from Δp at x_0 to zero at the inlet. The maximum pressure rise Δp cannot extend

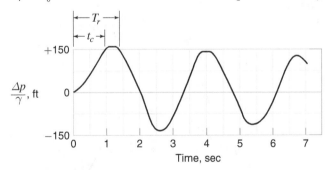

Figure 10.8 Rapid complete valve closure in time t_c (1 sec) less than T_r (1.40 sec). Actual measurements of pressure changes at the valve. (See footnote 3.)

---

[3] Figures 10.8 and 10.9 are from water-hammer studies made by the Southern California Edison Co. on an experimental pipe with the following data: $L = 3060$ ft, internal diameter $= 2.06$ in, $c_j = 4371$ fps, $V = 1.11$ fps, $\Delta p/\gamma = 150.7$ ft, $T_r = 1.40$ sec, static head $= 306.7$ ft, head before valve closure $= 301.6$ ft, $h_L = 5.1$ ft. In Fig. 10.8 the time of complete closure $t_c = 1$ sec and we see that the actual rise in pressure head is slightly more than 151 ft. In Fig. 10.9 the time of closure $= 3$ sec.

all the way to the reservoir intake. This modification of the transient hydraulic grade line is labeled "rapid closure" on Fig. 10.5. A linear variation of maximum transient pressure within x_0 is commonly assumed, and although this appears reasonable from Fig. 10.8 (for $t < t_c$), the actual shape of this curve depends on how the valve is operated.

Slow Closure ($t_c > T_r$)

The preceding discussion assumes a closure so rapid (or a pipe so long) that there is an insufficient time for a pressure wave to make the round trip before the valve is closed. We will define slow closure as one in which the time of valve movement is greater than $T_r = 2L/c_j$. In this case the maximum pressure rise will be less than in the preceding, because the wave of pressure unloading will reach the valve before the valve is completely closed. This will prevent any further increase in pressure.

Thus in Fig. 10.9 the pressure at the valve would continue to rise if it were not for the fact that at time T_r a return unloading pressure wave reaches the valve and stops the pressure rise at a value of about 53 ft as contrasted with nearly three times that value in Fig. 10.8.[4] Viewed another way, this is equivalent to saying that $x_0 > L$, or the valve is to the left of E in Fig. 10.5, so the entire pipeline only experiences maximum transient pressures which vary approximately linearly for $x < x_0$, and the maximum value given by Eqs. (10.12)–(10.14) is never achieved.

Tests have shown that for slow valve closure, i.e., in a time greater than T_r, the excess pressure produced decreases uniformly from the value at the valve to zero at the intake. The maximum water-hammer pressure developed by gradual closure of a valve when $t_c > T_r$ is given approximately by

$$\frac{\Delta p'}{\Delta p} \approx \frac{L}{x_0}$$

from which

Slow, partial or complete closure:

$$\Delta p' \approx \frac{L}{x_0}\Delta p = \frac{2L}{c_j t_c}\Delta p = \frac{L\rho c_j(V - V_2)}{x_0}$$
$$= \frac{2L\rho(V - V_2)}{t_c}$$

(10.16)

where t_c is the time of closure.

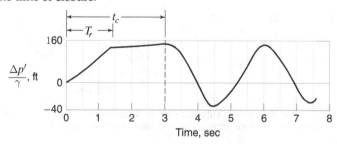

FIGURE 10.9   Slow complete valve closure in time t_c (3 sec) greater than T_r (1.40 sec). Actual measurements of pressure changes at the valve. (See footnote 3.)

[4] See footnote 3.

After an unloading pressure wave reaches the valve and the maximum pressure $\Delta p'$ is reached, elastic waves travel back and forth and pressure changes are very complex. They require a detailed step-by-step analysis that is beyond the scope of this text.[5] In brief, the method consists of assuming the valve movement to take place in a series of steps each of which produces a pressure Δp proportional to each ΔV.

Sample Problem 10.5

Assume that the elasticity and dimensions of the pipe in Fig. 10.5a are such that the celerity of the pressure wave is 3200 fps. Suppose that the pipe has a length of 2000 ft and a diameter of 4 ft. Find the maximum water hammer pressure at the valve, and the approximate maximum water hammer pressure at a point 300 ft from the reservoir, (a) if the valve is partially closed in 2.2 sec, thereby reducing the flow of water from 30 to 10 cfs; and (b) if water is initially flowing at 15 cfs and the valve is completely closed in 1.0 sec.

Solution

$$A = \pi(2)^2 = 12.57 \text{ ft}^2$$

Eq. (10.11): $$T_r = \frac{2(2000)}{3200} = 1.25 \text{ sec}$$

(a) $$t_c = 4.0 \text{ sec} > T_r, \text{ so closure is "slow."}$$

Eq. (10.15): $$x_0 = 0.5(3200)2.2 = 3520 \text{ ft}$$

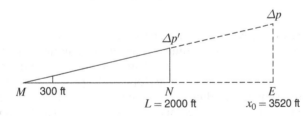

$$V = Q/A = 30/12.57 = 2.39 \text{ fps}; \quad V_2 = 10/12.57 = 0.796 \text{ fps}$$

Eq. (10.12): $$\Delta p = \frac{62.4}{32.2}(3200)(2.39 - 0.796) = 9870 \text{ psf}$$
$$= 68.5 \text{ psi}$$

At the valve N $(x < x_0)$,

Eq. (10.16): $$\Delta p' = \frac{2000}{3520}(68.5) = 38.9 \text{ psi} \qquad ANS$$

At the point $x = 300$ ft from the reservoir, assuming a linear variation,

$$(\Delta p)_x = \frac{300}{2000}(38.9) = 5.84 \text{ psi} \qquad ANS$$

[5] See, e.g., E. B. Wylie and V. L. Streeter, *Fluid Transients in Systems*, Prentice-Hall, Inc., Englewood Cliffs, New Jersey, 1993.

(b) $t_c = 1.0 \text{ sec} < T_r$, so closure is "rapid."

Eq. (10.15): $x_0 = 0.5(3200)1.0 = 1600 \text{ ft}$

M 300 ft E N

$x_0 = 1600 \text{ ft} \quad L = 2000 \text{ ft}$

$$V = Q/A = 15/12.57 = 1.194 \text{ fps}$$

At the valve N $(x > x_0)$,

Eq. (10.14): $\Delta p = \dfrac{62.4}{32.2}(3200)1.194 = 7400 \text{ psf}$

$$= 51.4 \text{ psi} \qquad \textbf{\textit{ANS}}$$

At the point $x = 300$ ft from the inlet, assuming a linear variation within x_0,

$$(\Delta p)_x = \frac{300}{1600}(51.4) = 9.64 \text{ psi} \qquad \textbf{\textit{ANS}}$$

Computer Techniques for Water Hammer

A stricter analysis of water hammer, based on Newton's second law of motion and continuity, results in two nonlinear partial differential equations each in two unknowns, V (or Q) and p (or p/γ), which are functions of x and t. These equations can take into account pipe friction and the slope of the pipe. Although there is no general solution to these equations, they can be solved by the *method of characteristics* using a finite difference solution on a computer. The two partial differential equations are combined in two particular ways to convert them into two ordinary differential equations, which we can write in finite difference form and solve simultaneously for the two unknowns. To do this, we must divide the pipe into a number of incremental lengths, assign initial values to all the nodes, and prescribe boundary conditions at each end of the pipe.

A major advantage of such a computerized solution method is that it can incorporate complex boundary conditions. We can specify these as variations in V or p, or relations between them, or connection to another pipeline or lines, or to a pump.

The details of these techniques are rather extensive, and can be found in more advanced texts.[6]

Protection from Water Hammer

A pipeline can be protected from damage by water hammer by connecting to it any of a variety of devices that keep the transient pressures within desired limits.

---

[6] See footnote 5.

Slow-closing valves can be used to control the flow, and by-pass valves can divert sudden changes in flow. Air valves can be connected to the line to admit air if the pressure falls too low. Similarly, automatic relief valves may be fitted to allow water to escape when pressure exceeds a certain value, but these are usually initiated by a pressure drop, and they only work if the reflection time T_r is long enough. Air chambers include so-called bladder accumulators and hydropneumatic surge arrestors. They contain compressed air or gas, which acts as a cushion to absorb pressure changes. Pump flywheels have been used to absorb energy by increasing the inertia of the rotating element, and they thus increase the time required for incremental changes in the flow rate.

Surge tanks are standpipes connected to the pipeline that store liquid when pressures are high and return it to the line when pressures fall. The simple surge tank, which allows diversion of the flow when the downstream valve is closed, is explained next in Sec. 10.7. Other surge tanks admit the liquid through an orifice, which in addition helps to dissipate energy. Yet others, known as differential surge tanks, have multiple chambers with interconnections; the different water levels that occur in the various chambers help to dampen out the oscillations.

10.7 Surge Tanks

In a hydroelectric plant the flow of water to a turbine must be decreased very rapidly whenever there is a sudden drop in load. This rapid decrease in flow will result in high water-hammer pressures and may result in the need for a very strong and hence expensive supply pipe called a ***penstock***. There are several ways to handle a situation of this sort; one is by use of a ***surge tank,*** or ***surge chamber.*** A simple surge tank is a vertical standpipe connected to the pipeline as shown in Fig. 10.10. With steady flow V_0 in the pipe, the initial water level z_0 in the surge tank is below the static (no flow) level $(z = 0)$, as shown in the figure. When the valve is suddenly closed, water rises in the surge tank. The water surface in the tank will then oscillate up and down until damped out by fluid friction. The section of pipe upstream of the surge tank is in effect afforded protection from the high water-hammer pressures that would exist on valve closure if there were no tank.

An approximate analysis for this simple surge tank may be performed as follows. Initially, for steady conditions, indicated by the subscript 0, the amount by which the water level in the surge tank is below the static level is the sum of the head loss due to

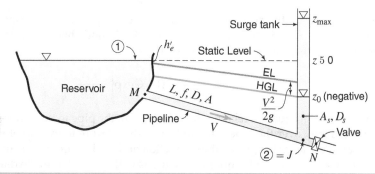

FIGURE 10.10 Definition sketch for surge tank analysis.

friction, the velocity head, and any minor losses in the line such as the entrance loss h_e' shown (see Sec. 7.22). Therefore, as z_0 is negative, we have

$$-z_0 = h_f + \frac{V_0^2}{2g} + \Sigma k \frac{V_0^2}{2g}$$

or

$$-z_0 = K \frac{V_0^2}{2g} \qquad \text{where} \qquad K = \frac{fL}{D} + 1 + \Sigma k \qquad (10.17)$$

where Σk is the sum of the coefficients for all minor losses in the line. For the condition soon after the valve is instantaneously closed, when the water is at some level z as it rises up the surge tank, we can write the energy equation given by Eq. (10.6) for unsteady flow between the surface of the reservoir (point 1 on Fig. 10.10) and the end of the pipeline at the surge tank (point 2 = J, on the centerlines). Neglecting fluid friction in the surge tank and inertial effects in the surge tank, and letting h be the vertical distance from point 2 to elevation $z = 0$, at point 2 we have $p_2/\gamma = h + z$ and $z_2 = -h$, so that $(p/\gamma + z)_2 = h + z - h = z$. Substituting this and conditions at point 1 into Eq. (10.6), we therefore obtain

$$0 - \left(z + \frac{V^2}{2g} \right) = \left(\Sigma k + \frac{fL}{D} \right) \frac{V^2}{2g} + \frac{L}{g} \frac{dV}{dt}$$

In this equation z represents the level of the water surface in the surge tank measured positively upward from the static water level where $z = 0$ and $(L/g)(dV/dt)$ represents the accelerative head h_a in the pipe between the reservoir and the surge tank (Sec. 10.3).

Using our definition of K from Eq. (10.17), the above equation can be written as

$$-z = K \frac{V^2}{2g} + \frac{L}{g} \frac{dV}{dz} \frac{dz}{dt} \qquad (10.18)$$

With the valve completely closed, since the flow velocity up the surge tank is $V_s = dz/dt$, the continuity equation is

$$AV = A_s \frac{dz}{dt} \qquad (10.19)$$

where A and A_s are the cross-sectional areas of the pipe and surge tank, respectively. Combining Eqs. (10.18) and (10.19) to eliminate dt, and then assuming f and K are constant while integrating and solving for V, yields

$$V^2 = \frac{2g}{K} \left(\frac{LA}{KA_s} - z \right) - C \exp \left(-\frac{KA_s}{LA} z \right) \qquad (10.20)$$

This is the general equation relating the velocity V in the pipe to the water-surface level z in the surge tank over the time interval from valve closure to the top of the first surge. After that, the water changes its flow direction and so this equation is no longer valid. However, we know that subsequent oscillations will be smaller due to damping by friction losses.

Before using this equation to find the maximum surge height z_{max}, we must first take care of the unknown constant of integration C. We can do this as follows. The initial conditions for the time interval considered are $V = V_0$ and $z = z_0$ [Eq. (10.17)]. Substituting these into Eq. (10.20) gives

$$V_0^2 = \frac{2g}{K}\left(\frac{LA}{KA_s} + K\frac{V_0^2}{2g}\right) - C\exp\left[\left(\frac{KA_s}{LA}\right)K\frac{V_0^2}{2g}\right]$$

which simplifies to

$$\frac{2gLA}{K^2A_s} = C\exp\left[\left(\frac{KA_s}{LA}\right)K\frac{V_0^2}{2g}\right] \tag{10.21}$$

The conditions at the end of the time interval, at the top of the first surge, are $V = 0$ and $z = z_{max}$. Substituting these into Eq. (10.20) and rearranging gives

$$\frac{2g}{K}\left(\frac{LA}{KA_s} - z_{max}\right) = C\exp\left(-\frac{KA_s}{LA}z_{max}\right) \tag{10.22}$$

Finally, dividing Eq. (10.22) by Eq. (10.21) to eliminate C results in

$$1 - \frac{KA_s}{LA}z_{max} = \exp\left[-\frac{KA_s}{LA}\left(K\frac{V_0^2}{2g} + z_{max}\right)\right] \tag{10.23}$$

Reviewing our result, we see that this equation is explicit for V_0, but is implicit for z_{max}, A_s, A, L, and K, and so therefore f (Appendix B). Also, we note that the right-hand side of this equation must always be positive because it is an exponential, so it follows that the left-hand side must always be positive also, leading to

$$\frac{KA_s}{LA}z_{max} < 1 \tag{10.24}$$

Note that in all cases we can conveniently replace A_s/A by $(D_s/D)^2$, where D_s is the diameter of the surge tank.

If we wish to find V_0, we can rearrange Eq. (10.23) into

$$V_0^2 = -\frac{2g}{K}\left[z_{max} + \frac{LA}{KA_s}\ln\left(1 - \frac{KA_s}{LA}z_{max}\right)\right] \tag{10.25}$$

where the right side will be positive because the ln term will be negative. However, if we wish to solve Eq. (10.23) for z_{max}, A_s, or K (in order to find f), we must either use an equation solver, or manually use trial and error. For the manual method, because the right side of Eq. (10.23) is usually quite small, it turns out that the limiting value given by Eq. (10.24) is a good indicator of the true answer. For example, when solving for z_{max}, Eq. (10.24) gives $z_{max} < LA/KA_s$, which serves as a good first estimate for trial and error.

Since this derivation neglected fluid friction and inertial effects in the surge tank, minor losses at the surge tank junction, and assumed constant f and instantaneous valve closure, the value of z_{max} as computed by Eq. (10.23) will be larger than the true value, and so the results provide a conservative estimate for preliminary design of simple surge tanks. For long pipelines, minor losses and the velocity head in the pipeline will be small compared with the friction head loss. If these are also neglected, according to Eq. (10.17) K must be replaced by fL/D. The results of doing this will provide a slightly more conservative estimate.

More accurate analysis of surge tank phenomena commonly uses computer programs. These use numerical methods to solve the differential and algebraic equations that represent the unsteady liquid motion, taking into account all the factors just mentioned, and including the actual rates of valve closure.

Surge tanks are usually open at the top and of sufficient height that they will not overflow. In some instances they are permitted to overflow if no damage will result. There are many types of surge tanks; other types were discussed at the end of Sec. 10.6.

The surge tank, in addition to providing protection against water-hammer pressures, fulfills another desirable function. That is, in the event of a sudden demand for increased flow, it can quickly provide some excess water, while the entire mass of water in a long pipeline is accelerating. The acceleration of masses of liquids in pipelines was discussed in Sec. 10.4.

Sample Problem 10.6

You are asked to design a surge tank for a 1.5-m-diameter steel pipeline ($f = 0.018$) 800 m long that supplies water to a small power plant. The surge tank will be connected to the pipeline at a point where the centerline is 35 m below the water surface in the reservoir. For a discharge of 12 m³/s, use only a basic scientific calculator to find the smallest surge tank diameter that will prevent surges from exceeding 18.5 m above the reservoir surface. Neglect inertial effects and fluid friction in the surge tank only. The pipeline inlet is square-edged.

Solution

Section 7.22 for square-edged entrance: $k_e = 0.5$

Eq. (10.17): $$K = \frac{0.018(800)}{1.5} + 1 + 0.5 = 11.10$$

$$V_0 = Q_0/A = 12/\pi(1.5/2)^2 = 6.79 \text{ m/s}$$

Eq. (10.17): $z_0 = -11.1(6.79)^2/2(9.81) = -26.09$ m, possible cf 35 m.

From Eq. (10.24): $$A_s < \frac{LA}{Kz_{max}} = \frac{800\pi(1.5/2)^2}{11.10(18.5)} \text{m}^2 = 6.88 \text{ m}^2$$

Solving Eq. (10.23) by trial and error (per Sample Problem 3.8):

First trial: Try $A_s = 6.88$

Last trial: $A_s = 6.06$ m² $= (\pi/4)D_s^2$; $D_s = 2.78$ m   *ANS*

Note: We can solve Eq. (10.23) without manual trial and error by using a programmable scientific calculator with an equation (root) solving capability, or by using equation solving software.

Programmed computing aids could help solve problems marked with this icon.

CHAPTER 11

Hydraulic Machinery— Pumps and Turbines[1]

There are various types of fluid machinery. Among them are those that transfer fluid energy (*torque converters*), those that convert mechanical energy to fluid energy (*pumps*), and those that convert fluid energy to mechanical energy (*turbines*). We convert mechanical energy to fluid energy by using pumps for incompressible fluids, and by using *blowers*, *fans*, and *compressors* for compressible fluids. In this chapter since pumps are much more prevalent than turbines, and because the analysis of turbines is similar to pumps, we discuss pumps primarily. We shall deal with *centrifugal and axial-flow pumps*, but shall not discuss positive displacement pumps such as reciprocating piston types.

In our discussion of hydraulic machinery, we will emphasize the selection of pumps for particular situations. To accomplish this, we shall first describe the various types of pumps and discuss their performance characteristics. We will show how pumps fit into the hydraulics of the systems in which they operate. Finally, we discuss turbines, highlighting the similarity in hydraulic analysis with pumps. We shall find that cavitation plays an important role in the selection of pumps and turbines.

11.1 Description of Centrifugal and Axial-Flow Pumps

We call the rotating element of a centrifugal pump the *impeller* (Fig. 11.1). The shape of an impeller may force water outward in a plane at right angles to its axis (*radial flow*), induce a spiral flow on coaxial cylinders in an axial direction (*axial flow*), or give the water an axial as well as radial velocity (*mixed flow*). We commonly refer to radial-flow and mixed-flow machines as centrifugal pumps, while we call axial-flow machines axial-flow pumps or *propeller pumps*. Radial- and mixed-flow impellers are either open or closed. The open impeller consists of a hub to which vanes are attached, while the closed impeller has plates (or *shrouds*) on each side of the vanes. The open impeller does not have as high an efficiency as the closed impeller, but it is less likely to become clogged and hence is suited to handling liquids containing solids.

---

[1] In this chapter, because of the difficulty of carrying everything in both sets of units, for the most part we shall deal with pumps and turbines using conventional English (BG) units in conformance with practice in the United States. However, a substantial number of sample problems, exercises and problems will be expressed in SI units.

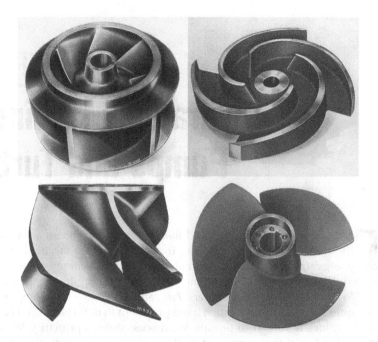

FIGURE 11.1   Types of pump impellers. Top left: closed or shrouded radial. Top right: open or unshrouded radial. Bottom left: mixed flow. Bottom right: axial flow propeller. (Courtesy of Flowserve Corporation)

Radial-flow pumps have a *spiral casing*, that we often call a *volute casing* (Fig. 11.2), which guides the flow from the impeller to the discharge pipe. The ever-increasing flow cross section around the casing tends to maintain a constant velocity within the casing. This helps to provide relatively smooth flow conditions at exit from the impeller. Some pumps have diffuser vanes instead of a volute casing. We call such pumps *turbine pumps*. Some radial pumps are of the double-suction type. They have identical, mirror-image impellers placed back to back. Water enters the pump from both sides and discharges into a volute casing or diffuser vanes. The advantage of the double-suction pump is the reduced mechanical friction that results because of the balanced thrust on the bearings.

Figure 11.3 shows typical centrifugal-flow and axial-flow pump installations. Pumps can be *single-stage* or *multistage*. A single-stage pump has only one impeller, while a multistage has two or more impellers arranged in such a way that the discharge from one impeller enters the eye of the next impeller. Deep-well pumps (Fig. 11.4), a type of turbine pump, are usually multistage, having several impellers on a vertical shaft suspended from a prime mover, usually an electric motor, located at the ground surface. Each impeller discharges into a fixed-vane diffuser, or bowl, coaxial with the drive shaft, which directs water to the next impeller.

We must properly arrange the suction and discharge piping for a centrifugal pump to operate at best efficiency. For economy, the diameter of the pump casing at suction and discharge is often smaller than that of the pipe to which it is attached. If there is a horizontal reducer between the suction and the pump, we should use an *eccentric reducer* (Fig. 11.3a) to prevent air accumulation. We can install a *foot valve* (check valve) in the suction pipe to prevent water from leaving the pump when it is stopped. We usually provide the discharge pipe with a check valve and a gate valve. The *check valve*

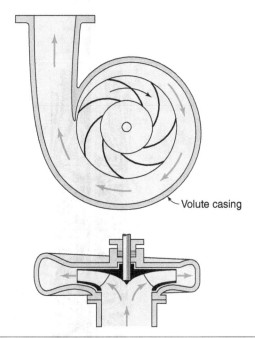

FIGURE 11.2   Radial-flow centrifugal pump with volute casing.

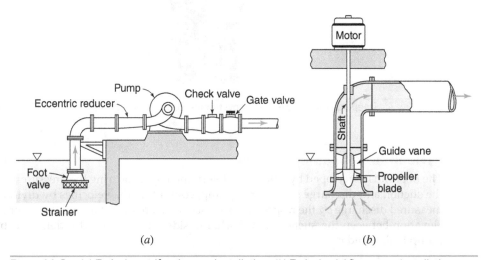

(a)   (b)

FIGURE 11.3   (a) Typical centrifugal pump installation. (b) Typical axial-flow pump installation.

prevents backflow through the pump if there is a power failure. And we usually provide suction pipes taking water from a sump or reservoir with a screen to prevent entrance of debris that might clog the pump.

Axial-flow pumps (Fig. 11.3b) usually have only two to four blades and, hence, have large unobstructed passages that permit handling of water containing debris without clogging. The blades of some large axial-flow pumps are adjustable to permit setting the pitch for the best efficiency under existing conditions.

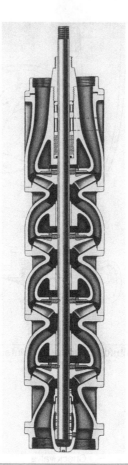

FIGURE 11.4   Deep-well multistage mixed-flow turbine pump. (Courtesy of Flowserve Corporation)

11.2   Head Developed by a Pump

The net head h developed by a pump is determined by measuring the pressures on both the suction and discharge sides of the pump, computing the velocities by dividing the measured discharge by the respective cross-sectional areas, and noting the difference in elevation between the suction and discharge sides. The net head h delivered by the pump to the fluid is

$$h = H_d - H_s = \left(\frac{p_d}{\gamma} + \frac{V_d^2}{2g} + z_d \right) - \left(\frac{p_s}{\gamma} + \frac{V_s^2}{2g} + z_s \right) \qquad (11.1)$$

where the subscripts d and s refer to the discharge and suction sides of the pump, as shown in Fig. 11.5. If the discharge and suction pipes are the same size, the velocity heads cancel out, but frequently the intake pipe is larger than the discharge pipe. Note that h, the net head put into the fluid by the pump, we previously designated h_M in Sec. 5.5 and h_p in Sec. 7.29.

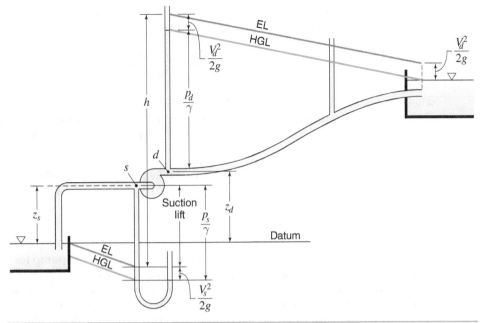

FIGURE 11.5   Net head developed by a pump. In this case p_s/γ is negative.

The official test code provides that the head developed by a pump be the difference between the total energy heads at the intake and discharge flanges. However, flow conditions at the discharge flange are usually too irregular for accurate pressure measurement, and we can more reliably measure the pressure at 10 or more pipe diameters away from the pump and add an estimated pipe friction head for that length of pipe. On the intake side, **prerotation** sometimes exists in the pipe near the pump, and this will cause the pressure reading on a gage to be different from the true average pressure at that section.

11.3   Pump Efficiency

As liquid flows through a pump, only part of the energy imparted to the shaft of the impeller transfers to the flowing liquid. There is friction in the bearings and packings; the impeller does not effectively act upon all liquid passing through the pump; and there is substantial loss of energy due to fluid friction which has a number of components including shock loss at entry to the impeller,[2] fluid friction as the fluid passes through the space between the vanes or blades, and head loss as the fluid leaves the impeller. The efficiency of a pump is quite sensitive to its operating conditions, as will be discussed in Sec. 11.5.

The power input to the pump, delivered to the pump shaft by the motor, is what we call the **shaft power** or the **brake power**. The power output from the pump, delivered to

---

[2] Shock loss occurs when the flow does not enter the impeller smoothly. This results in separation of the flow from the impeller blade.

the fluid, usually water, is what we call the *fluid power* or the *water power*. From Eq. (5.42) then, we may express the efficiency η (eta) of a pump by

$$\eta = \frac{\text{output power, } P_{\text{out}}}{\text{input power, } P_{\text{in}}} = \frac{\text{fluid power}}{\text{shaft (brake) power}} = \frac{\gamma Q h}{T \omega} \qquad (11.2a)$$

where γ, Q, and h are defined in the usual fashion; T is the torque exerted on the shaft of the pump by the motor that drives the shaft, and ω is the rate of rotation of the shaft in radians per second. From this, for convenience we may write Eq. (11.2a) as

$$\text{Fluid power} = P_{\text{out}} = \gamma Q h \quad \text{and} \quad \text{Brake power} = P_{\text{in}} = \frac{\gamma Q h}{\eta} \qquad (11.2b)$$

11.4   Similarity Laws for Pumps

Similarity laws enable us to predict the performance of a prototype pump (or turbine) from the test of a scaled model. Moreover, and of particular value in pump selection, these laws make it possible to predict the performance of a given machine under different conditions of operation from those under which it has been tested.

Similarity laws are based on the concept that two geometrically similar machines (i.e., same scale change, defined by a ratio of prototype to model length, in all three dimensions) with similar velocity diagrams at entrance to and exit from the rotating element are *homologous*. This means that their streamline patterns will be geometrically similar, i.e., that their behaviors will bear a resemblance to one another.

We can derive similarity laws by dimensional analysis. The most significant variables[3] affecting the operation of a turbomachine are the head h, the discharge Q, the rotative speed n, the diameter of the rotor D, and the acceleration due to gravity g. Thus, from the Buckingham Π-theorem (Sec. 7.2), since there are five dimensional variables and two fundamental dimensions (L and T), there will be three dimensionless groups. We have

$$f(h, Q, n, D, g) = 0$$

Upon grouping these variables into dimensionless quantities, we get

$$f'\left(\frac{Q}{nD^3}, \frac{g}{n^2 D}, \frac{h}{D}\right) = 0$$

Laboratory tests on turbomachines have demonstrated that the second dimensionless quantity is inversely proportional to the third. These can be combined to give

$$\frac{g}{n^2 D} = K \frac{D}{h} \quad \text{and} \quad K = \frac{gh}{n^2 D^2}$$

Thus

$$f''\left(\frac{Q}{nD^3}, \frac{gh}{n^2 D^2}\right) = 0 \qquad (11.3)$$

---

[3] If we wish to relate the operation of one pump to another with different fluids in each then kinematic viscosity is a significant variable.

From Eq. (11.3), we find

$$Q \propto nD^3 \quad \text{or} \quad Q = K_Q nD^3 \quad \text{or} \quad K_Q = \frac{Q}{nD^3} \tag{11.4}$$

and, assuming g is constant,[4]

$$h \propto n^2 D^2 \quad \text{or} \quad h = K_h n^2 D^2 \quad \text{or} \quad K_h = \frac{h}{n^2 D^2} \tag{11.5}$$

Since output power $P_{out} \propto Qh$ (Sec. 5.9 or end of Sec. 11.3), we get

$$P_{out} \propto n^3 D^5 \quad \text{or} \quad P_{out} = K_p n^3 D^5 \quad \text{or} \quad K_p = \frac{P_{out}}{n^3 D^5} \tag{11.6}$$

For any one design of a pump, K_Q, K_h, and K_p can be evaluated, preferably from test data, and then used to predict the performance of homologous pumps. Similarity laws like these for Q, h, and P_{out} are of great practical value, but care must be exercised when applying them. Thus, in comparing two machines of different sizes, the two must be homologous and the variation in the values of h, D, and n should not be too large. For example, a machine that operates satisfactorily at low speeds may cavitate at high speeds. The values of K in each of Eqs. (11.4)–(11.6) change somewhat as h, D, and n are varied, because the efficiencies of homologous machines are not identical. Large machines are usually more efficient than smaller ones, because their flow passages are larger. Also, efficiency usually increases with speed of rotation, because power output varies with the cube of the speed while mechanical losses increase only as the square of the speed. As a result, Eqs. (11.4)–(11.6) are approximate.

An empirical equation suggested by Moody,[5] originally intended for application to homologous reaction turbines, that gives fairly reasonable results for estimating the efficiency of a prototype pump from the test of a geometrically similar (model) pump is

$$\frac{1 - \eta_p}{1 - \eta_m} \approx \left(\frac{D_m}{D_p} \right)^{1/5} \tag{11.7}$$

The effects of rotation speed n on the performance (Q, h, P) of a given pump we call *affinity laws*. Equations (11.4)–(11.6) govern these effects, which generally change pump efficiency little. The effects of trimming the impeller outside diameter on the same performances we also call affinity laws. In this case the same equations do *not* govern, instead $Q \propto D$, $h \propto D^2$, $P \propto D^3$. Because trimming increases losses through larger clearances, it can reduce efficiency appreciably and thereby affect performance, so we should only use these relations for small diameter changes.

---

[4] If we were on the moon, or in some gravitational field different than that of the earth, we must include g in Eqs. (11.5) and (11.6).
[5] See J. H. T. Sun, "Hydraulic Machinery," in V. J. Zipparro and H. Hasen (Eds.), *Davis' Handbook of Applied Hydraulics*, 4th ed., 21–26, McGraw-Hill, New York, 1993.

Sample Problem 11.1

Develop a similarity equation for torque, assuming efficiency = 100% ($\eta = 1.0$), for geometrically similar pumps.

Solution

From Sec. 11.3 with $\eta = 1$: Output Power, $P_{out} = \gamma Qh = T\omega$

For water, with $g = 32.2$ ft/s^2, we can neglect γ because it has the same value for similar machines.

From Eqs. (11.4) and (11.5): $Q = K_Q nD^3$ and $h = K_h n^2 D^2$, so we get

$$(K_Q nD^3)(K_h n^2 D^2) = T\omega, \quad \text{but} \quad \omega = 2\pi n/60$$

Hence
$$T = K_Q K_h \left(\frac{60}{2\pi n}\right) n^3 D^5 = K_T n^2 D^5 \quad \textit{ANS}$$

Sample Problem 11.2

A small pump serving as a model has a diameter of 7.4 in. When tested in the laboratory at 3600 rpm, it delivered 3.0 cfs at a head of 125 ft. (a) If the efficiency of this model pump is 84%, what is the horsepower input to this pump? (b) Predict the speed, capacity, and horsepower input to the prototype pump if it is to develop the same head as the model pump and the model pump has a length scale ratio of 1:10 (model:prototype). Assume the efficiency of the prototype pump is 90%.

Solution

(a)   From Eq. (11.2b): $(P_m)_{in} = T_\omega = \gamma Qh/550\eta$ hp

$$(P_m)_{in} = 62.4(3.0)125/[550(0.84)] = 50.6 \text{ hp} \quad \textit{ANS}$$

(b)   From Eq. (11.5), with K_h constant for both model and prototype:

$$n_p = n_m (h_p/h_m)^{1/2}(D_m/D_p) = 3600(1)^{1/2}(1/10) = 360 \text{ rpm} \quad \textit{ANS}$$

From Eq. (11.4) with constant K_Q:

$$Q_p = Q_m (D_p/D_m)^3 n_p/n_m = 3.0(10/1)^3 \times 360/3600 = 300 \text{ cfs} \quad \textit{ANS}$$

Eq. (11.2b): $(P_p)_{in} = 62.4(300)125/[550(0.90)] = 4730 \text{ hp} \quad \textit{ANS}$

Note: To satisfy homologous conditions (streamlines geometrically similar) the larger prototype pump must operate at a *smaller* rotative speed than the model pump.

11.5   Performance Characteristics of Pumps at Constant Speed

The efficiency of a pump varies considerably, depending upon the conditions under which it must operate. Because of this, when selecting a pump for a given situation, it is important for the pump selector to have information regarding the performance of various pumps among which the selection is to be made. The pump manufacturer usually has information of this type, as determined by laboratory tests, for what are called shelf items, i.e., standard pumps. Large-capacity pumps, however, are sometimes custom-made. Often manufacturers make and test a model of such a pump before final design of the prototype.

Though some centrifugal pumps are driven by variable-speed motors, the usual mode of operation of a pump is at constant speed. Table 11.1 gives typical speeds of constant-speed electric motors. Figure 11.6 presents the ***pump characteristic curve*** (head versus capacity) and other performance curves for a typical mixed-flow centrifugal pump. This particular pump has a ***normal capacity*** or ***rated capacity*** of 10,500 gpm when developing a normal head of 60 ft at an operating speed of 1450 rpm. What we refer to as the "normal" or "rated" capacity corresponds to the ***point of optimum efficiency*** or ***BEP*** (best efficiency point); we shall denote rotative speeds at the BEP by a subscript e. Figure 11.7 gives similar curves for a typical axial-flow pump. Pump manufacturers usually determine curves like those in Figs. 11.6 and 11.7 through laboratory testing. By inspecting these two figures, we see the remarkable difference in the characteristics of these two pumps. We also observe that the efficiency of both pumps drops rather rapidly when the flow rate at which they are pumping exceeds the optimum.

The shape of the impellers and vanes and their relationship to the pump casing cause variations in the intensity of shock loss, fluid friction, and turbulence. These vary with head and flow rate, and cause the wide variation in pump characteristics. The ***shutoff head*** occurs when there is no flow. In the case of the mixed-flow centrifugal pump (Fig. 11.6), the shutoff head is usually about 10% greater than the normal head, which occurs at the point of optimum efficiency, while in the case of the axial-flow pump (Fig. 11.7), the shutoff head may be as much as three times the normal head.

	60-cycle		50-cycle	
Pairs of Poles	**Synchronous**	**Induction**[a]	**Synchronous**	**Induction**[a]
1	3600 rpm	3500 rpm	3000 rpm	2900 rpm
2	1800	1750	1500	1450
3	1200	1160	1000	960
4	900	870	750	720
5	720	695	600	575
6	600	580	500	480
8	450	435	375	360
10	360	350	300	290
12	300	290	250	240

[a] These values for the operating speed of induction motors are approximate. The speed of induction motors is usually 2–3% lower than that of synchronous motors.

TABLE 11.1   Operating speeds of constant-speed electric motors

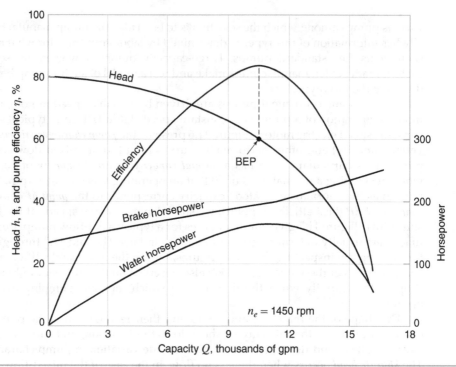

FIGURE 11.6 Characteristic curves for a typical mixed-flow centrifugal pump.

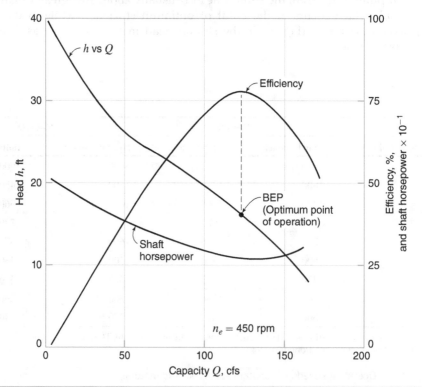

FIGURE 11.7 Characteristic curves for a typical axial-flow (propeller) pump.

11.6 Performance Characteristics at Different Speeds and Sizes

The choice of a pump for a given situation will depend on the rotative speed of the motor used to drive the pump. Knowing the characteristic curve for a pump at a given rotative speed, we can approximately derive the relation between head, h, and capacity, Q, at different rotative speeds through use of Eqs. (11.4) and (11.5). For example, in Fig. 11.8, assume we know the characteristic curve (Curve 1) of a pump when operating at n_1 rpm. We can approximately derive the characteristic curves at different *speeds* of operation such as n_2 and n_3 by transferring points on Curve 1 to corresponding points on Curves 2 and 3, respectively. Thus, from Eqs. (11.4) and (11.5), we have for Curve 2,

Different speeds, n_1 and n_2:

$$Q_2 = Q_1\left(\frac{n_2}{n_1}\right) \text{ and } h_2 = h_1\left(\frac{n_2}{n_1}\right)^2$$

We can use similar expressions to develop Curve 3. Lines of equal efficiency determined by testing are superimposed on Fig. 11.8. From this, we can see that efficiency drops off rather rapidly as one moves away from the BEP. We can also approximately

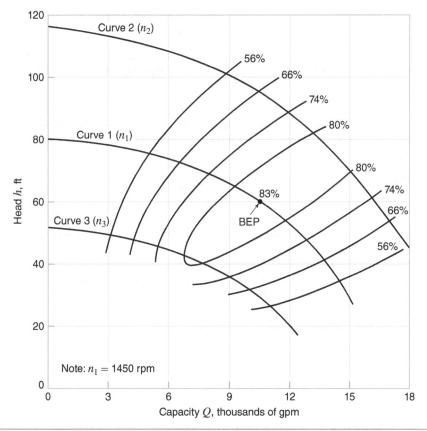

FIGURE 11.8 Characteristic performance curves of a typical mixed-flow centrifugal pump (Fig. 11.6) at various speeds of rotation with contours of equal efficiency.

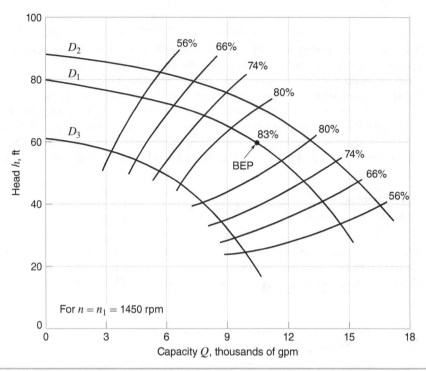

FIGURE 11.9 Characteristic performance curves with contours of equal efficiency for typical homologous mixed-flow centrifugal pumps having impellers of different size. D_1 is the same pump as in Fig. 11.6.

derive characteristic curves for pumps of different *size,* all operating at the same speed, using Eqs. (11.4) and (11.5). Fig. 11.9 shows such curves. The use of

Different sizes, D_1 and D_2:
$$Q_2 = Q_1 \left(\frac{D_2}{D_1}\right)^3 \quad \text{and} \quad h_2 = h_1 \left(\frac{D_2}{D_1}\right)^2$$

enables us to transfer corresponding points. Curves of equal efficiency were also super-imposed on Fig. 11.9.

We can approximately[6] develop characteristic curves for pumps having similar geometric shape, of both different *size* and operating at various constant *speeds,* using Eqs. (11.4) and (11.5), in which case

Both different sizes, D_1 and D_2 and speeds, n_1 and n_2:
$$Q_2 = Q_1 \left(\frac{n_2}{n_1}\right) \left(\frac{D_2}{D_1}\right)^3 \quad \text{and} \quad h_2 = h_1 \left(\frac{n_2}{n_1}\right)^2 \left(\frac{D_2}{D_1}\right)^2$$

---

[6] The accuracy drops off with large variations in D and n.

Sample Problem 11.3
In Fig. 11.8, $n_1 = 1450$ rpm. Estimate the rotative speeds n_2 and n_3.

Solution
For $Q = 0$ on Fig 11.8, we have $h_1 = 80$ ft. $h_2 \approx 117$ ft and $h_3 \approx 52$ ft.

From Eq. (11.5) with constant D: $h \propto n^2$

Thus:

$$\frac{80}{117} = \left(\frac{1450}{n_2}\right)^2, \quad \text{from which} \quad n_2 \approx 1450\sqrt{\frac{117}{80}} = 1753 \approx 1750 \text{ rpm} \qquad \textbf{\textit{ANS}}$$

and

$$\frac{52}{80} = \left(\frac{n_3}{1450}\right)^2, \quad \text{from which} \quad n_3 \approx 1450\sqrt{\frac{52}{80}} = 1169 \approx 1170 \text{ rpm} \qquad \textbf{\textit{ANS}}$$

11.7   Operating Point of a Pump

The manner in which a pump operates depends not only on the pump performance characteristics, but also on the characteristics of the system that it will operate in. In Fig. 11.10, for a particular pump under consideration, we show the pump operating characteristics (h versus Q) for a selected speed of operation, usually close to the speed that gives optimum efficiency. We also show the system characteristic curve (i.e., the required pumping head versus Q). In this case, the pump is delivering liquid through a piping system with a static lift of Δz. The head that the pump must develop is equal to the static lift plus the total head loss in the piping system (approximately proportional to Q^2). The intersection of the two curves determines the actual ***pump-operating head*** and flow rate.

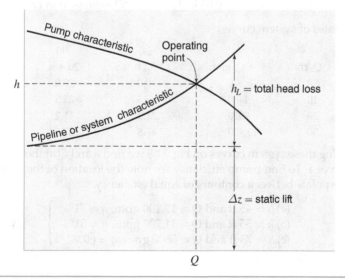

FIGURE 11.10   Graphical method for finding the operating point of a pump and pipeline.

The particular values of h and Q at this intersection may or may not be those for the maximum efficiency. If they are not, this means the pump is not exactly suited to the specific conditions. In Figs. 11.8 and 11.9 we see that efficiency drops off as one moves away from the BEP. Hence it is important to select a pump so that the pump performance and system characteristic curves intersect near the BEP. Changing the speed of the pump may help. In Secs. 11.12 and 11.13 we further discuss the behavior of pumps and their relationship to the systems in which they operate.

Sample Problem 11.4

The pump with the characteristic curve shown in Fig. 11.6 when operating at 1450 rpm pumps water from reservoir A to reservoir B through an 18-in-diameter pipe ($f = 0.032$) 500 ft long. Neglecting minor losses, find the flow rate for the following conditions: (a) reservoir water-surface elevations are identical; (b) water-surface elevation of reservoir B is 20 ft higher than that in reservoir A; (c) water-surface elevation of reservoir B is 65 ft higher than that in reservoir A. Assume f does not change with flow rate. Also, use Fig. 11.8 to find the three efficiencies.

Solution
Figure 11.6 gives the pump characteristic curve. Note that this curve is the same as Curve 1 of Fig. 11.8.

Energy equation: $$0 + h - h_L = \Delta z$$

from which we find the equation of the system characteristic:

$$h = \Delta z + h_L = \Delta z + f\frac{L}{D}\frac{V^2}{2g}$$

$$= \Delta z + (0.032)\frac{500}{1.5}\frac{Q^2}{[\pi(1.5)^2/4]^2 2(32.2)}$$

$$= \Delta z + 0.0530Q^2 \qquad (Q \text{ expressed in cfs})$$

Coordinates of system curves:

Q, cfs	Q, gpm	(a) h_L, ft	(b) $20 + h_L$	(c) $65 + h_L$
0	0	0.0	20.0	65.0
10	4480	5.3	25.3	70.3
20	8970	21.2	41.2	86.2
30	13,450	47.8	67.8	(112.8)[7]

By plotting these system curves on Fig. 11.8 we find h and Q at the points of intersection with Curve 1. To find pump efficiency we note the location of the points of intersection and interpolate between contours of equal efficiency.

(a) $h \approx 45$ ft and $Q \approx 13,200$ gpm, $\eta \approx 71\%$
(b) $h \approx 55$ ft and $Q \approx 11,700$ gpm, $\eta \approx 81\%$ *ANS*
(c) $h \approx 73$ ft and $Q \approx 5800$ gpm, $\eta \approx 60\%$

---

[7] For plotting purposes only. The pump cannot develop that much head when operating at 1450 rpm.

11.8   Specific Speed of Pumps

Specific speed is a number that defines the type of pump (radial-flow, mixed-flow, or axial-flow). Traditionally in the United States, we have expressed the specific speed N_s of a pump as

$$N_s = \left(\frac{n_e\sqrt{Q}}{h^{3/4}}\right)_{BEP} = \left(\frac{n_e\sqrt{gpm}}{h^{3/4}}\right)_{BEP} \qquad (11.8a)$$

where the values used for n_e (rpm), Q (gpm), and h (ft) are those that occur at the *point of optimum operating efficiency,* commonly referred to as the BEP (Sec. 11.5). Computed values of specific speed for a given pump throughout its entire operating range from zero discharge (shutoff head) to maximum discharge would give values from zero to some very large number. The only value that has any real significance is that corresponding to values of head, discharge, and speed at the BEP.

We derive Eq. (11.8a) by eliminating D from Eqs. (11.4) and (11.5) in such a way that n becomes a term in the numerator of the resulting expression. We leave it to the reader to work out this derivation. For large-capacity pumps, specific speed is sometimes calculated using Q expressed in cfs rather than gpm, in which case $(N_s)_{cfs} = 0.0472(N_s)_{gpm}$.

In the SI, specific speed of pumps is defined by

$$(N_s)_{SI} = \left(\frac{\omega_e\sqrt{Q}}{(gh)^{3/4}}\right)_{BEP} \qquad (11.8b)$$

where the values of ω_e (rad/s), Q (m³/s), and h (m) are those that occur at the BEP. The acceleration due to gravity g is 9.81 m/s². In this form, specific speed is dimensionless. However, rotative speed is expressed in radians per second, which is rather cumbersome. The relation between SI specific speed for pumps and the traditional mode of expressing specific speed is $(N_s)_{SI} = 0.000368(N_s)_{gpm}$.

Figure 11.11 shows several typical impellers in section and their corresponding specific speeds. Radial-flow impellers generally have specific speeds [Eq. (11.8a)] between 500 and 5000, mixed-flow between 4000 and 10,000 and axial-flow from 9000 to 15,000. In the SI the corresponding ranges of specific speed are approximately 0.2–2.0 for radial-flow, 1.5–3.7 for mixed-flow, and 3.3–5.5 for axial-flow pumps. Two impellers having the same geometric shape have the same specific speed though their sizes may differ. The curves of Fig. 11.11 show the variation of peak efficiency with specific speed. Note that pumps with specific speed below 800 tend to be inefficient.

Equations (11.8a) and (11.8b) indicate that pumping against high heads requires a low-specific speed pump. We will discuss this further in Sec. 11.10. For very high heads and low discharges, the required specific speed may fall below the values for normal design and result in a pump with a low efficiency. To overcome this problem, we can distribute the head among a number of pumps in series, or we can use a multistage unit. The head per stage is generally limited to about 400 ft, although some pumps in use develop more than 600 ft of head per stage. A multistage pump is usually less expensive than a series of individual pumps, but the very high pressures developed in the multistage pump can offset this. We can avoid excessive pressures in the system by spacing several pumps more or less uniformly along a pipeline.

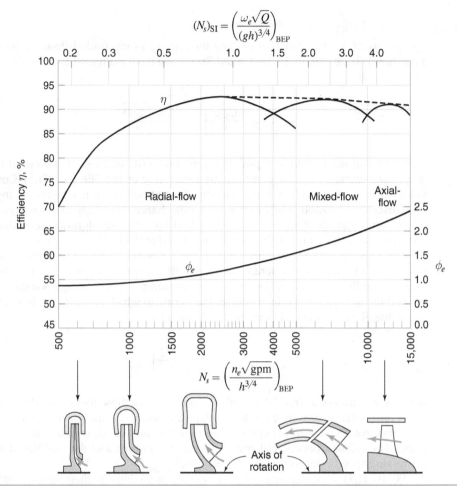

$$(N_s)_{\text{SI}} = \left(\frac{\omega_e \sqrt{Q}}{(gh)^{3/4}}\right)_{\text{BEP}}$$

$$N_s = \left(\frac{n_e \sqrt{\text{gpm}}}{h^{3/4}}\right)_{\text{BEP}}$$

FIGURE 11.11   Optimum efficiency and typical values of ϕ_e (Sec. 11.9) for water pumps as a function of specific speed.

Sample Problem 11.5

What is the specific speed of the pump whose performance characteristics are given in Fig. 11.6? What type of pump is this?

Solution

At its BEP this pump has a capacity of 10,500 gpm while developing a head of 60 ft at a rotative speed of 1450 rpm.

Hence [Eq. (11.8a)]: $N_s = \left(\dfrac{n_e \sqrt{\text{gpm}}}{h^{3/4}}\right)_{\text{BEP}} = \dfrac{1450\sqrt{10,500}}{60^{3/4}} = 6890$ *ANS*

From Fig. 11.11:   This is a mixed-flow centrifugal pump.   *ANS*

11.9 Peripheral-Velocity Factor

For a pump impeller (Fig. 11.1) or a turbine runner, the ratio of the peripheral velocity to $\sqrt{2gh}$ is referred to as the *peripheral-velocity factor*, denoted by ϕ (phi). Thus, for a pump,

$$u_2 = \phi\sqrt{2gh} \tag{11.9}$$

where u_2 is the peripheral speed of the impeller. For an axial-flow pump it is the vane-tip speed that is used in Eq. (11.9).

For any machine its peripheral velocity might be any value from zero up to some maximum under a given head, depending on the operating speed, and ϕ would consequently vary through a wide range. But the speed that is of most practical significance is that at which the efficiency is a maximum. We shall denote the value of ϕ at the speed of maximum efficiency by ϕ_e.

At maximum efficiency,

$$u_2 = \frac{2\pi r n_e}{60} = \frac{\pi D n_e}{60} = \phi_e\sqrt{2gh}$$

Thus,

$$D = \frac{60\sqrt{2g}\,\phi_e\sqrt{h}}{\pi n_e}$$

which in BG units reduces to

$$D = \frac{153.3\phi_e\sqrt{h}}{n_e} \tag{11.10}$$

Figure 11.11 presents typical values of ϕ_e for various types of pumps as a function of specific speed. We can use this curve, together with Eq. (11.10), to estimate the diameter of the impeller of a pump if we know the pump's specific speed.

Because many exercises and problems on this and subsequent sections depend on values read from graphs, numerical answers will probably vary somewhat.

Sample Problem 11.6

Find the value of ϕ_e for the Eagle Mountain pumping plant, part of the Colorado River aqueduct system, where each pump delivers 200 cfs at a head of 440 ft (134 m) with pumps equipped with 81.6-inch (2.07 m) diameter impellers. How does this computed value of ϕ_e compare with the value given in Fig. 11.11?

Solution

Eq. (11.10), for BG units:
$$D = 153.3\phi_e\sqrt{h}/n_e$$

$$(81.6/12) = 153.3\phi_e\sqrt{440}/450$$

From which:
$$\phi_e = 0.95 \quad ANS$$

Eq. (11.8a): Specific speed $N_s = \dfrac{450\sqrt{200(448.8)}}{440^{3/4}} = 1404$

From Fig. 11.11, $\phi_e \approx 1.0$, which is a close check. *ANS*

Sample Problem 11.7

Estimate the diameter of the impeller of the pump whose operating characteristic is shown in Fig. 11.6. See Sample Problem 11.5.

Solution

From Sample Problem 11.5:   Specific speed $N_s = 6890$

From Fig. 11.11 for $N_s = 6890$: $\phi_e = 1.7$

Eq. (11.10), for BG units: $D = 153.3\phi_e\sqrt{h}/n$

From Fig. 11.6 and Sample Problem 11.5: $n = 1450$ rpm, $h = 60$ ft.

So: $D = 153.3(1.7)\sqrt{60}/1450 = 1.392$ ft 16.70 in   *ANS*

11.10   Cavitation in Pumps

An important factor in the satisfactory operation of a pump is the avoidance of cavitation (Sec. 5.10), both for the sake of good efficiency and for the prevention of impeller damage. As liquid passes through the impeller of a pump, there is a change in pressure. If the absolute pressure of the liquid drops to the vapor pressure, cavitation will occur. The region of vaporization hinders the flow and places a limit on the capacity of the pump. As the fluid moves further into a region of higher pressure, the bubbles collapse and the implosion of the bubbles may cause pitting of the impeller. Cavitation is most likely to occur near the point of discharge (periphery) of radial-flow and mixed-flow impellers, where the velocities are highest. It may also occur on the suction side of the impeller, where the pressures are the lowest. In the case of an axial-flow pump, the blade-tip is the most vulnerable to cavitation.

For pumps, a *cavitation parameter* σ (sigma) is defined as

$$\sigma = \frac{(p_s)_{abs}/\gamma + V_s^2/2g - p_v/\gamma}{h} = \frac{\text{NPSH}}{h} \tag{11.11}$$

where subscript s refers to values at the pump intake (i.e., suction side of the pump), h is the head developed by the pump, and p_v is the vapor pressure. As the latter is normally given in absolute units, it follows that p_s must also be absolute pressure. NPSH, the numerator of Eq. (11.11), we refer to as the *net positive suction head* or sometimes as *available NPSH*.

There is a critical value of σ, called the *critical cavitation parameter* and denoted by σ_c, below which there is a drop in pump efficiency indicative of the onset of cavitation. The value of σ_c depends on the type of pump and the conditions of operation, including speed and flow rate. Pump manufacturers typically provide with the pump characteristic curve the *required NPSH* to avoid cavitation, or NPSH_R, where

$$\text{NPSH}_R = \sigma_c h$$

This is determined by the manufacturer based on factory testing of pumps under different operating conditions. Thus, to avoid cavitation, hydraulic analysis should ensure that $\sigma > \sigma_c$ or equivalently $\text{NPSH} > \text{NPSH}_R$.

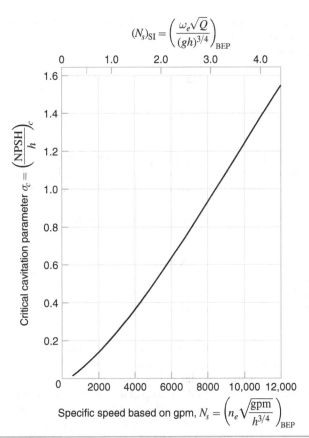

$$(N_s)_{\text{SI}} = \left(\frac{\omega_e \sqrt{Q}}{(gh)^{3/4}} \right)_{\text{BEP}}$$

FIGURE 11.12 Approximate values of critical cavitation parameter σ_c as a function of specific speed.

Figure 11.12 presents approximate values of σ_c for centrifugal pumps operating under normal conditions near optimum efficiency. In important installations the value of σ_c is usually determined experimentally in a model study. For modeling purposes, we define a useful parameter, the *suction specific speed S,*

$$S = \frac{n\sqrt{\text{gpm}}}{\text{NPSH}^{3/4}} \qquad (11.12)$$

If we operate a model and prototype at identical values of N_s and S, we achieve similarity of flow and cavitation provided the model and prototype are geometrically similar to one another.

We perform a hydraulic design to prevent cavitation (so $\sigma > \sigma_c$; NPSH > NPSH$_R$) by selecting the proper type and size of pump and speed of operation and by setting the pump at the proper point and elevation in the system. By inspecting Eq. (11.11), we can see that small values of σ result from large values of h, the head developed by the pump. Hence, for any given situation, there is a limiting value of h above which σ will be less than σ_c, resulting in cavitation.

By writing an energy equation for the situation depicted in Fig. 11.5 in terms of absolute pressure from the surface of the source reservoir to the suction side of the pump, we obtain

$$\frac{(p_0)_{abs}}{\gamma} - h_L = z_s + \frac{(p_s)_{abs}}{\gamma} + \frac{V_s^2}{2g}$$

or

$$\frac{(p_s)_{abs}}{\gamma} + \frac{V_s^2}{2g} = \frac{(p_0)_{abs}}{\gamma} - h_L - z_s$$

where $(p_0)_{abs}/\gamma$ is the absolute pressure on the surface of the source reservoir. If we draw the liquid from a closed reservoir, $(p_0)_{abs}/\gamma$ could be either greater or less than the atmospheric pressure. In the usual case the reservoir is open to the atmosphere, and $(p_0)_{abs}/\gamma = p_{atm}/\gamma$.

Substituting the right side of the preceding expression into Eq. (11.11), we get

For pumping from a reservoir (Fig. 11.5):

$$\sigma = \frac{NPSH}{h} = \frac{\dfrac{(p_0)_{abs}}{\gamma} - h_L - z_s - \dfrac{p_v}{\gamma}}{h}$$

$$\text{or } NPSH = \frac{(p_0)_{abs}}{\gamma} - h_L - z_s - \frac{p_v}{\gamma}$$

(11.13)

This expression indicates that σ will tend to be small (hence there will be a tendency toward cavitation) in the following situations: (a) high head; (b) low atmospheric pressure, i.e., high elevation; (c) large head loss between source reservoir and the pump; (d) large value of z_s, i.e., with the pump at a relatively high elevation compared with the elevation of the reservoir water surface; and (e) large value of vapor pressure, i.e., high temperature and/or a very volatile liquid such as gasoline being pumped. Conditions (b) through (e) also indicate low NPSH and increased cavitation risk. With (a), we can limit the required head by using multistage pumps. For a given liquid to be pumped at a certain elevation and temperature, we have no control over items (b) and (e). However, we can do something about (c) and (d). When we design the layout of a pumping system, we can reduce the tendency toward cavitation by minimizing h_L [item (c)] by placing the pump close to the source reservoir, and by setting the pump at a low elevation [item (e)] relative to the reservoir water surface.

In Eq. (11.13), when σ equals the critical value σ_c then $z_s = (z_s)_{max}$, the highest elevation at which we can safely set a pump and guard against cavitation. Rearranging, we get

For pumping from a reservoir (Fig. 11.5):

$$(z_s)_{max} = \frac{(p_0)_{abs}}{\gamma} - \frac{p_v}{\gamma} - \sigma_c h - h_L$$

$$\text{or } (z_s)_{max} = \frac{(p_0)_{abs}}{\gamma} - \frac{p_v}{\gamma} - NPSH_R - h_L$$

(11.14)

As long as z_s is less than $(z_s)_{max}$, there should be no problem with cavitation, assuming, of course, that we have an accurate value of σ_c or $NPSH_R$ for the given conditions of operation. Pump manufacturers often run tests on pumps to determine values of σ_c for different conditions of operation.

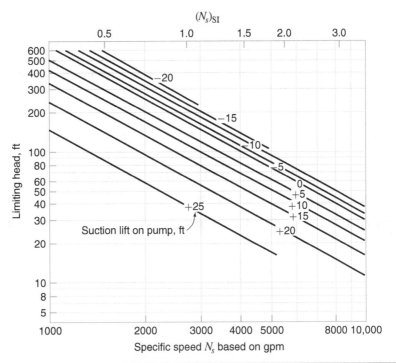

FIGURE 11.13 Recommended limiting heads for single-stage, single-suction pumps as a function of specific speed and suction lift, at sea level with water temperature of 80°F.

Based on experience with pumps, Fig. 11.13 depicts *limiting heads* for the prevention of cavitation in single-stage, single-suction pumps as a function of specific speed and *suction lift*, the elevation difference between the energy line at suction and the center of the impeller as indicated in Fig. 11.5. A positive suction lift indicates that the impeller is above the energy line. The curves of Fig. 11.13 are applicable to water at 80°F under sea-level atmospheric pressure. We can readily shift these curves for other conditions: different temperatures, different elevations, and different liquids.

Sample Problem 11.8

A pump with a value of $\sigma_c = 0.10$ is to pump against a head of 500 ft [NPSH$_R = \sigma_c h = 0.10(500) = 50$ ft]. The barometric pressure is 14.3 psia, and the vapor pressure is 0.5 psia. Assume friction losses in the intake piping are 5 ft. Find the maximum allowable elevation of the pump relative to the water surface at intake.

Solution

$$p_{atm}/\gamma = (14.3)144/62.4 = 33.0 \text{ ft}$$

$$\frac{p_v}{\gamma} = \frac{(0.5)144}{62.4} = 1.154 \text{ ft}$$

Eq. (11.14): $(z_s)_{max} = 33.0 - 1.154 - 50 - 5.0 = -23.2 \text{ ft}$

The pump should be placed at least 23.2 ft below the reservoir water surface.   ***ANS***

Sample Problem 11.9

A pump is delivering 7500 gpm at 140°F against a head of 240 ft when the barometric pressure is 13.8 psia. Determine the reading on a pressure gage in inches of mercury vacuum at the suction flange when cavitation is incipient if $\sigma_c = 0.085$. The suction pipe has a diameter of 2.0 ft.

Solution

Table A.1 at 140°F: $\gamma = 61.38$ pcf, and $\dfrac{p_v}{\gamma} = 6.67$ ft

$$V_s = Q/A_s = (7500/449)/(\pi 2^2/4) = 5.32 \text{ ft/sec}$$

Let p = gage pressure at suction flange: $(p_s)_{abs} = p_{atm} + p = (13.8 + p)$ psia

Eq. (11.11): $\sigma_c = 0.085 = [(13.8 + p)(144/61.38) + 5.32^2/2(32.2) - 6.67]/240$

From which $p = -2.45$ psi

$$-2.45(29.9 \text{ inHg}/14.7 \text{ psia}) = 4.98 \text{ inHg vacuum}   \textit{ANS}$$

11.11   Viscosity Effect

Centrifugal pumps are also used to pump liquids with viscosities different from that of water. Figure 11.14 shows actual test curves of performance for a very extreme range of viscosities, from water to an oil with a kinematic viscosity 3200 times that of water. We see

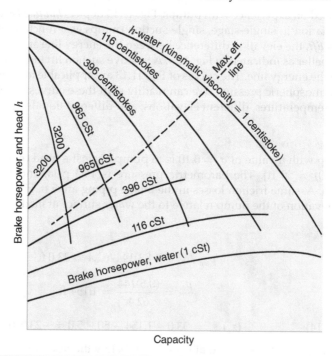

FIGURE 11.14   Centrifugal pump with viscous oils. (1 centistoke = 10^{-6} m²/s.)

that, as the viscosity increases, the head-capacity curve becomes steeper and the power required increases. The dashed line indicates the points of maximum efficiency for each curve. We see that both the head and capacity at the point of maximum efficiency decrease with increasing viscosity. As these are accompanied by an increase in brake horsepower, there is a marked decrease in efficiency. For example, based on the data underlying Fig. 11.14, if the optimum efficiency of the pump when pumping water is 0.85, its optimum efficiency is only 0.47 when pumping a liquid whose viscosity is 116 times that of water and only 0.18 when pumping a liquid whose viscosity is 396 times that of water.

11.12 Selection of Pumps

In selecting a pump for a given situation, we have a variety of pumps to choose among. We strive to select a pump that will operate at a relatively high efficiency for the given conditions of operation. Manufacturers provide pump performance information such as that in Figs. 11.6 and 11.7. The engineer's task is to select the pump or pumps that best fit in with the system characteristics. The scale of problems can vary widely. One of the world's largest pumping installations is the Edmonston Pumping Plant of the State of California water project. This plant lifts water from the California Aqueduct over the Tehachapi Mountains. At this plant there are 14 four-stage vertical-shaft centrifugal pumps, each capable of delivering 315 cfs (8.92 m³/s) against a head of 1970 ft (600 m) when rotating at 600 rpm. Their maximum efficiency is 92%. The maximum energy requirements for this plant are approximately 6×10^9 kWh per year. At the opposite end of the scale, axial pumps with impeller diameters as small as 6 mm have been developed for installing in human arteries to help the heart pump blood. They deliver up to 4 L/min against a pressure head of 80 mmHg.

Alternatives to be investigated for pump selection include specific speed N_s, size D of the impeller, and speed n of operation. Other alternatives include the use of multistage pumps, pumps in series, pumps in parallel, etc. Even, under certain conditions, throttling the flow in the system can result in energy savings.

In all cases we must carefully investigate the possibility of cavitation. We should avoid cavitation at all costs. We can eliminate problems with cavitation by limiting the head that the pump must develop, by selecting the proper type of pump, and by setting the pump at a low enough elevation.

The objective is to select a pump and its speed of operation such that the pump's performance characteristics relate to the system in which it operates in such a way that the operating point (Sec. 11.7) is close to the BEP (best operating point). This tends to optimize the pump efficiency, resulting in a minimization of energy expenditure.

We can shift the operating point by changing the pump characteristic curve, by changing the system characteristic curve, or by changing both curves. We can change the pump curve by changing the speed of operation of a given pump or by selecting a different pump with different performance characteristics (Fig. 11.15a). In some instances it is helpful to trim the impeller, i.e., reduce its diameter somewhat, perhaps 5% or so, by grinding it down (Sec. 11.4). We install the smaller impeller in the original casing. We can change the system characteristic curve by changing the pipe size or by throttling the flow (Fig. 11.15b).

A complication that often occurs is that the energy levels at the two ends of the system do not remain constant, such as that created by fluctuating reservoir levels. In such a case it is difficult to achieve a high efficiency for all modes of operation. In extreme cases a variable-speed motor is sometimes used.

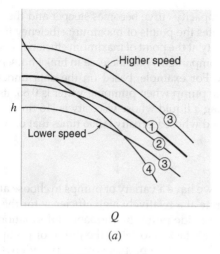

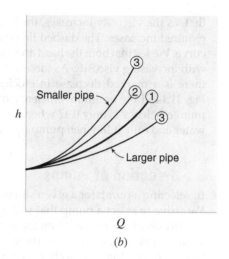

(a)   (b)

(1) Given pump
(2) Trim the impeller
(3) Change the speed
(4) Use a different pump

(1) Given system
(2) Use a throttling valve
(3) Use a different size pipe

FIGURE 11.15   Changing the operating point by (a) changing the pump characteristic curve or (b) changing the system characteristic curve.

Sample Problem 11.10

Water at 50°F will be pumped between reservoirs A and B whose water surfaces are at elevations 5910 ft and 6060 ft, respectively. The water will be delivered at 20 cfs through a 24-in-diameter pipe of length 2000 ft. Assume $f = 0.03$. (a) Find the specific speed of the most suitable pump if the speed of operation will be 600 rpm. (b) If this pump is installed at a point 300 ft from reservoir A at the same elevation as the water surface in reservoir A, will it be safe against cavitation?

Solution

(a) $$V = Q/A = 20/\pi(1)^2 = 6.37 \text{ ft/sec}$$

Find the total head delivered by the pump [Eq. (5.37)]:

$$h = \Delta z + h_L = (6060 - 5910) + 0.03\frac{2000}{2}\frac{(6.37)^2}{2(32.2)} = 168.9 \text{ ft}$$

Eq. (11.8a): $$N_s = \left(\frac{n_e\sqrt{\text{gpm}}}{h^{3/4}}\right)_{\text{BEP}} = \frac{600\sqrt{20 \times 449}}{(168.9)^{3/4}} = 1214 \quad \textbf{ANS}$$

(b)   From Table A.3 at mean elevation 5985 ft, by interpolation:

$$\frac{(p_0)_{\text{abs}}}{\gamma} = \frac{11.81(144)}{62.4} = 27.25 \text{ ft}$$

Table A.1 at 50°F: $$\frac{p_v}{\gamma} = 0.41 \text{ ft}$$

Fig. 11.12: $\sigma_c \approx 0.08$, $\sigma_c h = \text{NPSH}_R \approx 0.08\,(168.9) \approx 13.5\,\text{ft}$

Eq. (7.13): $h_L \approx h_f = f\dfrac{L}{D}\dfrac{V^2}{2g} = 0.03\dfrac{300}{2}\dfrac{(6.37)^2}{2(32.2)} = 2.83\,\text{ft}$

Eq. (11.14): $(z_s)_{max} = 27.25 - 0.41 - 13.5 - 2.83 = 10.5\,\text{ft}$

Yes, the pump will be safe from cavitation. It can be placed as much as 10.5 ft above the water-surface elevation in reservoir A.   ***ANS***

Sample Problem 11.11

Repeat Sample Problem 11.10 for a pump, where the rotative speed is 1200 rpm rather than 600 rpm, with all other data remaining the same.

Solution

Eq. (11.8a): $N_s \propto n_e$; hence $N_s = (1200/600)1214 = 2428$

From Fig. 11.12, when $N_s = 2428$: $\sigma_c \approx 0.18$

Thus: $\sigma_c h = \text{NPSH}_R \approx 0.18\,(168.9) \approx 30.4\,\text{ft}$

And $(z_s)_{max} = 27.3 - 0.4 - 30.4 - 2.8 \approx -6.4\,\text{ft}$

To prevent cavitation, the pump must be placed 6.4 ft or more below the water-surface elevation in reservoir A.   ***ANS***

Notes: By comparing the results of Sample Problems 11.10 and 11.11, we find that rotative speed can greatly influence the specific speed of the pump that should be selected at a given installation. We also observe that an increase in operating speed results in a pump with a higher specific speed that is more vulnerable to cavitation.

Sample Problem 11.12

Find the approximate diameters of the pumps of Sample Problems 11.10 and 11.11.

Solution

Fig. 11.11:   For $N_s = 1214$, $\phi_e \approx 0.9$, and for $N_s = 2428$, $\phi_e \approx 1.2$

Eq. (11.10), for BG units: $D = \dfrac{153.3\phi_e\sqrt{h}}{n_e}$

Sample Problem 11.10 with $N_s = 1214$: $D = \dfrac{153.3(0.9)\sqrt{168.9}}{600} = 2.99\,\text{ft} = 35.9\,\text{in}$ ***ANS***

Sample Problem 11.11 with $N_s = 2428$: $D = \dfrac{153.3(1.2)\sqrt{168.9}}{1200} = 1.99\,\text{ft} = 23.9\,\text{in}$ ***ANS***

Note: This and Sample Problems 11.10 and 11.11 show the faster speed permits us to use a smaller pump, but it is more vulnerable to cavitation (i.e., it must be set at a lower elevation).

11.13   Pumps Operating in Series and in Parallel

In Fig. 11.16 we summarize our discussion thus far by showing: (a) the general shape of the characteristic curves for the three different types of pumps, (b) the effect on performance of changing the rotative speed of a given pump, and (c) the effect on performance of different sizes of pump impellers. Figures 11.16d and 11.16e show the performance curves for two identical pumps in series and for two identical pumps in parallel. By examining Figs. 11.16d and 11.16e, we see that pumps in series tend to increase head, while pumps in parallel tend to increase flow rate.

When we install pumps in series or parallel, it is very important that they have reasonably similar, or better yet identical, head-capacity characteristics throughout their range of operation; otherwise, one pump will carry most of the load and, under certain conditions, all of the load, with the other pump acting as a hindrance rather than a help. In fact, in parallel, if the performance characteristics of the pumps are quite different, a condition of backflow can occur in one of the pumps. Finally, one must always be sure that the selected pump (or pumps) will not encounter cavitation problems over the full range of operating conditions.

We can best determine the mode of operation for any pumping system by plotting the pumping characteristics and the pipe system characteristics on the same diagram (Fig. 11.10). The point at which the two curves intersect indicates what will take place. The following sample problem demonstrates several aspects of the relationship between the pump and the system characteristics.

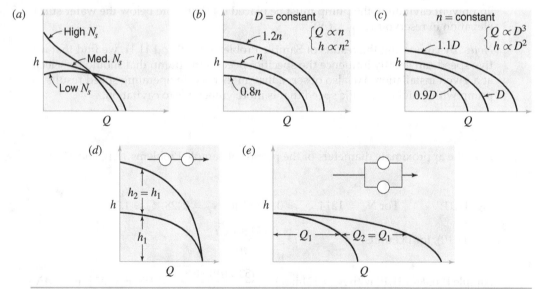

FIGURE 11.16   Pumping alternatives. (a) Different pumps with different characteristics. (b) A particular pump at different speeds. (c) Homologous pumps of different size. (d) Two identical pumps in series. (e) Two identical pumps in parallel. (Note: In series or parallel the pumps need not be identical, but their performance characteristics should be close to one another.)

Sample Problem 11.13

Two reservoirs A and B are connected with a long pipe that has characteristics such that the head loss through the pipe is expressed as $h_L = 20Q^2$, where h_L is in feet and Q is the flow rate in 100s of gpm. The water-surface elevation in reservoir B is 35 ft above that in reservoir A. Two identical pumps are available for use to pump the water from A to B. The following table gives the characteristic curve of each pump when operating at 1800 rpm.

Operation at 1800 rpm	
Head, ft	Flow rate, gpm
100	0
90	110
80	180
60	250
40	300
20	340

At the optimum point of operation, the pump delivers 200 gpm at a head of 75 ft, indicated by the point in Fig. S11.13. Determine the specific speed N_s of the pump and find the rate of flow under the following conditions: (a) A single pump operating at 1800 rpm; (b) two pumps in series, each operating at 1800 rpm; (c) two pumps in parallel, each operating at 1800 rpm.

Solution

Eq. (11.8a):
$$N_s = \left(\frac{n_e \sqrt{\text{gpm}}}{h^{3/4}} \right)_{\text{BEP}} = \frac{1800\,(200)^{1/2}}{75^{3/4}} = 999 \quad \textit{ANS}$$

We plot the head-capacity curves for the pumping alternatives (Fig. S11.13), and the h versus Q curve for the pipe system. In this case $h = \Delta z + h_L = 35 + 20Q^2$.

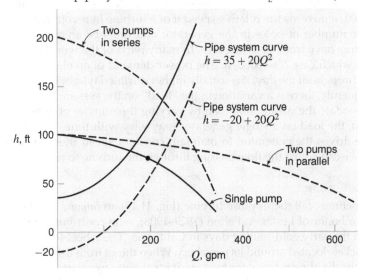

The answers are found at the points of intersection of the curves. They are as follows: (a) single pump, 156 gpm; (b) two pumps in series, 224 gpm; (c) two pumps in parallel, 170 gpm.   *ANS*

If Δz were greater than 100 ft, neither the single pump nor the two pumps in parallel would deliver any water. If Δz were −20 ft (i.e., with the water-surface elevation in reservoir B 20 ft below that in A), the flows would then be: (a) 212 gpm; (b) 258 gpm; (c) 232 gpm.

11.14   Turbines

We use water-driven turbines primarily for the development of hydroelectric energy. Turbines extract energy from flowing water and convert it to mechanical energy to drive electric generators. Called *hydroelectric power*, it is an important source of energy. The United States derives about 6% of its electric energy from hydropower, though in Washington state hydropower constitutes over 70% of production. In Norway nearly 90% of the electric energy is from hydroelectric plants.

There are two basic types of hydraulic turbines. In the *impulse turbine* a free jet of water impinges at atmospheric pressure on the revolving element of the machine. In a *reaction turbine*, flow takes place under pressure in a closed chamber. Although the energy delivered to an impulse turbine is all kinetic, while the reaction turbine utilizes pressure energy as well as kinetic energy, the action of both turbines depends on a change in the momentum of the water so that a dynamic force is exerted on the rotating element, or *runner*. The runner of a reaction turbine is similar in design, but not identical, to a pump impeller. By rotating either rotor (i.e., centrifugal pump or reaction turbine) in the reverse direction, the machine assumes the function of its counterpart. The efficiency, however, will be small in the reverse mode because the geometric shape of the rotating element and its surrounding casing will not be optimum.

We operate turbines at constant speed. In the United States, 60-cycle (cycles/sec or Hz) electric current is most common, and under such conditions the expression $N = 7200/n$ governs the rotative speed n of a turbine in revolutions per minute, where N is the number of poles in the generator and must be an even integer. Most 60-Hz generators have from 12 to 96 poles. In many parts of the world they use 50-cycle current, in which case $N = 6000/n$. The power demand of an electric distribution system varies throughout the day; it is usually higher during daylight hours and lower at night. Consequently, there is a variation in the "load" on the system. In a large system we can accommodate the variation in load by varying the number of generators in operation. Even so, the load on a single generator may vary with time. Therefore, if a hydraulic turbine drives the generator, to maintain constant speed there must be some way in which we can adjust the flow passing through the turbine to regulate its power output.

Impulse Turbines

We sometimes call the impulse turbine (Fig. 11.17) an *impulse wheel* or *Pelton wheel*, so-called in honor of Lester A. Pelton (1829–1908), who contributed much to its development in the early gold-mining days in California. The wheel (or runner) has a series of split buckets located around its periphery. When the jet from the nozzle (at B in Fig. 11.17) strikes the dividing ridge of the buckets, it is split into two parts that discharge from both sides of the bucket. Each split bucket has a notch that enables the bucket to attain a position nearly tangent to the direction of the jet before the bucket lip intercepts the jet.

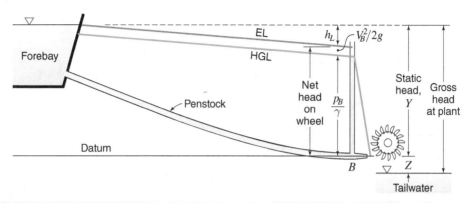

FIGURE 11.17   Definition sketch for an impulse turbine.

The **net head**, or **effective head** h on an impulse turbine (including nozzle), i.e., the head at the base B of the nozzle in Fig. 11.17, is the **static head**, Y, minus the pipe friction losses. Thus, for net head, we have

Impulse turbine:
$$h = \frac{p_B}{\gamma} + \frac{V_B^2}{2g} = Y - h_L \qquad (11.15)$$

There are four ways the impulse turbine uses the energy, or head, available at the base of the nozzle. Fluid friction in the nozzle consumes some energy, known as the nozzle loss [Sec. 7.24 and Eq. (9.14)]; fluid friction over the buckets uses another portion; the water discharged from the buckets carries away kinetic energy ($V_2^2/2g$); and the buckets receive the rest.

Reaction Turbines

A reaction turbine is one in which flow takes place in a closed chamber under pressure. The flow through a reaction turbine may be radially inward, axial, or mixed (partially radial and partially axial). There are two types of reaction turbines in general use, the *Francis turbine* and the *axial-flow* (or *propeller*) *turbine*. The Francis turbine is named after James B. Francis (1815–1892), an eminent American hydraulic engineer who designed, built, and tested the first efficient inward-flow turbine in 1849.

To operate properly, reaction turbines must have a submerged discharge. The water, after passing through the runner, enters the *draft tube* (Fig. 11.18), which directs the water to the point of discharge (point 2 in Fig. 11.18). The draft tube is an integral part of a reaction turbine, and the turbine manufacturer usually specifies its design criteria. A function of the draft tube is to reduce the head loss at submerged discharge and thereby increase the net head available to the turbine runner. A gradually diverging tube whose cross-sectional area at discharge is considerably larger than that at the tube entrance accomplishes this.

For a reaction turbine the **net head** h is the difference between the energy level just upstream of the turbine and that of the tailrace. Thus in Fig. 11.18 the net head on the turbine $h = H_B - H_C$, and $H_B = H_A - h_L$, or

Reaction turbine:
$$h = \left(z_B + \frac{p_B}{\gamma} + \frac{V_B^2}{2g} \right) - \frac{V_C^2}{2g} = z_A - h_L - \frac{V_C^2}{2g} \qquad (11.16)$$

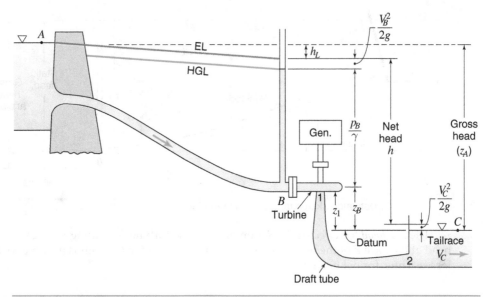

FIGURE 11.18   Definition sketch for a reaction turbine. h is the net head; z_B is the draft head.

where z_B is defined as the **draft head**, z_A is the **gross head** measured as the elevation difference between the reservoir at A and that of the tailrace at C, and V_C is the velocity in the tailrace. In most instances we may neglect $V_C^2/2g$ because it is very small. By comparing Fig. 11.18 with Fig. 11.17, we can see that, for the same setting, the net head on a reaction turbine will be greater than that on a Pelton wheel. The difference is of small importance in a high-head plant, but it is important for a low-head plant. Because we consider the draft tube an integral part of a reaction turbine, the **effective head**, h', available to act on the runner of a reaction turbine must subtract the head losses in the draft tube and discharge, $h_{L(draft)}$.

Reaction turbine: $h' = h - h_{L(draft)}$ (11.17)

Turbine Design Considerations

We calculate the efficiency of impulse turbines and reaction turbines in the same way. In general, the efficiency η (eta) of turbines is defined by:

$$\eta = \frac{\text{power delivered to the shaft (brake power)}}{\text{power taken from the water}} = \frac{T\omega}{\gamma Qh}$$ (11.18)

where T is the torque delivered to the shaft by the turbine, ω is the rotative speed in radians per second, Q is the flow rate, and h is the net head on an impulse turbine [Eq. (11.15)] or the effective head for reaction turbines [Eq. (11.17)].

As with pumps, design considerations for turbines include assessing specific speed to assist in turbine selection, and designing to avoid cavitation. Hydropower stations have used impulse wheels for heads as low as 50 ft if the capacity is small, but they are more commonly employed for heads greater than 500 or 1000 ft. The limiting head for Francis turbines is about 1500 ft because of possible cavitation and the difficulty of building casings to withstand such high pressures. Generally good practice requires us to have at least two turbines at an installation so that the plant can continue operation while we shut down one of the turbines for repairs or inspection. Topography largely governs the head h, and the hydrology of the watershed and characteristics of the reservoir primarily determine the flow Q.

The water stored in a reservoir can have an immense potential energy. To take advantage of this, the ***pump-turbine hydraulic machine*** was developed, effectively allowing a reservoir to act as a battery, storing and releasing energy as needed. A pump-turbine is very similar in design and construction to the Francis turbine. When water enters the rotor at the periphery and flows inward the machine acts as a turbine. With water entering at the center (or *eye*) and flowing outward, the machine acts as a pump. The direction of rotation is, of course, opposite in the two cases. The pump turbine connects to a motor generator, which acts as either a motor or generator depending on the direction of rotation. Used at pumped-storage hydroelectric plants, pump turbines pump water from a lower reservoir to an upper reservoir during off-peak load periods to make water available to drive the machine as a turbine during the time that peak power generation is needed.

Sample Problem 11.14

Refer to Fig. 11.18. A reaction turbine is supplied by a reservoir at an elevation 150 ft above the tailrace, with 130 cfs delivered via a 600-ft-long 4-ft-diameter pipe ($f = 0.025$). A draft tube leading from the discharge side of the turbine to the submerged discharge into the tailrace consists of a 40-ft-long pipe ($f = 0.03$) of constant diameter 3 ft. Determine the power delivered by the turbine if it operates at 88% efficiency. Neglect minor losses and assume the velocity V_C in the tailrace is negligible.

Solution

Head loss in intake pipe,

Eq. (7.13): $h_f = f \dfrac{L}{D} \dfrac{V^2}{2g}$,   where $V = \dfrac{Q}{A} = \dfrac{130}{\pi (2.0)^2} = 10.35 \text{ ft/sec}$

Thus: $h_f = 0.025 \dfrac{600}{4} \dfrac{(10.35)^2}{2(32.2)} = 6.24 \text{ ft}$

Head loss due to draft tube,

Eq. (7.13): $h_f = f \dfrac{L}{D} \dfrac{V^2}{2g}$,   where $V = \dfrac{Q}{A} = \dfrac{130}{\pi (1.5)^2} = 18.4 \text{ ft/sec}$

Thus: $h_f = 0.03 \dfrac{40}{3} \dfrac{(18.4)^2}{2(32.2)} = 2.10 \text{ ft}$

Head loss at discharge,

Eq. (7.74): $h_d^{'} = \dfrac{V_2^2}{2g} - \dfrac{V_C^2}{2g} = \dfrac{18.4^2}{2(32.2)} - 0 = 5.26 \text{ ft}$

Total draft tube head losses,

$$h_{L(\text{draft})} = h_f + h_d' = 2.10 \text{ ft} + 5.26 \text{ ft} = 7.36 \text{ ft}$$

Combining Eqs. (11.16) and (11.17) for effective head on the reaction turbine,

$$h' = z_A - h_L - h_{L(\text{draft})} - \frac{V_C^2}{2g} = 150 - 6.24 - 7.36 - 0 = 136.4 \text{ ft}$$

Power generation is then,

Eq. (5.40): $\text{Water power} = \dfrac{\gamma Q h}{550} = \dfrac{62.4(130)(136.4)}{550} = 2012 \text{ hp}$

Eq. (11.18): $\text{Shaft power} = \eta(\text{Water power}) = 0.88 \times 2012 \text{ hp} = 1770 \text{ hp}$ *ANS*

It is typical to express power generation in SI units. Using values from Appendix D,

$$P = (1770 \text{ hp})(0.7457 \text{ kW}/\text{hp}) = 1320 \text{ kW} = 1.320 \text{ MW}.$$

Fluid and Geometric Properties, and Pipe Diameters

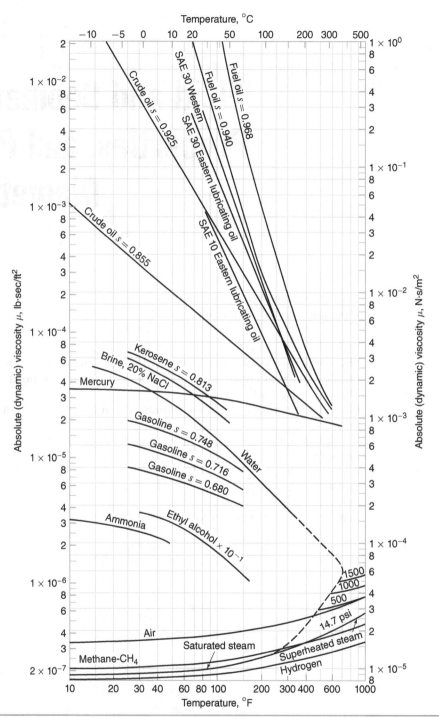

FIGURE A.1   Absolute viscosity μ of fluids. (s = specific gravity at 60°F relative to water at 60°F. For water and air, see also Tables A.1 and A.2, respectively.)

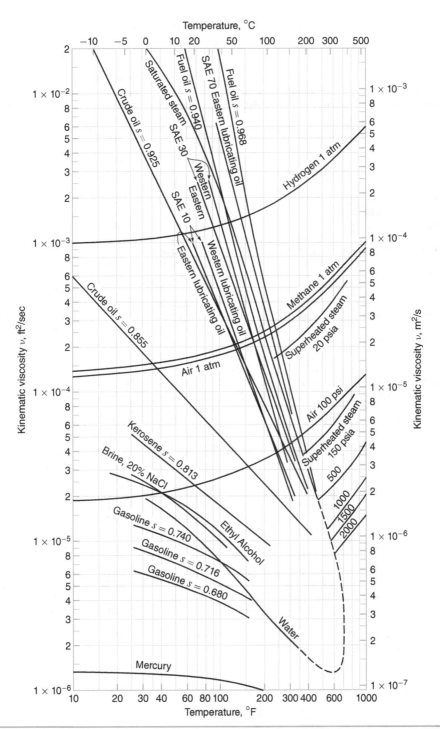

FIGURE A.2 Kinematic viscosity ν of fluids. (s = specific gravity at 60°F relative to water at 60°F. For water and air, see also Tables A.1 and A.2, respectively.)

Table A.1 Physical properties of water at standard sea-level atmospheric pressure[a]

Tem-pera-ture, T	Specific weight, γ	Density, ρ	Absolute viscosity,[b] μ	Kinematic viscosity,[b] ν	Surface tension, σ	Satura-tion vapor pressure, p_v	Satura-tion vapor pressure head, p_v/γ	Bulk modulus of elasticity, E_v
°F	lb/ft³	slugs/ft³	10^{-6} lb·sec/ft²	10^{-6} ft²/sec	lb/ft	psia	ft abs	psi
32°F	62.42	1.940	37.46	19.31	0.005 18	0.0885	0.204	293,000
40°F	62.43	1.940	32.29	16.64	0.005 14	0.122	0.281	294,000
50°F	62.41	1.940	27.35	14.10	0.005 09	0.178	0.411	305,000
60°F	62.37	1.938	23.59	12.17	0.005 04	0.256	0.592	311,000
70°F	62.30	1.936	20.50	10.59	0.004 98	0.363	0.839	320,000
80°F	62.22	1.934	17.99	9.30	0.004 92	0.507	1.173	322,000
90°F	62.11	1.931	15.95	8.26	0.004 86	0.698	1.618	323,000
100°F	62.00	1.927	14.24	7.39	0.004 80	0.949	2.20	327,000
110°F	61.86	1.923	12.84	6.67	0.004 73	1.275	2.97	331,000
120°F	61.71	1.918	11.68	6.09	0.004 67	1.692	3.95	333,000
130°F	61.55	1.913	10.69	5.58	0.004 60	2.22	5.19	334,000
140°F	61.38	1.908	9.81	5.14	0.004 54	2.89	6.78	330,000
150°F	61.20	1.902	9.05	4.76	0.004 47	3.72	8.75	328,000
160°F	61.00	1.896	8.38	4.42	0.004 41	4.74	11.18	326,000
170°F	60.80	1.890	7.80	4.13	0.004 34	5.99	14.19	322,000
180°F	60.58	1.883	7.26	3.85	0.004 27	7.51	17.84	318,000
190°F	60.36	1.876	6.78	3.62	0.004 20	9.34	22.28	313,000
200°F	60.12	1.868	6.37	3.41	0.004 13	11.52	27.59	308,000
212°F	59.83	1.860	5.93	3.19	0.004 04	14.69	35.36	300,000
°C	kN/m³	kg/m³	N·s/m²	10^{-6} m²/s	N/m	kN/m² abs	m abs	10^6 kN/m²
0°C	9.805	999.8	0.001 781	1.785	0.0756	0.611	0.0623	2.02
5°C	9.807	1000.0	0.001 518	1.519	0.0749	0.872	0.0889	2.06
10°C	9.804	999.7	0.001 307	1.306	0.0742	1.230	0.1255	2.10
15°C	9.798	999.1	0.001 139	1.139	0.0735	1.710	0.1745	2.14
20°C	9.789	998.2	0.001 002	1.003	0.0728	2.34	0.239	2.18
25°C	9.777	997.0	0.000 890	0.893	0.0720	3.17	0.324	2.22
30°C	9.765	995.7	0.000 798	0.800	0.0712	4.24	0.434	2.25
40°C	9.731	992.2	0.000 653	0.658	0.0696	7.38	0.758	2.28
50°C	9.690	988.0	0.000 547	0.553	0.0679	12.33	1.272	2.29
60°C	9.642	983.2	0.000 466	0.474	0.0662	19.92	2.07	2.28
70°C	9.589	977.8	0.000 404	0.413	0.0644	31.16	3.25	2.25
80°C	9.530	971.8	0.000 354	0.364	0.0626	47.34	4.97	2.20
90°C	9.467	965.3	0.000 315	0.326	0.0608	70.10	7.40	2.14
100°C	9.399	958.4	0.000 282	0.294	0.0589	101.33	10.78	2.07

[a] In these tables, if (e.g., at 32°F) μ is given as 37.46 and the units are 10^{-6} lb·sec/ft² then $\mu = 37.46 \times 10^{-6}$ lb·sec/ft².
[b] For viscosity, see also Figs. A.1 and A.2.

Table A.2   Physical properties of air at standard sea-level atmospheric pressure[a]

Temperature, T	Density, ρ	Specific weight, γ	Absolute viscosity,[b] μ	Kinematic viscosity,[b] ν
°F	slug/ft³	lb/ft³	10^{-6} lb·sec/ft²	10^{-3} ft²/sec
−40°F	0.002940	0.09460	0.312	0.106
−20°F	0.002807	0.09030	0.325	0.116
0°F	0.002684	0.08637	0.338	0.126
10°F	0.002627	0.08453	0.345	0.131
20°F	0.002572	0.08277	0.350	0.136
30°F	0.002520	0.08108	0.358	0.142
40°F	0.002470	0.07945	0.362	0.146
50°F	0.002421	0.07790	0.368	0.152
60°F	0.002374	0.07640	0.374	0.158
70°F	0.002330	0.07495	0.382	0.164
80°F	0.002286	0.07357	0.385	0.169
90°F	0.002245	0.07223	0.390	0.174
100°F	0.002205	0.07094	0.396	0.180
120°F	0.002129	0.06849	0.407	0.189
140°F	0.002058	0.06620	0.414	0.201
160°F	0.001991	0.06407	0.422	0.212
180°F	0.001929	0.06206	0.434	0.225
200°F	0.001871	0.06018	0.449	0.240
250°F	0.001739	0.05594	0.487	0.280
°C	kg/m³	N/m³	10^{-6} N·s/m²	10^{-6} m²/s
−40°C	1.515	14.86	14.9	9.8
−20°C	1.395	13.68	16.1	11.5
0°C	1.293	12.68	17.1	13.2
10°C	1.248	12.24	17.6	14.1
20°C	1.205	11.82	18.1	15.0
30°C	1.165	11.43	18.6	16.0
40°C	1.128	11.06	19.0	16.8
60°C	1.060	10.40	20.0	18.7
80°C	1.000	9.81	20.9	20.9
100°C	0.946	9.28	21.8	23.1
200°C	0.747	7.33	25.8	34.5

[a] In these tables, if (e.g., at −40°F) μ is given as 0.312 and the units are 10^{-6} lb·sec/ft² then $\mu = 0.312 \times 10^{-6}$ lb·sec/ft².
[b] For viscosity, see also Figs. A.1 and A.2. Absolute viscosity μ is virtually independent of pressure, whereas kinematic viscosity ν varies with pressure (density) (Sec. 2.11).

Table A.3   The ICAO[a] standard atmosphere[b]

Elevation above sea level	Temperature, T	Absolute pressure, p	Specific weight, γ	Density, ρ	Absolute viscosity, μ	Kinematic viscosity, ν	Speed of sound, c	Gravitational acceleration, g
ft	°F	psia	lb/ft³	slug/ft³	10^{-6} lb·sec/ft²	10^{-3} ft²/sec	ft/sec	ft/sec²
0	59.000	14.695 9	0.076 472	0.002 376 8	0.373 72	0.157 24	1116.45	32.1740
5,000	41.173	12.228 3	0.065 864	0.002 048 1	0.363 66	0.177 56	1097.08	32.158
10,000	23.355	10.108 3	0.056 424	0.001 755 5	0.353 43	0.201 33	1077.40	32.142
15,000	5.545	8.297 0	0.048 068	0.001 496 1	0.343 02	0.229 28	1057.35	32.129
20,000	−12.255	6.758 8	0.040 694	0.001 267 2	0.332 44	0.262 34	1036.94	32.113
25,000	−30.048	5.460 7	0.034 224	0.001 066 3	0.321 66	0.301 67	1016.11	32.097
30,000	−47.832	4.372 6	0.028 573	0.000 890 65	0.310 69	0.348 84	994.85	32.081
35,000	−65.607	3.467 6	0.023 672	0.000 738 19	0.299 52	0.405 75	973.13	32.068
40,000	−69.700	2.730 0	0.018 823	0.000 587 26	0.296 91	0.505 59	968.08	32.052
45,000	−69.700	2.148 9	0.014 809	0.000 462 27	0.296 91	0.642 30	968.08	32.036
50,000	−69.700	1.691 7	0.011 652	0.000 363 91	0.296 91	0.815 89	968.08	32.020
60,000	−69.700	1.048 8	0.007 217 5	0.000 225 61	0.296 91	1.316 0	968.08	31.991
70,000	−67.425	0.650 87	0.004 448 5	0.000 139 20	0.298 36	2.143 4	970.90	31.958
80,000	−61.976	0.406 32	0.002 736 6	0.000 085 707	0.301 82	3.521 5	977.62	31.930
90,000	−56.535	0.255 40	0.001 695 2	0.000 053 145	0.305 25	5.743 6	984.28	31.897
100,000	−51.099	0.161 60	0.001 057 5	0.000 033 182	0.308 65	9.301 8	990.91	31.868
km	°C	kPa abs	N/m³	kg/m³	10^{-6} N·s/m²	10^{-6} m²/s	m/s	m/s²
0	15.000	101.325	12.013 1	1.225 0	17.894	14.607	340.294	9.806 65
1	8.501	89.876	10.898 7	1.111 7	17.579	15.813	336.43	9.803 6
2	2.004	79.501	9.865 2	1.006 6	17.260	17.147	332.53	9.800 5
3	−4.500	70.121	8.908 3	0.909 25	16.938	18.628	328.58	9.797 4
4	−10.984	61.660	8.025 0	0.819 35	16.612	20.275	324.59	9.794 3
5	−17.474	54.048	7.210 5	0.736 43	16.282	22.110	320.55	9.791 2
6	−23.963	47.217	6.461 3	0.660 11	15.949	24.161	316.45	9.788 2
8	−36.935	35.651	5.143 3	0.525 79	15.271	29.044	308.11	9.782 0
10	−49.898	26.499	4.042 4	0.413 51	14.577	35.251	299.53	9.775 9
12	−56.500	19.399	3.047 6	0.311 94	14.216	45.574	295.07	9.769 7
14	−56.500	14.170	2.224 7	0.227 86	14.216	62.391	295.07	9.763 6
16	−56.500	10.352	1.624 3	0.166 47	14.216	85.397	295.07	9.757 5
18	−56.500	7.565	1.186 2	0.121 65	14.216	116.86	295.07	9.751 3
20	−56.500	5.529	0.866 4	0.088 91	14.216	159.89	295.07	9.745 2
25	−51.598	2.549	0.390 0	0.040 08	14.484	361.35	298.39	9.730 0
30	−46.641	1.197	0.178 8	0.018 41	14.753	801.34	301.71	9.714 7

[a] International Civil Aviation Organization; see Sec. 2.9.

[b] In these tables, if (e.g., at 0 ft) μ is given as 0.373 72 and the units are 10^{-6} lb·sec/ft² then $\mu = 0.373\ 72 \times 10^{-6}$ lb·sec/ft².

Table A.4   Physical properties of common liquids at standard sea-level atmospheric pressure[a]

Liquid	Temper-ature, T	Density, ρ	Specific gravity,[b] s	Absolute viscosity,[c] μ	Surface tension, σ	Vapor pressure, p_v	Bulk mod-ulus of elasticity, E_v	Specific heat, c
	°F	slug/ft³	—	10⁻⁶ lb·sec/ft²	lb/ft	psia	psi	ft·lb/(slug·°R) = ft²/(sec²·°R)
Benzene	68°F	1.70	0.88	14.37	0.002 0	1.45	150,000	10,290
Carbon tetrachloride	68°F	3.08	1.594	20.35	0.001 8	1.90	160,000	5,035
Crude oil	68°F	1.66	0.86	150	0.002	—	—	—
Gasoline	68°F	1.32	0.68	6.1	—	8.0	—	12,500
Glycerin	68°F	2.44	1.26	31,200	0.004 3	0.000 002	630,000	14,270
Hydrogen	−430°F	0.143	0.074	0.435	0.000 2	3.1	—	—
Kerosene	68°F	1.57	0.81	40	0.001 7	0.46	—	12,000
Mercury	68°F	26.3	13.56	33	0.032	0.000 025	3,800,000	834
Oxygen	−320°F	2.34	1.21	5.8	0.001	3.1	—	~5,760
SAE 10 oil	68°F	1.78	0.92	1,700	0.002 5	—	—	—
SAE 30 oil	68°F	1.78	0.92	9,200	0.002 4	—	—	—
Fresh water	68°F	1.936	0.999	21.0	0.005 0	0.34	318,000	25,000
Seawater	68°F	1.985	1.024	22.5	0.005 0	0.34	336,000	23,500
	°C	kg/m³	—	10⁻³ N·s/m²	N/m	kN/m² abs	10⁶ N/m²	N·m/(kg·K) = m²/(s²·K)
Benzene	20°C	876	0.88	0.65	0.029	10.0	1 030	1720
Carbon tetrachloride	20°C	1 588	1.594	0.97	0.026	13.1	1 100	842
Crude oil	20°C	856	0.86	7.2	0.03	—	—	—
Gasoline	20°C	680	0.68	0.29	—	55.2	—	2100
Glycerin	20°C	1 258	1.26	1494	0.063	0.000 014	4 344	2386
Hydrogen	−257°C	73.7	0.074	0.021	0.002 9	21.4	—	—
Kerosene	20°C	808	0.81	1.92	0.025	3.20	—	2000
Mercury	20°C	13 550	13.56	1.56	0.51	0.000 17	26 200	139.4
Oxygen	−195°C	1 206	1.21	0.278	0.015	21.4	—	~964
SAE 10 oil	20°C	918	0.92	82	0.037	—	—	—
SAE 30 oil	20°C	918	0.92	440	0.036	—	—	—
Fresh water	20°C	998	0.999	1.00	0.073	2.34	2171	4187
Seawater	20°C	1 023	1.024	1.07	0.073	2.34	2 300	3933

[a] In these tables, if (e.g., for benzene at 68°F) μ is given as 1.437 and the units are 10⁻⁶ lb·sec/ft² then $\mu = 1.437 \times 10^{-6}$ lb·sec/ft².
[b] Relative to pure water at 60°F.
[c] For viscosity, see also Figs. A.1 and A.2.

Table A.5 **Physical properties of common gases at standard sea-level atmospheric pressure[a]**

Gas	Chemical formula	Molar mass, M	Density, ρ	Absolute viscosity,[b] μ	Gas constant, R	Specific heat, c_p	c_v	Specific heat ratio, $k = c_p/c_v$
at 68°F	—	slug/ slug-mol	slug/ft³	10^{-6} lb·sec/ft²	ft·lb/(slug·°R) $= $ ft²/(sec²·°R)	ft·lb/(slug·°R) $= $ ft²/(sec²·°R)		—
Air		28.96	0.00231	0.376	1,715	6,000	4,285	1.40
Carbon dioxide	CO_2	44.01	0.00354	0.310	1,123	5,132	4,009	1.28
Carbon monoxide	CO	28.01	0.00226	0.380	1,778	6,218	4,440	1.40
Helium	He	4.003	0.000323	0.411	12,420	31,230	18,810	1.66
Hydrogen	H_2	2.016	0.000162	0.189	24,680	86,390	61,710	1.40
Methane	CH_4	16.04	0.00129	0.280	3,100	13,400	10,300	1.30
Nitrogen	N_2	28.02	0.00226	0.368	1,773	6,210	4,437	1.40
Oxygen	O_2	32.00	0.00258	0.418	1,554	5,437	3,883	1.40
Water vapor	H_2O	18.02	0.00145	0.212	2,760	11,110	8,350	1.33
at 20°C	—	kg/ kg-mol	kg/m³	10^{-6} N· s/m²	N·m/(kg·K) $= $ m²/(s²·K)	N·m/(kg·K) $= $ m²/(s²·K)		—
Air		28.96	1.205	18.0	287	1003	716	1.40
Carbon dioxide	CO_2	44.01	1.84	14.8	188	858	670	1.28
Carbon monoxide	CO	28.01	1.16	18.2	297	1040	743	1.40
Helium	He	4.003	0.166	19.7	2077	5220	3143	1.66
Hydrogen	H_2	2.016	0.0839	9.0	4120	14450	10330	1.40
Methane	CH_4	16.04	0.668	13.4	520	2250	1730	1.30
Nitrogen	N_2	28.02	1.16	17.6	297	1040	743	1.40
Oxygen	O_2	32.00	1.33	20.0	260	909	649	1.40
Water vapor	H_2O	18.02	0.747	10.1	462	1862	1400	1.33

[a] In these tables, if (e.g., for air at 68°F) μ is given as 0.376 and the units are 10^{-6} lb·sec/ft² then $\mu = 0.376 \times 10^{-6}$ lb·sec/ft².

[b] For viscosity, see also Figs. A.1 and A.2. Absolute viscosity μ is virtually independent of pressure, whereas kinematic viscosity ν varies with pressure (density) (Sec. 2.11).

Table A.6   Areas of circles

Diameter	Area	
in	in²	ft²
0.25	0.049087	0.00034088
0.5	0.19635	0.0013635
1.0	0.78540	0.0054542
2.0	3.1416	0.021817
3.0	7.0686	0.049087
4.0	12.566	0.087266
6.0	28.274	0.19635
8.0	50.265	0.34907
9.0	63.617	0.44179
10.0	78.540	0.54542
12.0	113.097	0.78540
m	m²	
0.05	0.0019635	
0.10	0.0078540	
0.15	0.017671	
0.20	0.031416	
0.25	0.049087	
0.30	0.070686	
0.50	0.19635	
1.00	0.78540	
1.50	1.76715	
2.00	3.14159	

Table A.7   Properties of areas

Shape	Sketch	Area	Location of centroid	I_c or $I = I_c + Ay_c^2$
Rectangle		bh	$y_c = \dfrac{h}{2}$	$I_c = \dfrac{bh^3}{12}$
Triangle		$\dfrac{bh}{2}$	$y_c = \dfrac{h}{3}$	$I_c = \dfrac{bh^3}{36}$
Circle		$\dfrac{\pi D^2}{4}$	$y_c = \dfrac{D}{2}$	$I_c = \dfrac{\pi D^4}{64}$
Semicircle		$\dfrac{\pi D^2}{8}$	$y_c = \dfrac{4r}{3\pi}$	$I = \dfrac{\pi D^4}{128}$
Circular sector		$\dfrac{\theta r^2}{2}$	$y_c = \dfrac{4r}{3\theta}\sin\dfrac{\theta}{2}$	$I = \dfrac{r^4}{8}(\theta + \sin\theta)$
Ellipse		$\dfrac{\pi bh}{4}$	$y_c = \dfrac{h}{2}$	$I_c = \dfrac{\pi bh^3}{64}$
Semiellipse		$\dfrac{\pi bh}{4}$	$y_c = \dfrac{4h}{3\pi}$	$I = \dfrac{\pi bh^3}{16}$
Parabola		$\dfrac{2bh}{3}$	$x_c = \dfrac{3b}{8}$ $y_c = \dfrac{3h}{5}$	$I = \dfrac{2bh^3}{7}$

Table A.8   Properties of solid bodies

Body	Sketch	Volume	Surface area	Location of centroid
Cylinder		$\dfrac{\pi D^2 h}{4}$	$\pi D h + \dfrac{\pi D^2}{2}$	$y_c = \dfrac{h}{2}$
Cone		$\dfrac{1}{3}\left(\dfrac{\pi D^2 h}{4}\right)$	$\dfrac{\pi D h}{2}\sqrt{1+\dfrac{D^2}{4h^2}}+\dfrac{\pi D^2}{4}$	$y_c = \dfrac{h}{4}$
Sphere		$\dfrac{\pi D^3}{6}$	πD^2	$y_c = \dfrac{D}{2}$
Hemisphere		$\dfrac{\pi D^3}{12}$	$\dfrac{\pi D^2}{2}+\dfrac{\pi D^2}{4}=\dfrac{3\pi D^2}{4}$	$y_c = \dfrac{3r}{8}$
Paraboloid		$\dfrac{1}{2}\left(\dfrac{\pi D^2 h}{4}\right)$		$y_c = \dfrac{h}{3}$
Half ellipsoid		$\dfrac{2}{3}\left(\dfrac{\pi D^2 h}{4}\right)$		$y_c = \dfrac{3h}{8}$

Table A.9 Typical pipe diameters and wall thicknesses[a]

Nominal pipe size	Nominal pipe size[b]	Outside diameter (OD)	Wall thickness	Inside diameter (ID)	Nominal pipe size	Nominal pipe size[b]	Outside diameter (OD)	Wall thickness	Inside diameter (ID)
in	mm	in	in	in	in	mm	in	in	in
Polyvinyl chloride pipe (PVC)					Steel pipe				
4	100	4.50	0.34	3.83	4	100	4.50	0.34	3.83
5	125	5.56	0.38	4.81	5	125	5.56	0.38	4.81
6	150	6.63	0.43	5.76	6	150	6.63	0.43	5.76
8	200	8.63	0.50	7.63	8	200	8.63	0.50	7.63
10	250	10.75	0.59	9.56	10	250	10.75	0.59	9.56
12	300	12.75	0.69	11.38	12	300	12.75	0.69	11.37
14	350	14.00	0.75	12.50	14	350	14.00	0.75	12.50
16	400	16.00	0.84	14.31	16	400	16.00	0.84	14.31
18	450	18.00	0.94	16.12	18	450	18.00	0.94	16.12
20	500	20.00	1.03	17.94	20	500	20.00	1.03	17.94
24	600	24.00	1.22	21.56	24	600	24.00	1.22	21.56
Ductile iron pipe (DIP)[c]					Reinforced concrete pipe (RCP)				
6	150	6.9	0.34	6.2	15	375	19.5	2.25	15
8	200	9.1	0.36	8.3	18	450	23	2.5	18
10	250	11.1	0.38	10.3	24	600	30	3	24
12	300	13.2	0.40	12.4	30	750	37	3.5	30
14	350	15.3	0.42	14.5	36	900	44	4	36
16	400	17.4	0.43	16.5	42	1050	51	4.5	42
18	450	19.5	0.44	18.6	48	1200	58	5	48
20	500	21.6	0.45	20.7	54	1350	65	5.5	54
24	600	25.8	0.47	24.9	60	1500	72	6	60
30	750	32.0	0.51	31.0	72	1800	86	7	72
36	900	38.3	0.58	37.1					
42	1050	44.5	0.65	43.2					
48	1200	50.8	0.72	49.4					
54	1350	57.1	0.81	55.5					

[a] Wall thicknesses are mid-range standards (RCP B-wall, DIP Class 53, PVC and Steel Schedule 80) and may be thicker or thinner based on the design specifications. For different wall thickness, RCP retains the specified ID and changes OD, where DIP, PVC, and steel retain OD.

[b] Outside of the United States, metric nominal sizes differ from those shown here, since they are based on outside diameters. For example, a 5-inch PVC may be considered a nominal 140 mm pipe and a nominal 250 mm pipe would fall between an 8 and 10-inch pipe in the U.S.

[c] DIP and cast iron are available in the same nominal diameters, though cast iron has greater wall thickness.

Equations in Fluid Mechanics[1]

Equations are a vital part of the language of fluid mechanics. Therefore, below we summarize the characteristics of the principal types and forms of equations used in fluid mechanics, and we present some of their useful properties.

All equations are either *identities* or *conditional equations.*

Identities are equations that are true for *all* values of the quantities involved. For example

$$\sin^2\theta + \cos^2\theta = 1 \quad \text{and} \quad x^2 - y^2 = (x+y)(x-y)$$

are identities. $y = f(x)$, where a function is being *defined* by an equation, is another type of identity. The equality sign is sometimes written $\equiv$ for identities.

Conditional equations are true only for certain values of the unknown quantities involved. For example,

$$x^2 - 1 = 0$$

is true for only $x = \pm 1$. More generally, $f(x) = 0$ is conditional.

Rational equations are derived from fundamental laws of physics; they are all dimensionally homogeneous.

Theoretical (or analytical) equations result from analysis. They are usually dimensionally homogeneous. All equations are either theoretical (the majority in this book), empirical, or a combination of the two.

Empirical equations are based on experimental data. We discussed these in Sec. 1.1 and presented some classical examples from fluid mechanics in Sec. 7.19. It is interesting that some of these, of fundamental importance to the subject, are dimensionally nonhomogeneous (their dimensions do not "agree").

Explicit equations. When the variable of interest is isolated on one side of an equation, as in $y = f(x, z)$, mathematicians say this equation has *explicit form.* Any equation

---

[1] The authors gratefully acknowledge the suggestions for this appendix provided by Dr. Peter Ross.

that we can arrange into this form we shall call an explicit equation in y. Thus we can compute explicitly the value of the unknown. For example,

$$pv^n = p_1 v_1^n = \text{constant} \tag{2.6a}$$

is explicit in v, and

$$\frac{1}{\sqrt{f}} = -2\log\left(\frac{e/D}{3.7} + \frac{2.51}{\mathbf{R}\sqrt{f}}\right) \tag{7.51}$$

is explicit in e.

Implicit equations. If an equation cannot be rearranged to have a variable isolated on one side, in mathematics we say the equation has *implicit form*. Some equations that have implicit form, such as Eq. (2.6), we can rearrange into explicit form, and others we cannot. Any equation that we *cannot* rearrange into explicit form to isolate the variable of interest we shall call an implicit equation in that variable. Such equations only "imply" the value of the unknown variable of interest, and so we usually find there is more challenge in solving for it. A famous example in fluid mechanics is Eq. (7.51), which is implicit in f. Equation (7.81) is implicit in Q, L, D, h_L, and Σk. An equation is either explicit or implicit in a particular variable; but it may be explicit in some variables and implicit in others, as in Eq. (7.51).

Algebraic equations contain only variables raised to various constant powers, not necessarily integers. They do not include transcendental or differential equations, etc., discussed here later. An example of an algebraic equation is Eq. (2.6) given earlier here.

Transcendental equations are those in which the unknown variable or variables occur within a nonalgebraic function such as a trigonometric, logarithmic, or exponential function. Equations (7.46), (7.51), and (10.23) are examples. It follows that transcendental equations cannot be algebraic or polynomial.

Polynomial equations are algebraic equations containing only terms with variables raised to positive integral powers and with constant coefficients. The equation

$$y = ax^n + bx^{n-1} + \cdots + dx^2 + ex + f = 0 \tag{B.1}$$

is a polynomial equation in x. All terms need not occur, i.e., some coefficients may be zero. If $a \neq 0$ we say that Eq. (B.1) is of *degree* n, from its largest exponent, and it has n *roots*, i.e., n real or complex values of x that cause y to be zero. Some hand calculators will display the values of all the roots when given the values of all the coefficients.

Linear equations are first-degree polynomial equations, e.g.,

$$y = mx + c \tag{B.2}$$

Quadratic equations are second-degree polynomial equations, e.g.,

$$y = ax^2 + bx + c \tag{B.3}$$

Their name includes "quad" because we can use second-order equations to compute areas of (four-sided) rectangles. Equations (5.44) and (8.45) are examples of quadratic equations.

The two roots of the equation

$$ax^2 + bx + c = 0$$

we know well are

$$x = \frac{-b \pm \sqrt{b^2 - 4ac}}{2a} \tag{B.4}$$

Cubic equations are third-degree polynomial equations, e.g.,

$$ax^3 + bx^2 + cx + d = 0 \tag{B.5}$$

They occur quite frequently in fluid mechanics, as in Sample Problems 3.8, 7.13, 8.7, and 8.9. Formulas to find the three roots from a, b, c, and d are quite involved and require a series of substitutions. As we noted earlier, some handheld calculators will display all the roots; others will solve for the root closest to an estimate provided by the user.

Sometimes we know one of the nonzero roots, say $x = r$, as in Sample Problem 8.7; if not, we can find one by trials (see Sample Problem 3.8). Then, dividing Eq. (B.5) by $(x - r)$, we obtain the quadratic equation

$$ax^2 + (b + ar)x - \frac{d}{r} = 0 \tag{B.6}$$

from which we can easily find the other two roots; they are

$$x = -\left(\frac{b}{2a} + \frac{r}{2}\right) \pm \sqrt{\left(\frac{b}{2a} + \frac{r}{2}\right)^2 + \frac{d}{ar}} \tag{B.7}$$

A particular form of equation, which occurs frequently in Chap. 8, is the specific energy equation given by Eq. (8.17), which we can write

$$E = y + \frac{G}{y^2}$$

where E and G are known positive constants. We can rearrange this into the cubic equation

$$y^3 - Ey^2 + G = 0$$

If we know one nonzero root, $y = r$, then dividing through by $(y - r)$ yields the quadratic equation

$$y^2 - \frac{G}{r^2}y - \frac{G}{r} = 0 \tag{B.8}$$

from which the other two roots are

$$y = \frac{G}{2r^2} \pm \sqrt{\left(\frac{G}{2r^2}\right)^2 + \frac{G}{r}} \tag{B.9}$$

Quartic equations are fourth-degree polynomial equations.

Infinite series equations are those including an infinite number of terms, such as Eq. (5.18). Such equations may "converge" (to a specific total value) or "diverge" (the sum of the terms is increasing without bound).

Nonlinear equations are all equations other than linear (first-order polynomial) equations. They may be higher degree polynomial equations, transcendental equations, or equations with nonintegral exponents such as

$$pv^n = \text{constant} \qquad (\text{for } n \neq 1) \tag{2.6a}$$

Differential equations, both ordinary and partial, have their own special terminology.

Ordinary differential equations (ODEs) contain ordinary derivatives, like dy/dx, i.e., derivatives of one or more unknown functions (dependent variables, y) of a *single* (independent) variable (x). Examples are

$$\tau = \mu \frac{du}{dy} \tag{2.9}$$

and

$$\frac{dp}{\gamma} + dz + d\left(\frac{V^2}{2g}\right) = 0 \tag{5.6}$$

The highest derivative in a differential equation determines its *order;* if the highest derivative is $d^n y/dx^n$ we say the equation is of the nth order.

Partial differential equations (PDEs) contain partial derivatives of an unknown function (or functions) involving *more* than one independent variable; an example is

$$(a_x)_{st} = u\frac{\partial u}{\partial x} + v\frac{\partial u}{\partial y} + w\frac{\partial u}{\partial z} \tag{4.23a}$$

A celebrated *second-order* PDE is Laplace's equation,

$$\frac{\partial^2 \phi}{\partial x^2} + \frac{\partial^2 \phi}{\partial y^2} = 0$$

This equation is *linear* in ϕ, i.e., if ϕ_1 and ϕ_2 are each solutions, then so is $\phi_3 = a_1\phi_1 + a_2\phi_2$. An example of a *vector* PDE is

$$\mathbf{a}_{st} = u\frac{\partial \mathbf{V}}{\partial x} + v\frac{\partial \mathbf{V}}{\partial y} + w\frac{\partial \mathbf{V}}{\partial z} \tag{4.22}$$

where the boldface symbols are vectors.

Some equations governing fluid phenomena occur in the form of *systems* of simultaneous differential equations, such as the Navier-Stokes equations given by Eqs. (6.8).

Difference equations include differences between two quantities, usually indicated by Δ. An example is

$$\frac{\Delta v}{v} \approx -\frac{\Delta p}{E_v} \tag{2.3a}$$

Equations with integrals occur in fluid mechanics, such as

$$\alpha = \frac{1}{V^2} \frac{\int u^3 dA}{\int u\,dA} \tag{5.4}$$

Some of these include *definite* integrals (with limits), like

$$\bar{\tau}_0 = \frac{1}{P} \int_0^P \tau_0\,dP \tag{7.6}$$

Note that these are not the same as *integral equations*, in which the unknown function $y(x)$ appears within an integral sign.

Simultaneous equations or *systems of equations* occur when more than one equation with the same unknown variables holds at the same time; all the equations are satisfied by the same values of the variables. Generally, the number of equations must equal the number of unknowns; otherwise, the system may be underdetermined or overdetermined. Examples include the system of four equations governing flow in a single pipe, as summarized in Sec. 7.15. Larger such systems govern all multipipe flow problems (Secs. 7.30–7.33).

Vector equations are those involving vector quantities; an example is

$$\sum \mathbf{F} = \dot{m}_2 \mathbf{V}_2 - \dot{m}_1 \mathbf{V}_1 \tag{6.5}$$

where the boldface symbols are vectors.

References

There is a great volume of literature available on the various aspects of fluid mechanics and hydraulics. The results of original research are usually reported in papers published in peer-reviewed technical journals. We list here selected highly ranked journals that have published numerous articles dealing with fluids, their properties, and applications; some of them have changed their names over time:

Environmental Fluid Mechanics, 2001–

Experiments in Fluids, 1983–

Geophysical and Astrophysical Fluid Dynamics, 1977–

Geophysical Fluid Dynamics, 1970–1976

International Journal of Multiphase Flow, 1973–

International Journal for Numerical Methods in Fluids, 1981–

International Journal of Numerical Methods for Heat & Fluid Flow, 1991–

Journal of Applied Mechanics, ASME, 1935–

Journal of Computational Physics, 1966–

Journal of Fluid Mechanics, 1956–

Journal of Fluids and Structures, 1987–

Journal of Fluids Engineering, ASME, 1973–

Journal of Geophysical Research, Atmospheres, 1984–

Journal of Geophysical Research, Oceans, 1984–

Journal of Hydraulic Research, IAHR/AIHR, 1964–

Journal of the Hydraulic Division, Proceedings of the ASCE, 1956–1982

Journal of Hydraulic Engineering, ASCE, 1983–

Journal of Non-Newtonian Fluid Mechanics, 1976–

Journal of Turbomachinery, ASME, 1986–

Physical Review Fluids. 2016–

Journal & Proceedings, Institution of Mechanical Engineers (London), 1847–

Physics of Fluids, 1958–1988, 1994–

Physics of Fluids, A: Fluid Dynamics, 1989–1993

Proceedings of the ASCE, 1873–1955

Proceedings of the Institution of Civil Engineers (GB), 1952–

Recent Advances in Numerical Methods in Fluids, 1980–

Theoretical and Computational Fluid Dynamics, 1989–

Transactions of the ASCE, 1867–

Transactions of the ASME, 1880–

Transactions of the Institution of Civil Engineers (GB), 1836–1842

For the convenience of students, we next include a list of books covering many different topics of fluid mechanics and its engineering applications. This list by no means includes all the important books that have been written; we intend merely to provide a representative list. We encourage students to "probe deeper" and to widen their horizons by further reading.

1. Ackers, P., W. R. White, A. J. Harrison, and J. A. Perkins. *Weirs and Flumes for Flow Measurement.* Wiley, New York, 1978.
2. Anderson, J. D. *Computational Fluid Dynamics: The Basics with Applications.* McGraw-Hill, New York, 1995.
3. Anderson, J. D. *Fundamentals of Aerodynamics,* 2d ed. McGraw-Hill, New York, 1991.
4. Anderson, J. D. *Modern Compressible Flow: With Historical Perspective,* 2d ed. McGraw-Hill, New York, 1989 & 1990.
5. Barenblatt, G. I. *Dimensional Analysis,* translated from Russian by P. Makinen. Gordon and Breach, New York, 1987.
6. Batchelor, G. K. *An Introduction to Fluid Dynamics.* Cambridge University Press, Cambridge, England, 2000.
7. Bear, J. *Dynamics of Fluids in Porous Media.* Dover, New York, 1988.
8. Bertin, J. J., and M. L. Smith. *Aerodynamics for Engineers,* 3d ed. Prentice Hall, Upper Saddle River, NJ, 1998.
9. Böhme, G. *Non-Newtonian Fluid Mechanics.* Elsevier, New York, 1987.
10. Bos, M. G., J. A. Replogle, and A. J. Clemmens. *Flow Measuring Flumes for Open Channel Systems.* Wiley, New York, 1984.
11. Brater, E. F., H. W. King, J. E. Lindell, and C. Y. Wei. *Handbook of Hydraulics,* 7th ed. McGraw-Hill, New York, 1996.
12. Cengel, Y. A., M. A. Boles, and M. Kanoglu. *Thermodynamics: An Engineering Approach,* 10th ed. McGraw-Hill, New York, 2024.
13. Chadwick, A. J., J. Morfett, and M. Borthwick *Hydraulics in Civil and Environmental Engineering,* 6th ed. CRC Press, New York, 2021.
14. Chaudhry, M. H. *Applied Hydraulic Transients,* 3rd ed. Springer, New York, 2014.
15. Chaudhry, M. H. *Open Channel Flow,* 2nd ed., Springer, New York, 2007.
16. Cheremisinoff, N. P. *Fundamentals of Wind Energy.* Ann Arbor Science, Ann Arbor, MI, 1978.
17. Chin, D. A. *Water-Resources Engineering,* 4th ed., Pearson, New York, 2021.
18. Chorin, A. J., and J. E. Marsden. *A Mathematical Introduction to Fluid Mechanics,* 3d ed. Springer-Verlag, New York, 1994.
19. Chow, C.-Y. *An Introduction to Computational Fluid Mechanics.* Seminole, Boulder, CO, 1983.
20. Chow, V. T. *Open-Channel Hydraulics.* McGraw-Hill, New York, 1959.
21. Churchill, S. W., and H. Brenner. *Viscous Flows: The Practical Use of Theory.* Butterworth-Heinemann, Boston, 1988.

22. Clift, R., J. R. Grace, and M. E. Weber. *Bubbles, Drops, and Particles.* Academic Press, New York, 1978.

23. Colt Industries, Inc. *Hydraulic Handbook: Fundamental Hydraulics and Data Useful in the Solution of Pump Application Problems,* 11th ed. Fairbanks Morse Pump Division, Kansas City, KS, 1979.

24. Crowe, C. T., D. F. Elger, and J. A. Roberson. *Engineering Fluid Mechanics,* 9th ed. Wiley, New York, 2008.

25. Eggleston, D. M., and F. S. Stoddard. *Wind Turbine Engineering Design.* Van Nostrand Reinhold, New York, 1987.

26. Evett, J. B., and C. Liu. *Fundamentals of Fluid Mechanics.* McGraw-Hill, New York, 1987.

27. Fischer, H. B., E. J. List, R. C. Y. Koh, J. Imberger, and N. H. Brooks. *Mixing in Inland and Coastal Waters.* Academic Press, New York, 1979.

28. Fox, R. W., P. J. Pritchard, and A. T. McDonald. *Introduction to Fluid Mechanics,* 7th ed. Wiley, New York, 2008.

29. Freeze, R. A., and J. A. Cherry. *Groundwater.* Prentice-Hall, Englewood Cliffs, NJ, 1979.

30. French, R. H. *Open-Channel Hydraulics.* McGraw-Hill, New York, 1985.

31. *Friction Factors for Large Conduits Flowing Full,* Engineering Monograph No. 7. U.S. Department of the Interior, Bureau of Reclamation, Denver, CO, 1977.

32. Garay, P. N. *Pump Application Desk Book.* Fairmont Press, Liburn, GA, 1990.

33. Goldstein, R. J. (ed.). *Fluid Mechanics Measurements,* 2d ed. Taylor & Francis, Washington, DC, 1996.

34. Graf, W. H. *Hydraulics of Sediment Transport.* McGraw-Hill, New York, 1971, reprinted by Water Resource Publications, Littleton, CO, 1984.

35. Granger, R. A. *Experiments in Fluid Mechanics.* Holt, Rinehart and Winston, New York, 1988.

36. Gulliver, J. S., and R. E. A. Arndt (eds.). *Hydropower Engineering Handbook.* McGraw-Hill, New York, 1991.

37. Hamrock, B. J. *Fundamentals of Fluid Film Lubrication.* McGraw-Hill, New York, 1994.

38. Henderson, F. M. *Open Channel Flow.* Macmillan, New York, 1966.

39. Herring, J. R., and J. C. McWilliams. *Lecture Notes on Turbulence.* World Scientific, Singapore, 1989.

40. Hinze, J. O. *Turbulence,* 2d ed. McGraw-Hill, New York, 1975.

41. Holton, J. R., and G. J. Hakim, *An Introduction to Dynamic Meteorology,* 5th ed., Academic Press, San Diego, CA, 2012.

42. Horváth, I. *Hydraulics in Water and Waste-Water Treatment Technology.* Wiley, New York, 1994.

43. Hucho, W.-H. (ed.). *Aerodynamics of Road Vehicles: From Fluid Mechanics to Vehicle Engineering,* 4th ed. Society of Automotive Engineers, Warrendale, PA, 1998.

44. Hydraulic Institute. *Engineering Data Book,* 2d ed. Hydraulic Institute, Cleveland, OH, 1990.

45. Hydraulic Institute. *Hydraulic Institute Standards for Centrifugal, Rotary, & Reciprocating Pumps,* 15th ed. Hydraulic Institute, Parsippany, NJ, 1994.

46. Hydraulic Institute. *Pipe Friction Manual,* 3d ed. Hydraulic Institute, Cleveland, OH, 1975.

47. Hydraulic Institute. *Standards of Hydraulic Institute,* 14th ed. Hydraulic Institute, New York, 1983.

48. Idel'chik, I. E. *Handbook of Hydraulic Resistance,* 3d ed. CRC Press, Boca Raton, FL, 1994.

49. Johnson, R. W. (ed.). *The Handbook of Fluid Dynamics*. CRC Press, Boca Raton, FL, 1998.
50. Kirchhoff, R. H. *Potential Flows: Computer Graphic Solutions*. M. Dekker, New York, 1985.
51. Kline, S. J. *Similitude and Approximation Theory*. Springer-Verlag, New York, 1986.
52. Knapp, R. T., J. W. Daily, and F. G. Hammitt. *Cavitation*. McGraw-Hill, New York, 1970, reprinted by Institute of Hydraulic Research, University of Iowa, Iowa City, IA, 1979.
53. Kuethe, A. M., and C.-Y. Chow. *Foundations of Aerodynamics: Bases of Aerodynamic Design*, 5th ed. Wiley, New York, 1998.
54. Lamb, H. *Hydrodynamics*, 6th ed. Cambridge University Press, Cambridge, England, 1932, reprinted 1993.
55. Langhaar, H. L. *Dimensional Analysis and the Theory of Models*. Wiley, New York 1951, reprinted by Krieger, Huntington, NY, 1980.
56. Liggett, J. A. *Fluid Mechanics*. McGraw-Hill, New York, 1994.
57. Lighthill, M. J. *Waves in Fluids*. Cambridge University Press, London, 1978.
58. Linsley, R. K., J. B. Franzini, D. L. Freyberg, and G. Tchobanoglous. *Water Resources Engineering*, 4th ed. McGraw-Hill, New York, 1992.
59. Lobanoff, V. S., and R. R. Ross. *Centrifugal Pumps: Design and Application*, 2d ed. Gulf Publishing, Houston, TX, 1992.
60. Merzkirch, W. *Flow Visualization*, 2d ed. Academic Press, Orlando, FL, 1987.
61. Miller, R. W. *Flow Measurement Engineering Handbook*, 3d ed. McGraw-Hill, New York, 1996.
62. Moran, M. J., H. N. Shapiro, D. D. Boettner, and M. B. Bailey. *Fundamentals of Engineering Thermodynamics*, 8th ed. Wiley, New York, 2014.
63. Munson, B. R., A. P. Rothmayer, T. H. Okiishi, and W. W. Huebsch. *Fundamentals of Fluid Mechanics*, 7th ed. Wiley, New York, 2012.
64. National Committee for Fluid Mechanics Films. *Illustrated Experiments in Fluid Mechanics*. MIT Press, Cambridge, MA, 1972.
65. Novak, P., and J. Cábelka. *Models in Hydraulic Engineering: Physical Principles and Design Applications*. Pitman, Boston, 1981.
66. Olfe, D. B. *Fluid Mechanics Programs for the IBM PC*. McGraw-Hill, New York, 1987.
67. Pai, S.-I., and S. Luo. *Theoretical and Computational Dynamics of a Compressible Flow*. Van Nostrand Reinhold, New York, 1991.
68. Parker, S. P. (ed.). *Fluid Mechanics Source Book*. McGraw-Hill, New York, 1988.
69. Peyret, R., and T. D. Taylor. *Computational Methods for Fluid Flow*. Springer-Verlag, New York, 1990.
70. Potter, M. C., D. C. Wiggert, and B. H. Ramadan. *Mechanics of Fluids*, 4th ed. Prentice Hall, Upper Saddle River, NJ, 2011.
71. Rahman, M. *Water Waves: Relating Modern Theory to Advanced Engineering Applications*. Oxford University Press, Oxford, 1995.
72. Raudkivi, A. J. *Loose Boundary Hydraulics*, 4th ed. Balkema, Rotterdam, 1998.
73. Roberson, J. A., J. J. Cassidy, and M. H. Chaudhry. *Hydraulic Engineering*, 2d ed. Wiley, New York, 1998.
74. Rouse, H., and S. Ince. *History of Hydraulics*. Dover, New York, 1963.
75. Saad, M. A. *Compressible Fluid Flow*, 2d ed. Prentice Hall, Englewood Cliffs, NJ, 1993.
76. Saad, M. A. *Thermodynamics Principles and Practices*. Prentice Hall, Upper Saddle River, NJ, 1997.
77. Sabersky, R. H., A. J. Acosta, E. G. Hauptmann, and E. M. Gates. *Fluid Flow: A First Course in Fluid Mechanics*, 4th ed. Prentice Hall, Upper Saddle River, NJ, 1999.

78. Schetz, J. A. *Boundary Layer Analysis*. Prentice-Hall, Englewood Cliffs, NJ, 1993.
79. Schlichting, H., and K. Gersten. *Boundary Layer Theory*, 9th ed. Springer, New York, 2017.
80. Shames, I. H. *Mechanics of Fluids*, 4th ed. McGraw-Hill, New York, 2002.
81. Sharpe, G. J. *Solving Problems in Fluid Dynamics*. Longman Scientific and Technical, Harlow, Essex, England, 1994.
82. Sherman, F. S. *Viscous Flow*. McGraw-Hill, New York, 1990.
83. Sonntag, R. E., and C. Borgnake. *Fundamentals of Thermodynamics*, 10th ed. Wiley, New York, 2020.
84. Soo, S. L. *Particulates and Continuum: Multiphase Fluid Dynamics*. Hemisphere, New York, 1989.
85. Stepanoff, A. J. *Centrifugal and Axial Flow Pumps: Theory, Design, and Application*, 2d ed. Wiley, New York, 1957, reprinted by Krieger, Malabar, FL, 1993.
86. Stewart, H. L. *Pumps*, 5th ed. Macmillan, New York, 1991.
87. Stoker, J. J. *Water Waves: The Mathematical Theory with Applications*. Wiley, New York, 1992.
88. Street, R. L., G. Z. Watters, and J. K. Vennard. *Elementary Fluid Mechanics*, 7th ed. Wiley, New York, 1996.
89. Streeter, V. L., E. B. Wylie, and K. W. Bedford. *Fluid Mechanics*, 9th ed. WCB/McGraw-Hill, New York, 1998.
90. Sullivan, J. A. *Fluid Power: Theory and Applications*, 3d ed. Prentice-Hall, Englewood Cliffs, NJ, 1989.
91. Sutton, G. P., and O. Biblarz. *Rocket Propulsion Elements*, 9th ed. Wiley, New York, 2016.
92. Tannehill, J. C., and R. H. Pletcher. *Computational Fluid Mechanics and Heat Transfer*, 3rd ed. CRC Press, Boca Raton, FL, 2011.
93. Task Committee on Hydraulic Modeling (R. Ettema, chmn.). *Hydraulic Modeling: Concepts and Practice*. ASCE Manuals and Reports on Engineering Practice No. 97, American Society of Civil Engineers, Reston, VA, 2000.
94. Tokaty, G. A. *A History and Philosophy of Fluid Mechanics*. Foulis, Henley-on-Thames, England, 1971.
95. Tullis, J. P. *Hydraulics of Pipelines: Pumps, Valves, Cavitation, Transients*. Wiley, New York, 1989.
96. Turton, R. K. *Rotodynamic Pump Design*. Cambridge University Press, Cambridge, England, 1994.
97. Upp, E. L. *Fluid Flow Measurement: A Practical Guide to Accurate Flow Measurement*. Gulf Publishing, Houston, 1993.
98. U. S. Committee on Extension to the Standard Atmosphere. *U.S. Standard Atmosphere, 1976*. Government Printing Office, Washington, DC, 1976. Also produced by the National Oceanic and Atmospheric Administration.
99. Vallentine, H. R. *Applied Hydrodynamics*, 2d ed. Butterworth, London, and Plenum Press, New York, 1967.
100. Walski, T. M. *Advanced Water Distribution Modeling and Management*. Haestad Press, Waterbury, CT, 2003.
101. Warnick, C. C. *Hydropower Engineering*. Prentice-Hall, Englewood Cliffs, NJ, 1984.
102. Watters, G. Z. *Analysis and Control of Unsteady Flow in Pipelines*, 2d ed. Butterworths, Boston, 1984.
103. White, F. M. *Fluid Mechanics*, 8th ed. WCB/McGraw-Hill, New York, 2015.
104. White, F. M. *Viscous Fluid Flow*, 3rd ed. McGraw-Hill, New York, 2005.

105. Wylie, E. B., and V. L. Streeter. *Fluid Transients in Systems*. Prentice-Hall, Englewood Cliffs, NJ, 1993.
106. Yalin, M. S. *River Mechanics*. Pergamon, Oxford, 1992.
107. Yang, W.-J. (ed.). *Handbook of Flow Visualization.*, 2nd ed., Hemisphere, New York, 2001.
108. Yaws, C. L. *Handbook of Viscosity*, 3 vols. Gulf Publishing, Houston, TX, 1995–1997.
109. Zipparro, V. J., and H. Hasen (eds.). *Davis' Handbook of Applied Hydraulics*, 4th ed. McGraw-Hill, New York, 1993.
110. Záruba, J. *Water Hammer in Pipe-Line Systems*. Elsevier, Amsterdam, 1993.

Common Conversions and Constants

Table D.1 Conversion of BG (English) units to SI (metric) units[a]

	Multiply the number of	By[b]	To obtain the number of
Acceleration	ft/sec²	0.3048*	m/s²
Area	in²	645.16*	mm²
	ft²	0.092 903 04*	m²
	acre = 43,560* ft²	0.404 686	hectare (ha) = (100 m)²
Density, ρ	slug/ft³	515.379	kg/m³
Energy (work or quantity of heat)	ft·lb	1.355 818	joule (J) = N·m = 10⁷ erg
	ft·lb	3.766 16 × 10⁻⁷	kWh
	Btu = 777.649 ft·lb	1054.350	J = N·m = 0.239 006 cal
Flow rate, Q (of volume)	cfs = ft³/sec = 448.831 USgpm	0.028 316 8	m³/s = 10³ L/s
	mgd = 1.547 229 cfs	0.043 812 6	m³/s = 10³ L/s
	USgpm = 0.002 228 01 cfs	0.063 090 2	L/s = 10⁻³ m³/s
Force	lb [≡ slug·ft/sec²] = 16* oz	4.448 22	newton (N) [≡ kg·m/s²] = 10⁵ dyne
Kinematic viscosity, ν	ft²/sec	0.092 903 04*	m²/s = 10⁴ stokes (St)
Length	in = 1000* mil	25.4*	mm
	ft = 12* in	0.3048*	m = 10⁶* micron
	yard = 3* ft	0.9144*	m
	mile = 5280* ft	1.609 344*	km
	nautical mile = 6076.12 ft	1.852*	km
Mass	slug ≡ lb·sec²/ft = mass of 32.1740 lb weight	14.593 90	kg = mass of 9.806 65 N weight on earth
Power	ft·lb/sec	1.355 818	watt (W) = J/s = N·m/s
	hp = 550* ft·lb/sec = 2546.14 Btu/hr	0.745 700	kW

[a] To convert from SI to BG units, *divide* by the factor shown.

[b] For example, 1 ft/sec² = 0.3048* m/s².

[c] For the kgf definition, see Sec. 1.5.

* Exact conversion.

Table D.1 Conversion of BG (English) units to SI (metric) units[a] (*Continued*)

	Multiply the number of	By[b]	To obtain the number of
Pressure	psi = lb/in²	6894.76	pascal (Pa) = N/m²
	psf = lb/ft²	47.8803	Pa = N/m² = 10⁻⁵ bar
	psi	6.89476	kPa = kN/m² = 10 mb
	psi = lb/in²	6894.76	pascal (Pa) = N/m²
Specific heat	ft·lb/(slug·°R) [≡ ft²/(sec²·°R)]	0.1672255	N·m/(kg·K) [≡ m²/(s²·K)]
Specific weight	lb/ft³ = pcf	157.0875	N/m³
Velocity	fps = ft/sec	0.3048*	m/s
	mph = mile/hr = 1.466667 fps (30 mph = 44* fps)	1.609334*	km/h = 0.277778 m/s
	mph	0.44704*	m/s
	knot (nautical mph) = 1.15078 mph	1.852*	km/h
	knot = 1.687820 fps	0.514444	m/s
	fps = ft/sec	0.3048*	m/s
Viscosity, absolute, μ (dynamic)	lb·sec/ft² [≡ slug/(ft·sec)]	47.8803	N·s/m² [≡ kg/(m·s)] = 10 poise (P)
Volume	ft³ = 7.48052 U.S. gal	0.0283168	m³
	U.S. gal = 0.133681 ft³ = 231* in³ = 4* quarts = 8* pints	3.785 41	liter (L) = 10⁻³ m³
	pint = 2* cups = 16* fluid ounces	0.473176	liter (L) = 10⁻³ m³
	fluid ounce = 2* Tbsp = 6* tsp	29.5735	milliliter (mL) = 10⁻³ L
	imperial gal = 277.417 in³ = 1.20094 U.S. gal	4.5461	liter (L) = 10⁻³ m³
	acre-ft	1233.48183755*	m³
Weight (see Force)	U.S. (short) ton = 2000* lb	0.90718486	metric ton = 1000* kgf[c]
	British (long) ton = 2240* lb	1.0160470	metric ton = 1000* kgf

[a] To convert from SI to BG units, *divide* by the factor shown.
[b] For example, 1 ft/sec² = 0.3048* m/s².
[c] For the kgf definition, see Sec. 1.5.
* Exact conversion.

Table D.2 Other conversions

Engineering gas constant, R	1 ft·lb/(slug·°R) = 0.167226 N·m/(kg·K)
Heat	
SI	1 cal = 4.184* J (heat required to raise 1.0 g of water 1.0 K)
BG	1 Btu = 251.996 cal (heat required to raise 1.0 lb of water 1.0°R)
Temperature	
SI	K = 273.15° + °C (273° + °C for most calculations)
BG	°R = 459.67° + °F (460° + °F for most calculations)
	ΔT of 1°C = ΔT of 1 K = ΔT of 1.8°F = ΔT of 1.8°R

* Exact conversion.

Table D.3   Relationships between temperatures

[celsius (°C), fahrenheit (°F), kelvin (K), and rankine (°R)]

°C = (5/9)(°F − 32); °F = 32 + (9/5)(°C); K = (5/9)(°R); °R = (9/5)(K); K = °C + 273; °R = °F + 460

°C	−30	−20	−10	0	10	20	30	40	50	60	70	80	90	100
°F	−22	−4	14	32	50	68	86	104	122	140	158	176	194	212
K	243	253	263	273	283	293	303	313	323	333	343	353	363	373
°R	438	456	474	492	510	528	546	564	582	600	618	636	654	672

Table D.4   Definition of metric quantities

Hectare (ha)	area	10^4 m^2 = (100 m square)
Joule (J)	energy (or work)	newton·meter (N·m)
Liter (L)	volume	10^{-3} m^3
Newton (N)	force (or weight)	1 kg × 1 m/s^2
Pascal (Pa)	pressure	newton/meter^2 (N/m^2)
Poise (P)	absolute viscosity	10^{-1} N·s/m^2
Stoke (St)	kinematic viscosity	10^{-4} m^2/s
Watt (W)	power	newton·meter/second (N·m/s)

Table D.5   Commonly used prefixes for SI units

Factor by which unit is multiplied	Prefix	Symbol
10^9	giga	G
10^6	mega	M
10^3	kilo	k
10^{-2}	centi	c**
10^{-3}	milli	m
10^{-6}	micro	μ
10^{-9}	nano	n

** Avoid if possible (see Sec. 1.5).

Table D.6   Important quantities

	BG (English) unit	SI (metric) unit
Acceleration due to gravity (at sea level), g	32.1740 ft/sec^2	9.80665 m/s^2
for most calculations, use	32.2 ft/sec^2	9.81 m/s^2
Density of water (39.2°F, 4°C), ρ	1.940 slug/ft^3 = 1.940 lb·sec^2/ft^4	1000 kg/m^3 = 1.0 g/cm^3 = 1.0 Mg/m^3
Specific weight of water (40°F, 5°C), γ	62.43 lb/ft^3	9807 N/m^3 or 9.807 kN/m^3
for most calculations, use	62.4 lb/ft^3	9.81 kN/m^3
Standard atmosphere at sea level (59°F, 15°C)	14.696 psia	101.325 kPa abs
	29.92 inHg	760 mmHg
	33.91 ft H$_2$O	10.34 m H$_2$O
	2116.2 psfa	1013.25 millibars abs

Index

Page locators in **"bold"** refer to figures and locators in *"Italics"* refer to tables. Page numbers followed by an *n* refer to footnotes on those pages.